"Alternative data has become a hot topic in finance. New kinds of data, new data sources, and of course new tools for processing such data offer the possibility of new and previously unsuspected signals. In short alternative data lead to the promise of enhanced predictive power. But such advance does not come without its challenges - in terms of the quality of the data, the length of its history, reliable data capture, the development of appropriate statistical, AI, machine learning, and data mining tools, and, of course, the ethical challenges in the face of increasingly tough data protection regimes. Gautam Mitra and his colleagues have put together a superb collection of chapters discussing these topics, and more, to show how alternative data, used with care and expertise, can reveal the bigger picture."
– *Professor David J. Hand, Emeritus Professor of Mathematics and Senior Research Investigator, Imperial College, London*

"Digital capital is now so important that it can rightly be viewed as a factor of production, especially in the financial sector. This handbook does for the field of alternative data what vendors of alternative data do for data itself; and that is to provide structure, filter noise, and bring clarity. It is an indispensable work which every financial professional can consult, be it for an overview of the field or for specific details about alternative data."
– *Professor Hersh Shefrin, Mario L. Belotti Professor of Finance, Santa Clara University*

An impressive and timely contribution to the fast developing discipline of data driven decisions in the trading and management of financial risk. Automated data collection, organization, and dissemination is part and parcel of Data Science and the Handbook covers the current breadth of these activities, their risks, rewards, and costs. A welcome addition to the landscape of quantitative finance.
–*Professor Dilip Madan, Professor of Finance, Robert H. Smith School of Business*

"The Handbook of Alternative Data in Finance is the most comprehensive guide to alternative data I have seen. It could be called the Encyclopaedia of Alternative Data. It belongs to the desktop, not the bookshelf, of every investor."
– *Ernest Chan, Respected Academic, Author, Practicing Fund Manager, Entrepreneur and Founder of PredictNow.AI*

"Professor Gautam Mitra and his team unpack the topic of alternative data in finance, an ambitious endeavor given the fast-expanding nature of this new and exciting space. Alternative data powered by Natural Language Processing and Machine Learning has emerged as a new source of insights that can help investors make more informed decisions, stay ahead of competition and mitigate emerging risks. This handbook provides a strong validation of the substantial added value that alternative data brings. It also helps promote the idea that data driven decisions are better and more sustainable – something we, at RavenPack, firmly believe."
– *Armando Gonzalez, CEO and Founder of RavenPack*

"As the 1st Duke of Marlborough, John Churchill, wrote in 1715: 'No war can be conducted successfully without early and good intelligence.' The same can be said for successful trading. In that light, the Handbook of Alternative Data in Finance contains vital insights about how to gather and use alternative data —in short, intelligence —to facilitate successful trading."
– *Professor Steve H. Hanke, Professor of Applied Economics, The Johns Hopkins University, Baltimore, USA*

"*The Handbook of Alternative Data in Finance* is cutting edge and it bridges a huge gap in the representative studies on emerging areas of finance where alternative data can be profitably utilised for better informed decisions. The practical insights in the book would come very handy to both investors and researchers who look for fresh ideas."
– *Ashok Banerjee, Director, Indian Institute of Management Udaipur, Formerly Dean, and Faculty-in-charge of the Finance Lab at Indian Institute of Management Calcutta*

Handbook of Alternative Data in Finance
Volume I

Handbook of Alternative Data in Finance, Volume I motivates and challenges the reader to explore and apply Alternative Data in finance. The book provides a robust and in-depth overview of Alternative Data, including its definition, characteristics, difference from conventional data, categories of Alternative Data, Alternative Data providers, and more. The book also offers a rigorous and detailed exploration of process, application and delivery that should be practically useful to researchers and practitioners alike.

Features

- Includes cutting-edge applications in machine learning, fintech, and more
- Suitable for professional quantitative analysts, and as a resource for postgraduates and researchers in financial mathematics
- Features chapters from many leading researchers and practitioners.

CRC Press/OptiRisk Series in Finance

Series Editors:
Gautam Mitra
OptiRisk Systems

Forthcoming books:

Handbook of Alternative Data in Finance, Volume I
Edited by Gautam Mitra, Christina Erlwein-Sayer, Kieu Thi Hoang, Diana Roman, and Zryan Sadik

For more information about the series, please visit: https://www.routledge.com/CRC-PressOptiRisk-Series-in-Finance/book-series/ORSF

Handbook of Alternative Data in Finance

Volume I

Edited by
Gautam Mitra
Christina Erlwein-Sayer
Kieu Thi Hoang
Diana Roman
Zryan Sadik

CRC Press
Taylor & Francis Group
Boca Raton London New York

CRC Press is an imprint of the
Taylor & Francis Group, an **informa** business

A CHAPMAN & HALL BOOK

Designed cover image: ©ShutterStock Images

First edition published 2023
by CRC Press
6000 Broken Sound Parkway NW, Suite 300, Boca Raton, FL 33487-2742

and by CRC Press
4 Park Square, Milton Park, Abingdon, Oxon, OX14 4RN

CRC Press is an imprint of Taylor & Francis Group, LLC

Library of Congress Cataloging-in-Publication Data

Names: Mitra, Gautam, editor. | Erlwein-Sayer, Christina, editor. | Hoang, Kieu Thi, editor. | Roman, Diana, editor. | Sadik, Zryan, editor.
Title: Handbook of alternative data in finance / edited by Gautam Mitra, Christina Erlwein-Sayer, Kieu Thi Hoang, Diana Roman, Zryan Sadik.
Description: Boca Raton : C&H/CRC Press, 2023. | Series: CRC Press/OptiRisk series in finance | "A Chapman & Hall book." | Includes bibliographical references and index.
Identifiers: LCCN 2022060251 (print) | LCCN 2022060252 (ebook) | ISBN 9781032276489 (hardback) | ISBN 9781032276700 (paperback) | ISBN 9781003293644 (ebook)
Subjects: LCSH: Finance--Data processing. | Finance--Mathematical models.
Classification: LCC HG101 .H34 2023 (print) | LCC HG101 (ebook) | DDC 332.028541--dc23/eng/20230417
LC record available at https://lccn.loc.gov/2022060251
LC ebook record available at https://lccn.loc.gov/2022060252

ISBN: 978-1-032-27648-9 (hbk)
ISBN: 978-1-032-27670-0 (pbk)
ISBN: 978-1-003-29364-4 (ebk)

DOI: 10.1201/9781003293644

Typeset in Latin Modern font
by KnowledgeWorks Global Ltd.

Publisher's note: This book has been prepared from camera-ready copy provided by the authors

Contents

(PART IV.C) Case Studies on ESG

PART (V) DIRECTORY OF ALTERNATIVE DATA VENDORS

Preface

SETTING THE SCENE: BACKGROUND

We at OptiRisk have set out to research, acquire knowledge and be experts in the domain of News Analytics, Sentiment Analysis and Alternative Data in Finance. Our journey started with The Handbook of News Analytics in Finance (2011), which was followed by The Handbook of Sentiment Analysis in Finance (2016). We have stayed the course and as a Financial Analytics company continued to research in the domain of trading and fund management; this has culminated in our latest work: The Handbook of Alternative Data in Finance (2022). We wish to set a context for this multiauthor volume of the Handbook. In a Tome by a single author, one might get a fair depth of analysis. We refer the readers to (Denev & Amen, 2020),[1] Mahjouri (Mahjouri, 2023),[2] and Neptune Knowledge (Knowledge, 2022).[3] All these books discuss how Alternative Data is used to predict market movements make smarter investment decisions and risk control decisions. But these also invariably include the author's biases and a single perspective. In a multiauthor book, however, you get a 360-degree view and a wider perspective. In applied finance, an area of empirical research, this provides a better value to the market participants; specially, to traders, analysts, fund managers and others (Mitra et al., 2020).[4] The target audience of the Handbook is the latter group of financial market professionals as well as academics who are interested in the financial markets.

RECENT PROGRESS

Over the last 50 years, economists, operations researchers and management science and business analysts have researched extensively the topic of "models for decision making". In the early days, the role of data in such models were not even considered. Later from 1990s role of data was acknowledged but it

---

[1]Denev, A. & Amen, S., 2020. *The Book of Alternative Data: A Guide for Investors, Traders and Risk Managers.* s.l.:Wiley.

[2]Mahjouri, M., 2020. *Alternative Data: Capturing the Predictive Power of Big Data for Investment Success.* s.l.:Willey.

[3]Knowledge, N., 2022. *Introduction to Alternative Data: The power of data; a book for company executives and investment professionals,* s.l: Amazon

[4]Mitra, G., Baigurmanova, D. and Jhunjhunwala, S. (2020). The new landscape of financial markets: Participants and asset classes. *Available at SSRN 4104690.*

appeared as the "kid brother" of sophisticated modelling paradigms. Then at the beginning of new millennium data entered the modeling scene and started gaining importance. In the next phase, it appeared as "data driven decision making". This was then followed by the era of "big data". Soon, thereafter, "big data" became the "big brother" of decision modeling. We at OptiRisk have kept abreast of the emergence of big data, data analytics and data science and have observed their growing importance in financial decision making. Today data is "the king" the preponderance of data in almost all aspects of Political, Economic, Social, and Technological in short (PEST) is tipping the scale in favour of "data" as opposed to models for decision making. This paradigm shift of [data and models]; from "kid brother" to "big brother" has ushered in the era of "Alternative Data". As early as 2017 Greenwich Associates' research report[5] "Alternative Data for Alpha" pointed out, "on the trading floors of the world's largest investment managers, the use of Alternative Data to make institutional investments is still in its infancy".

ORGANIZATION OF THE HANDBOOK

An Overview of Alternative Data is presented in **Chapter 1**. This chapter is a joint effort of the OptiRisk's editorial team together with Dr Keith Black and Dr Ganesh Mani of the FDPInstitute.

Dr Black and Dr Mani bring their knowledge of the finance industry to this chapter, namely, **Chapter 1- Alternative Data: Overview**. This chapter is followed by five parts.

In **Part I ALTERNATIVE DATA: PROCESSING AND IM-PACT**, there are two chapters. **Chapter 2** is by David Jessop who presents his Contemplation and Reflection on using alternative data for trading and fund management. The second chapter, namely, **Chapter 3**: "Global Economy and Markets Sentiment Model" is by the research team of Northwestern Mutual. The authors [Jacob Gelfand, Kamilla Kasymova, Seamus M. O'Shea and Weijie Tan] introduce the novel data source "GDELT" and discuss its impact on financial market.

The **COUPLING** of **MODELS WITH ALTERNATIVE DATA FOR FINANCIAL ANALYTICS** is discussed in **Chapter 4** appearing in **Part II** of the Handbook. In the SenRisk project (see http://senrisk.eu/): a project funded by EU, a consortium of OptiRisk, Fraunhofer ITWM, and others investigated how "Macro News" can be exploited in sovereign and corporate bond modeling. AI, Machine Learning & Quantitative Models form the back-drop of research directed towards the application AI/ML and neo-classical Quant Models in Finance. In **Chapter 5**, we consider this topic and argue that these two modeling paradigms interact and are closely coupled.

---

[5]McPartland, K., 2017. *Alternative Data for Alpha*, s.l.: Greenwich Associates.

Our thesis is illustrated with an exemplar problem of predicting short-term market movement.

PART III HANDLING DIFFERENT ALTERNATIVE DATASETS

This part is made up of five chapters which cover handling different alternative datasets like news, micro-blog, company filings, earning calls and sensors data. The first two chapters discuss Asset Allocation Strategies Enhanced by Micro-blog and Enhanced by News. In the next two chapters the authors describe several methods to Extract Structured Datasets from Textual Sources like company filings and earning calls. The fifth chapter concern various kinds of Sensors Data.

How researchers have used News (meta) data, Sentiment (meta) data and other alternative data sources is presented in **Part III**. Two OptiRisk whitepapers may interest the readers as we first introduce the concept of filters, RSI filters to be specific. The (RSI) filters use only market data. We then proceed to describe two models in which these filters are enhanced using news (meta) data and micro-blog (meta) data respectively; see **Chapter 6** and **Chapter 7**, respectively.

PART IV ALTERNATIVE DATA USE CASES IN FINANCE

This part presents several case studies of applying alternative data in finance. We focus on three major applications of alternative data in finance, which include application in Trading and Fund Management to Find new Alpha, application in Risk Control and application in ESG.

Application in Trading and Fund Management (Finding new Alpha) includes three chapters:

- Media Sentiment Momentum: Global Equity Media-Driven Price Drift

- Defining Market States with Media Sentiment

Application in Risk Control includes three chapters:

- A Quantitative Metric for Corporate Sustainability

- Hot off the Press: Predicting Intraday Risk and Liquidity with News Analytics

- Exogenous Risks Alternative Data Implications for Strategic Asset Allocation—Multi-Subordination Levy Processes Approach

Case Studies on ESG there are three chapters:

- ESG Controversies and Stock Returns

- Oil and Gas Drilling Waste—A Material Externality

- ESG Scores and Price Momentum Are Compatible: Revisited

A list of alternative data providers is supplied in the last part of the Handbook:

PART V DIRECTORY OF ALTERNATIVE DATA VENDORS

In the directory fourteen (14) Alternative Data Vendors are listed. The information about these vendors and their products and services are available in the public domain and listed in their respective web portals. But having the information in a compact format and in one place provides a convenient and comparative summary.

Suggested Reading Sequence: After reading the overview chapter the rest of this book can be read in any order, depending on the reader's interest and job role, as the chapters have all been written independently. There is some referencing between the chapters but this does not hinder the understanding of content within each chapter. The purpose of this handbook is to invigorate and instigate readers to become active and pursue the exploration of sentiment analysis in finance. There are many applications to be explored, techniques and methods to be tested. Our mission is to motivate and excite the researchers and equally practitioners to participate in the research and development or exploitation of research knowledge in their respective areas of interest.

We have made a start in compiling the **Volume II** of the Handbook. We continue with the same theme and introduce (i) recent progress in theory and (ii) more use cases.

Gautam Mitra and Kieu Thi Hoang,
OptiRisk Systems Ltd, London

Acknowledgments

We first of all thank the members of the editorial panel for their painstaking effort in (i) the thorough review of the chapter contributions and (ii) their helpful discussions in determining the Parts and the Chapter structure of the Handbook.

We further thank our distinguished contributors who have given due care and effort in interacting with the reviewers and have revised their chapters in a timely fashion.

We thank Callum Fraser, the Editor of Mathematics Books CRC Press/Taylor & Francis Group, for his constructive suggestions in consummating the publication agreement. Further thanks are due to him and Mansi Kabra of the editorial team for working closely with us and speeding up the publication process.

Finally, we thank the members of our Handbook Team (Shradha Berry, Shrey Jhunjhunwala and Akshita Porwal), who have given their best and made the task of compilation of the Handbook go forward as clockwork.

Editors and Contributors

EDITORS

Gautam Mitra is founder and MD of Optirisk Systems. He is internationally renowned research scientist in the field of Operational Research in general and computational optimization and modeling in particular. He is an alumni of UCL and currently a visiting professor at UCL. In 2004 he was awarded the title of "distinguished professor" by Brunel University in recognition of his contributions in the domain of computational optimization, risk analytics and modeling. Professor Mitra is also the founder and chairman of the sister company UNICOM seminars.

Christina Erlwein-Sayer is a consultant and associate researcher at OptiRisk Systems. Her research interests lie in financial analytics, portfolio optimisation and risk management with sentiment analysis, involving time series modelling and machine learning techniques. She holds a professorship in Financial Mathematics and Statistics at HTW University of Applied Sciences, Berlin. She completed her PhD in Mathematics at Brunel University, London in 2008. She was then a researcher and consultant in the Financial Mathematics Department at Fraunhofer ITWM, Kaiserslautern, Germany. Between 2015 and 2018, prior to joining HTW Berlin in 2019, she was a full-time senior quantitative analyst and researcher at OptiRisk Systems, London, UK. She teaches modules on statistics, machine learning and financial mathematics and is part of the CSAF faculty. Christina is an experienced presenter at conferences and workshops: among others, she presented at workshops in London, IIM Calcutta in Kolkata and Mumbai and in Washington to World Bank.

Kieu Thi Hoang is a Financial Analyst and Relationship Manager at OptiRisk Systems. Kieu has a bachelor's degree (with high distinction) in International Economics from Foreign Trade University, Hanoi, Vietnam. She was among the top 10% of all the global candidates in her CFA level 2 examination (December 2020). Kieu has a strong foundation in advanced financial analysis and work experience in the finance industry. She has years of experience working at different renowned BFSI firms in Vietnam. Joining OptiRisk Systems as a Financial Analyst and Relationship Manager, she has done a lot of thorough research on alternative data in company projects. She also works with a variety of alternative data providers who are partners of her firm.

Diana Roman is a Consultant and Research Associate at OptiRisk Systems. After completing her PhD at Brunel University under late Professor Darby-Dowman and Professor Mitra, Dr Roman joined OptiRisk Systems as a software developer. She had designed the scenario-generator library which was used inSPInEthe first version of the SP Tool developed by OptiRisk Systems. Together with Professor Mitra she has written a few seminal papers on the topic of portfolio construction with downside risk control in general and use of Second Order Stochastic Dominance (SSD) in particular. Dr Roman is a senior lecturer in the Department of Mathematics at Brunel University London.

Zryan Sadik is a senior Quantitative Analyst and Researcher at OptiRisk Systems. Dr Sadik has a bachelor's degree in Mathematics from Salahaddin University – Erbil in the Kurdistan region of Iraq. After working as an IT technician, he pursued an MSc Degree in Computational Mathematics with Modelling at Brunel University, London (2012). Dr Sadik completed his PhD in Applied Mathematics with a thesis on the "Asset Price and Volatility Forecasting Using News Sentiment" at Brunel University, London (2018). His research interests include news sentiment analysis, macroeconomic sentiment analysis, stochastic volatility models, filtering in linear and nonlinear time series applying Kalman filters, volatility forecasting as well as optimization and risk assessment. His current research interests lie in the areas of empirical finance and quantitative methods, and the role of Alternative data in financial markets. He has been involved in developing predictive models of sentiment analysis, and sentiment-based trading strategies for the last seven years. These models and strategies are developed in C, C++, MATLAB, Python and R as appropriate. His prior studies include the impact of macroeconomic news on the spot and futures prices of crude oil, and the impact of firm-specific news on the movement of asset prices and on the volatility of asset price returns. Dr Sadik is fluent in Kurdish (his native language), as well as in English and Arabic.

EDITORIAL PANEL

Keith Black is the managing director and program director of the FDP Institute. Previously, he served as the managing director of content strategy at the CAIA Association, where he was a co-author of the CAIA curriculum. During a prior role, Keith advised foundations, endowments and pension funds on their asset allocation and manager selection strategies in hedge funds, commodities and managed futures. He has also traded commodity and equity derivatives and built quantitative stock selection models. Dr. Black earned a BA from Whittier College, an MBA from Carnegie Mellon University and a PhD from the Illinois Institute of Technology. He is a CFA, CAIA and FDP charterholder.

Arkaja Chakraverty is a Senior Research Associate at OptiRisk Systems. Arkaja is a dynamic academic actively engaged in research in the domain of corporate finance and financial markets. She has the experience of investing in Indian equity derivatives, namely, futures and options. Through a number of consulting projects involving small-scale enterprises (start ups companies such as — Man Capital, Clark & Kent Inc), she had been active in the industry. She received her PhD from Indian School of Business in Financial Economics in 2017; she is affiliated with Higher School of Economics, Moscow. Currently she is based out of Melbourne and is working on a series of research papers. As of April 2021, Arkaja has joined the OptiRisk team as an Associate (Consultant).

Yuanqi Chu joined OptiRisk Systems as an industry sponsored PhD student in the Department of Mathematics at Brunel University. Yuanqi is supervised by Professor Keming Yu and her research interests lie primarily in developing new methods on regression analysis of skewness, over-dispersion and heavy-tailedness. She obtained her master's degree in Statistics (with Distinction) from University of Nottingham in 2019. Yuanqi has a strong programming background and is well versed in R, MATLAB and SPSS.

Alexander Gladilin is the CEO of a start-up company Transolved and a research associate of OptiRisk. He is also a Doctoral Researcher at the Mathematics Department of Brunel University, actively involved in the research projects funded by OptiRisk. Alexander has work experience and master's degrees in finance and Data Analytics. His research focuses on applications of implied volatility and alternative data in financial modelling as applied to the BFSI sector.

Ganesh Mani is on the adjunct faculty of Carnegie Mellon University and is considered a global thought leader, helping scale human expertise via AI in many application areas, particularly financial services. He has been a pioneer in applying innovative techniques to multiple asset classes and has worked with large investment management firms incl. hedge funds, after having sold one of the earliest AI/ML-based asset management boutiques into SSgA (in the late nineties), nucleating the Advanced Research Center there. Ganesh has been featured on ABC Nightline and in a Barron's cover story titled "New Brains on the Block". He has an MBA in Finance and a PhD specializing in Artificial Intelligence from the University of Wisconsin-Madison; as well as an undergraduate degree in Computer Science from the Indian Institute of Technology, Bombay. Ganesh is often invited to speak and moderate panels at industry conferences around "The Art of Data Science" theme.

CONTRIBUTORS

Shradha Berry joined OptiRisk Systems in October 2019 as a Data Scientist and Research Analyst. She has over 5 years' experience in software development with a multinational consulting company in India. Shradha has a Bachelor's degree in Information Technology from Meghnad Saha Institute of technology, Kolkata, India and a Master's degree (with Distinction) in Data Science and Analytics from Brunel University, London. In OptiRisk her research was focused on AI & ML applied to quant finance models for the BFSI sector. Shradha left OptiRisk in June 2022.

Zhixin Cai was a Quantitative Research Intern at OptiRisk from 2018 to 2019. He has a Bachelor's Degree in Economics and Finance in The College of Economics, Henan University, China. He engaged in the exchange programme to Cardiff University from 2014 to 2016 and has done the BSc Econ Economics and Finance and BSc Econ Banking and Finance with first class result. He also holds a MSc Financial Risk Management in University College London.

Matteo Campellone is the founder and Executive Chairman of Brain, a company focused on the development of algorithms for trading strategies and investment decisions. He holds a Ph.D. in Physics and a Master in Business Administration. Matteo's past activities included Financial Modelling for financial institutions and Corporate Risk and Value Based Management for industrial companies. As a Theoretical Physicist he worked in the field of statistical mechanics of complex systems and of non-linear stochastic equations.

Joshua Clark-Bell is a Quantitative Researcher and has been working at MarketPsych for over a year. He holds a degree in econometrics and has an interest in applications of NLP towards sustainable finance.

Francesco Cricchio is the founder and Chief Executive Officer of Brain, a company focused on the development of algorithms for trading strategies and investment decisions. Francesco obtained his Ph.D. in Computational Methods applied to Quantum Physics from Uppsala University in 2010. He focused his career in solving complex computational problems in different sectors using a wide range of techniques, from density functional theory in Solid State

Physics to the application of Machine Learning in different industrial sectors with focus on Finance.

Giuliano De Rossi is the head of Portfolio Analytics in Prime Services at Goldman Sachs. Prior to this, he worked at Macquarie and PIMCO. He also spent six years in the Quant research team at UBS. Giuliano has a PhD in economics from Cambridge University and worked for three years as a college lecturer in economics at Cambridge before joining the finance industry on a full-time basis. Giuliano's Masters degree is from the LSE and his first degree is from Bocconi University in Milan. He has worked on a wide range of topics, including pairs trading, low volatility, the tracking error of global ETFs, cross asset strategies, downside risk and applications of machine learning to finance. His academic research has been published in the Journal of Econometrics and the Journal of Empirical Finance.

Dan diBartolomeo is President and founder of Northfield Information Services, Inc. Based in Boston since 1986, Northfield develops quantitative models of financial markets. He sits on boards of numerous industry organizations include IAQF and CQA, and is a director and past president of the Boston Economic Club. His publication record includes fifty books, book chapters and research journal articles. In addition, Dan spent several years as a Visiting Professor at Brunel University and has been admitted as an expert witness in litigation matters regarding investment management practices and derivatives in both US federal and state courts. He became editor in chief of the *Journal of Asset Management* at the start of 2019.

Jacob Gelfand, CFA, Director of Quantitative Strategy and Research, Investment Risk Management, is responsible for macro research and analysis for the fixed income, currencies, equity and cross-asset class portfolios. His focus areas include global markets risk analysis, asset allocation, investment style and risk management, relative value analysis and other quantitative and global macroaspects of risk and portfolio management.

Mr. Gelfand joined Northwestern Mutual in 2004 as a member of the Fixed Income Department at Mason Street Advisors (MSA). In 2015, he joined the Investment Strategy Division and in 2016 became a member of the Investment Risk Management Division. Before joining Northwestern Mutual, Mr. Gelfand was a Senior Consultant with Cap Gemini, serving clients in investment management and financial industries. Mr. Gelfand is an adjunct faculty in the Lubar School of Business, University of Wisconsin- Milwaukee, where he teaches classes on Global Investments, and Options and Derivatives. Mr. Gelfand serves on the board of the CFA Society in Milwaukee, Wisconsin.

Mr. Gelfand received a MS in Computer Science with Honors from the Moscow State University of Civil Engineering (MSUCE), Russia, and an MBA in Finance and Strategy with Honors from the University of Chicago Booth School of Business. He holds the Chartered Financial Analyst (CFA) designation.

Ryoko Ito works on equities execution research and content generation at Goldman Sachs. Prior to this, she worked at UBS, during which she managed a team of quants responsible for designing and running econometric models used in firm-wide regulatory stress tests. She was also a postdoctoral researcher at Oxford University, during which she lectured financial econometrics to graduate students at Saïd Business School. She has published academic papers on time series analysis and financial econometrics. Her recent work on modeling time series with zero-valued observations appeared in the Journal of Econometrics in 2019. Ryoko has a PhD in economics from Cambridge University. She undertook research projects with her PhD sponsors; one on liquidity prediction and order-book analysis in high-frequency FX at Morgan Stanley, and another on modeling systemic risk in the European banking system at the International Monetary Fund. Ryoko also completed Part III of the Mathematical Tripos at Cambridge University.

David Jessop is Head of Investment Risk at Columbia Threadneedle Investments EMEA APAC. He has responsibility for overseeing the independent investment risk management process for all portfolios managed in the EMEA region.

Before joining Columbia Threadneedle Investments, David was the Global Head of Quantitative Research at UBS. Over his 17 years at UBS, his research covered many topics but in particular he concentrated on risk analysis, portfolio construction and more recently cross asset factor investing / the application of machine learning and Bayesian techniques in investment management. Prior to this he was Head of Quantitative Marketing at Citigroup. David started his career at Morgan Grenfell, initially as a derivative analyst and then as a quantitative portfolio manager.

David has an MA in Mathematics from Trinity College, Cambridge.

Dan Joldzic, CFA, FRM is CEO of Alexandria Technology, Inc, which develops artificial intelligence to analyze financial news. Prior to joining Alexandria, Dan served dual roles as an equity portfolio manager and quantitative research analyst at Alliance Bernstein where he performed factor research to enhance the performance of equity portfolios.

Christopher Kantos is a Managing Director at Alexandria Technology. In this role, he focuses on maintaining and growing new business in EMEA and also co-heads research efforts at Alexandria, focusing on exploring ways in which natural language processing and machine learning can be applied in the financial domain. Prior, he spent 15 years working in financial risk at Northfield Information Services as a Director and Senior Equity Risk Analyst. Mr. Kantos earned a BS in computer engineering from Tufts University.

Kamilla Kasymova is a Quantitative Research and Analytics Director in the Investment Risk Management of Northwestern Mutual Life Insurance Company. She joined the company in 2014. Kamilla has extensive experience using analytical and research techniques (time series analysis and forecasting, econometrics, financial mathematics) in developing internal capital market assumptions and economic scenario generator used by Actuarial and Investment departments across the company. Prior to joining Northwestern Mutual, Kamilla taught various undergraduate mathematics and economics classes during her graduate studies.

Kamilla has a BSc in Economics from Moscow State University, MSc in Finance from University of Ulm, MSc in Mathematics from University of Wisconsin-Milwaukee and PhD in Economics from University of Wisconsin-Milwaukee.

Richard L. Peterson is CEO of MarketPsych Data which produces psychological and macroeconomic data derived from text analytics of news and social media. MarketPsych's data is consumed by the world's largest hedge funds. Dr. Peterson is an award-winning financial writer, an associate editor of the Journal of Behavioral Finance, has published widely in academia and performed postdoctoral neuroeconomics research at Stanford University.

Changjie Liu is Head of Research at MarketPsych. He has more than fifteen years of experience in investigating sentiment data, including ten years in MarketPsych. He leads the quant and web development team in planning and developing sentiment-based investment strategies and products. He holds a MFE from UC Berkeley and a Computer Science BSc with honors from NUS.

Anthony Luciani, MSc is a Quantitative Researcher at MarketPsych. He constructs and validates sentiment-based signals, with a focus on relative value strategies and using machine learning techniques. His research is built upon an education in mathematics and information theory, and his prior work in behavioral financial studies.

Seamus M. O'Shea, CFA is the Managing Director of the Public Investment in Northwestern Mutual. He joined Northwestern Mutual in 2018 after spending two years with a specialty finance start-up based outside of Boston. In his current role, he is responsible for the sector allocation process for the $105bn+ Public Markets fixed income portfolio, which includes corporate, mortgage-backed and municipal bonds. In addition, he has direct portfolio management oversight for the internally managed EM assets.

Prior to trying his hand in the start-up world, he co-managed $50bn in U.S. Credit-oriented separate accounts and mutual funds at Wellington Management, where he was a Managing Director and the lead risk-taker for the Emerging Markets, High Yield and ABS segments of client accounts. In addition to his portfolio management duties there, he maintained extensive client-facing responsibilities and was regularly sought out to present to current and prospective clients and consultancies both in the US as well as overseas.

Prior to joining Wellington, Seamus was a Global Bond portfolio manager at Pacific Investment Management Company (PIMCO) in Newport Beach, CA, where he helped manage over $60bn in Global-Agg benchmarked accounts. He has 15+ years of investment experience, and began his career at Fidelity Investments.

Seamus holds a MBA from MIT's Sloan School of Management and a BA in Mathematics from the College of the Holy Cross. He is a CFA Charterholder, and a citizen of the United States, Canada and Ireland.

Matus Padysak has a strong background in probability and statistics of which he holds a master´s degree. Currently, he is a Ph.D. candidate with a research interest in pension funds investments. Apart from the research, his responsibilities also lie in teaching several undergraduate courses. He is always keen to utilize all his knowledge in finances, investing or trading. To escape from the research and clear the head, Matus likes to swim, bike and run.

Tiago Quevedo Teodoro is a Quantitative Researcher at MarketPsych. He specialises in applying statistical and machine learning methods toward sentiment usage in timing and event studies. He holds a Ph.D. in Physical Chemistry as well as an MSc in Finance & Technology, with 16 peer-reviewed publications in top scientific journals.

Boryana Racheva-Iotova is Senior Vice President, Senior Director of Buy-Side Strategy at Factset, driving the strategy and vision for FactSets's solutions for institutional asset managers and asset owners. Prior to that, she led the Risk and Quantitative Analytics Business Line for FactSet, responsible for all risk and quantitative solutions, which includes quantitative research, product development and business strategy. She is a co-founder of

FinAnalytica and the former Global Head of Risk at BISAM and has over 20 years of experience in building risk and quant portfolio management software solutions. Before founding FinAnalytica, Ms. Racheva-Iotova led the implementation of a Monte-Carlo based VaR calculation to meet the Basel II requirements at SGZ Bank, as well as the development of six patented methodologies for FinAnalytica. In 2018, she received the Risk Professional of the Year Award from Waters Technology based on her achievements in building risk management software solutions and translating the latest academic advancements in practical applications to meet the needs of financial industry practitioners. Ms. Racheva-Iotova earned a Master of Science in Probability and Statistics at Sofia University and a Doctor of Science from Ludwig Maximilian University of Munich.

J. Blake Scott is the President of Waste Analytics LLC. Blake has been a leader in the oil and gas sector for more than 25 years. As founder and president of Scott Environmental Services and Scott Energy Technologies, Blake pioneered groundbreaking research and development in the drilling-waste recycling industry. Blake has received multiple awards and recognitions for his work, and has presented to organizations such as the Interstate Oil and Gas Compact Commission, has served on ITRC's Solidification/Stabilization Team and was an expert for GRI's Oil and Gas Sector Standard. Blake founded Waste Analytics in 2018 in order to provide unique data on oil and gas companies for current insight and financial prediction. Blake has a BA and MBA from Texas Christian University.

Michael Steliaros is global head of Quantitative Execution Services at Goldman Sachs. He is responsible for the research, development and implementation of quantitative processes for portfolio and electronic trading, as well as the management of the firm's relationships with the quantitative client-base across regions. Michael manages a variety of teams globally, spanning algorithmic research, portfolio quants, client solutions, analytics and quantitative content generation. He joined the firm as a managing director in 2017.

Weijie Tan is Associate Quantitative Analyst in the Investment Risk Management of Northwestern Mutual. He joined Northwestern Mutual in 2020. In his current role, he focuses on Capital Market Assumptions, which produces long-term return, volatility and correlation projection of a variety of asset classes. Besides, he builds and maintains indicators based on sentiment analysis to support asset allocation. Weijie holds a Ph.D. in Engineering from the University of Wisconsin - Milwaukee and a bachelor's degree in Economics from the Shanghai University of Finance and Economics.

Ziwen Tan was a Quantitative Research Intern at OptiRisk from 2018 to 2019. Before that, she was a Research Assistant for two years at the University of Toronto. She became a Senior Data Analyst at PHD Canada in 2020. Ziwen holds a Bachelor's Degree with Distinction in Statistics and Linguistics from the University of Toronto. She is currently pursuing her Master's Degree in Library and Information Science at the University of Illinois Urbana-Champaign.

Gareth Williams is a Data Scientist at Transolved Ltd. He had previously completed an undergraduate degree in Physics with Mathematics at Keele University before obtaining a PhD in Theoretical Physics at the University of Liverpool. His research interests include applied Bayesian statistics and computational physics, in particular the calculation of marginalized likelihoods and the use of the Bayes factor as a tool for model comparison.

William Zieff is Director at Northfield Information Services, a leading provider of financial risk analytics. He has extensive experience in finance and investing. Prior to Northfield, he was Chief Investment Officer of Global Strategic Products, Wells Fargo Asset Management; Partner and Co-Chief Investment Officer of Global Asset Allocation at Putnam Investments; and Director of Asset Allocation at Grantham, Mayo, Van Otterloo. He has taught master's and undergraduate courses in finance at Northeastern University and Boston College. He holds an A.B. degree in Economics and Mathematics from Brown University and MBA from Harvard Business School.

Abbreviations

ABBREVIATION	EXPLANATION
ACOG	Association of Central Oklahoma Governments
ADF	Augmented Dickey–Fuller test
AHT	After-Hours Trading
AGG	Aggregated bond index ETF
AI	Artificial Intelligence
AIC	Akaike Information Criterion
ANN	Artificial Neural Network
API	American Petroleum Institute
API	Application Programming Interface
AQ	Analysts' Questions
ARCH	Auto-regressive Conditional Heteroskedastic
ARIMAX	Autoregressive–Moving-Average Model with Exogenous Inputs Model
AUM	Assets under Management
BERT	Bidirectional Encoder Representations from Transformers
BLMCF	Brain Language Metrics on Company Filings
BOW	Bag of Words
BSI	Brain Sentiment Indicator
CAC	Central America and the Caribbean
CAGR	Compound Annual Growth Rate
CAPM	capital asset pricing model
CDS	Credit Default Swap
CFP	Corporate Financial performance
CIS	Commonwealth of Independent States
CLOB	Central Limit Order Book
CNN	Convolutional Neural Network
DM	Diebold-Mariano
DNS	Domain Name System
DRSI	Derived RSI
ECG	Electric Company of Ghana
ECT	Earnings Calls Transcripts
EEL	Estimated Environmental Liability

EMA	Exponential Moving Average
EMD	empirical modal decomposition
EMH	Efficient Market Hypothesis
ESG	Environmental, Social and Governance
ESPP	Employee Stock Purchase Plan
ETL	Expected Tail Loss
EU	European Union
EVSI	External Vulnerability Sentiment Index
FAA	Federal Aviation Administration
FASB	Financial Accounting Standards Board
FN	False-Negative
FP	False-Positive
FSD	First-Order Stochastic Dominance
FT	Financial Times
FTRI	Financial Technology Research Institute
FX	Foreign Exchange
GARCH	Generalized ARCH
GCAM	Global Content Analysis Measures
GDELT	Global Database of Events, Location and Tone
GDPR	Global Data Protection Regulation
GEMS	Global Economy and Markets Sentiment
GHG	Greenhouse Gas Emissions
GMM	Gaussian mixture model
GPS	Global Positioning System
GRI	Global Reporting Initiative
GUI	Graphical User Interface
HFT	High-Frequency Trading
ICBDAC	International Conference on Big Data Analytics and Computational
IG	Investment Grade
IIRC	International Integrated Reporting Council
IMF	International Monetary Fund
IoT	Internet of Things
IPIECA	International Petroleum Industry Environmental Conservation Association
IRR	Internal Rate of Return
IRS	Interest Rate Swap
IS	Impact Scores
IVSI	Internal Vulnerability Sentiment Index
LM	Loughran McDonald
LP	Linear Programming
LR	logistic regression

LSTM	long short-term memory
LT VSI	Long-Term Vulnerability Sentiment Index
MA	Moving Average
MAAQ	Management Answers to Analysts' Questions
MAE	Mean Absolute Error
MAPE	Mean Absolute Percent Error
MASE	Mean Absolute Scaled Error
MD	Management Discussion
MIP	Mixed Integer Programming
ML	Machine Learning
MLE	maximum likelihood estimation
MOM	Momentum
MPT	Modern Portfolio Theory
MRG	Mergers and Acquisitions
MRSI	Micro-blog Relative Strength Index
NCQ	Neo-Classsical Quant
NETL	National Energy Technology Lab
NIH	National Institutes of Health
NLP	Natural Language Processing
NORM	Naturally Occurring Radioactive Material
NRSI	News Relative Strength Index
NYSE	New York Stock Exchange
OCC	Oklahoma Corporation Commission
OHLC	Open, High, Low and Close
OLS	ordinary least square
PADEP	Pennsylvania Department of Protection
PAH	Polycyclic Aromatic Hydrocarbons
PEAD	Post-Earnings Announcement Drift
PII	Personally Identifiable Information
QA	Questions and Answers
RCRA	Resource Conservation and Recovery Act
RF	Random Forest
RFF	Resources for the Future
RFQ	Request for Quote
RFR	Risk Free Rate
RIC	Reuters Instrument Code
RMA	Refinitiv MarketPsych Analytics
RM-ESG	Refinitiv MarketPsych ESG
RMSE	Root Mean Square Error
RND	Risk Neutral Density
ROP	Rate of Penetration
RPNA	RavenPack News Analytics

RSI	Relative Strength Index
SASB	Sustainability Accounting Standards Board
SCOTUS	Supreme Court of the United States
SD	Stochastic Dominance
SDG	Sustainable Development Goal
SEC	Securities and Exchange Commission
SPE	Society of Petroleum Engineers
SSD	Second Order Stochastic Dominance
SSE	Shanghai Stock Exchange
ST VSI	Short-Term Vulnerability Sentiment Index
SVM	support vector machine
TB	Terabytes
TN	True-Negative
TP	True-Positive
TRMA	Thomson Reuters MarketPsych Analytics
TSD	Third-order Stochastic Dominance
UAT	User Acceptance Training
UK	United Kingdom
UN	United Nations
UNPRI	United Nations Principles on Responsible Investing
USEPA	United States Environmental Protection Agency
USSEC	United States Securities and Exchange Commission
UTC	Coordinated Universal Time
VIX	Volatility Index
VSI	Vulnerability Sentiment Indices
VSW	Velocity Solar Wind
XGB	XGBoost eXtreme Gradient Boosting

CHAPTER 1

Alternative Data: Overview

Gautam Mitra

Research Director, OptiRisk Systems and UCL Department of Computer Science, London, United Kingdom

Kieu Thi Hoang

Senior Financial Analyst, OptiRisk Systems

Alexander Gladilin

Research Associate, OptiRisk Systems

Yuanqi Chu

Sponsored PhD Candidate and Intern, OptiRisk Systems

Keith Black

Managing Director and Program Director, FDP Institute

Ganesh Mani

Adjunct Faculty, Carnegie Mellon University

CONTENTS

DOI: 10.1201/9781003293644-1

1.1 INTRODUCTION

The finance industry has always employed data. The digital era has generated exponential growth in various types of data, and the big data phenomenon has revolutionized the modern age. A growing number of researchers and investors have embraced big data advances to enhance investment performance. What set the current evolution apart are large volumes of alternative data, which have the potential to profoundly reshape financial landscape in the foreseeable future. Alternative data can be conceptualized as the data derived from outside of the standard repository of financial data. [Quinlan and Cheng, 2017] refer to the alternative data sources as "any useable information or data that is not from a financial statement or report". Different from traditional repertoire of financial information, such datasets are granular, real time and speed centric, including transaction activities, social-media streams, satellite images, clickstream data, digital footprints and other emerging data sensed via the Internet of Things (IoT).

The exploitation of alternative data has been catalyzed by the boom of the data vending industry [Kolanovic and Krishnamachari, 2017] and the belief that integrating alternative data can reduce the cost of obtaining information [Goldfarb and Tucker, 2019]. The application of alternative data has attracted great interest in various perspectives within today's financial landscape, including trading strategy generation [Xiao and Chen, 2018], enhancing credit scoring [Cheney, 2008]; [Djeundje et al., 2021], portfolio construction and allocation [Henriksson et al., 2019]; [Bertolotti, 2020]; [De Spiegeleer et al., 2021], among others. Using contemporary data science and Artificial Intelligence (AI) techniques, economic researchers and market practitioners can exploit key alternative data to interpret and condense increasing amounts of traditional information, extracting pithy insights for cost reduction, revenue enhancement and maintaining a competitive edge.

While the intertwinement between alternative data and finance systems is broadening the possibilities for innovation and value creation, various challenges and issues also arise through the exploration of the data. As [Dourish and Gomez Cruz, 2018] point out, "Data must be narrated – put to work in particular contexts, sunk into narratives that give them shape and meaning, and mobilized as part of broader processes of interpretation and meaning-making", the lack of an eligible narration and characterization paradigm makes alternative data employment more vulnerable to issues like data ownership, security and privacy than traditional data sources. The challenges and potential pitfalls of managing, using and analyzing alternative data sources should be highlighted by researchers and financial practitioners in the process of extracting improved analytics-driven insights.

In this chapter we describe the current landscape of alternative data with a view to provide sufficient information and guidance for analysts to exploit such datasets. The rest of this chapter is organized as follows. In Section 1.2 we discuss the increasing availability of data in the financial arena. In Section 1.3 we outline the role and use of traditional data. In Section 1.4 we formally define alternative data and highlight the differences between traditional and alternative data. In this section we bring out the differences between alternative data and big data and briefly discuss a few use cases and consider some ethical challenges. In Section 1.5 we consider the phenomenal growth of alternative data; our discussions cover recent history, the drivers of growth and the positive impact experienced in the financial markets. Alternative data providers collect, aggregate, structure and act as a financial intermediary for the buyers of alternative data. In Section 1.6 we describe the services provided by these vendors; some examples of established data vendors are cited and their services are discussed. In Section 1.7 we consider the challenges faced by diverse users and financial market participants who want to exploit alternative data and we end the chapter with a summary discussion and conclusion in Section 1.8.

1.2   DATA

In recent times in the realm of analytics, data has gained substantial importance. The growth of data analytics and data science has vindicated this; [Donoho et al., 2000] have articulated well via: "The coming century is surely the century of data". Recent technological innovations are based on this central role of data and the proliferation of globally produced data. There has been an explosion in the volume, velocity and variety of the data which are collected by scraping web portals; and from mobile phone usage, social media activities, online payment records, customer service records and embedded sensors. As the OECD [The Organization for Economic Cooperation and Development, 2013] suggests, "In business, the exploitation of data, promises to create added value in a variety of operations, ranging from optimizing the value chain, manufacturing production to an efficient use of labor and improved customer relationships". The technologies are constantly evolving; they intertwine and embrace "Big Data", the "Internet of Things" (IoT) and the "Internet of Signs" [O'Leary, 2013]. These achieve the desirable goals of knowledge management, knowledge sharing and effective decision-making.

According to [Aldridge, 2015] a decade ago, finance was a small-data discipline. Because data was a scarce resource it naturally relied upon small-data paradigm. To most investors, exchanges offered only four prices per stock per day: Open, High, Low and Close (OHLC). Data at higher frequency was not stored by even the largest market makers [Aldridge, 2015]. [Wang, 2008] reinforces this paradigm shift and observes: "Financial econometrics is only made possible by the availability of vast economic and financial data". In the new landscape of the financial markets, see [Mitra et al., 2020], data takes the central role and has led to the transformation of financial service sectors and institutions. In discussing applications of "Big Data in Finance", [Goldstein et al., 2021] recently pointed out that three properties, namely, large size, high dimension and complex structure which taken together underpin the role of "Big Data in Finance" research.

"According to IBM, companies have captured more data in the last two years than in the previous 2000 years" [Syed et al., 2013]. "ONE MEASURE OF PROGRESS in empirical econometrics is the frequency of data used" [Engle, 2000]. In contrast to the availability of OHLC data reported the following day (on the T+1 basis) in the small-data age, in current financial markets, a stock can experience 500 quote changes and 150 trades in one microsecond [Lewis, 2014]. High-frequency trading (HFT) has grown tremendously and is becoming increasingly important in the functioning of the financial markets. In the world of "Algo Trading", massive amounts of high-frequency data are consumed by sophisticated algorithms. Algo Traders take advantage of HFT to spot profitable opportunities that may only be open for milliseconds.

In financial portfolio construction and optimization, high dimensionality is encountered when asset managers estimate the covariance matrix or its inverse matrix of the returns of a large pool of assets [Fan et al., 2011]. High-dimensional time series such as the country-level quarterly macroeconomic data [Stock and Watson, 2009], simultaneously observed return series for a large number of stocks [Barigozzi and Hallin, 2017] and large series of company financials based on their quarterly reports [Wang et al., 2019] also abound. These data sources have gained considerable attention in modern finance and economics. The "curse of dimensionality" that arises in financial data, presents new challenges from a computational and statistical perspective. As for financial market participants, they have to delve into real-time analytics and decision making to effectively exploit the value hidden in datasets.

Another important characteristic of finance-related big data is complexity. Unstructured data refers to information that is not organized and does not fall into a pre-determined model [Fang and Zhang, 2016]. So the data that is gathered from various sources, such as news articles, blogs, forums, twitters, emails or images, audio, video, is unstructured. Unstructured data is often characterized by high dimensionality [Goldstein et al., 2021]. Recent research findings have shed light on the prominent role of social media data in the analysis of financial dynamics.

From the era of Big Data, we have now moved on to alternative data ("alt-data"). Alternative data sources can be characterized as "any useable information or data that is not from a financial statement or report" [Quinlan and Cheng, 2017]. These are different from traditional sources of financial information; typically, asset prices, corporate annual reports and consumer spending. Such datasets are granular, real time and speed centric, including transactional activities, social-media streams, satellite images, clickstream data, digital footprints and other emerging data from the IoT. In financial landscape, market participants are embracing and unlocking alternative data to extract competitive edges and identify trading opportunities: [Jame et al., 2016] assert the value of crowdsourced forecasts in forecasting earnings and measuring the market's expectations of earnings; [Cong et al., 2021] highlight the promising utilization of different categories of alt-data in economics and business-related fields. [In et al., 2019] show the environmental, social and governance (ESG) data, when it is of high quality, is capable of mapping onto the investment decision-making processes. [Monk et al., 2019] emphasize that with efficient organizing and processing, alt-data can be employed to delve deep and gain insights of risk and generate operational alpha, [Berg et al., 2020] suggest that even the simple, easily accessible users' digital footprints are highly informative in predicting consumer default, to name only a few.

Today alternative data has assumed a central role in financial analytics and thereby in financial markets; it is well set to drive future innovations in financial systems. We observe that many challenges such as data acquisition, information extraction, analytical technique as well as data security and privacy have emerged and need to be suitably addressed through research and innovation.

1.3 TRADITIONAL DATA

Traditional data has been an input into quantitative investment decision making for over 40 years. In most countries, publicly traded companies make substantial disclosures on a quarterly basis. In addition, stock exchanges publish data on stock prices and trading volume, while investment banks and brokerage firms publish estimates for futures earnings releases.

Analysis of traditional investment information has focused on ratio-based measures from data disclosed on quarterly corporate income statements and balance sheets. Using the income statement, growth-oriented investors focus on revenue and earnings growth measures, comparing the current quarter and current year revenue and earnings relative to prior periods. Value investors use the market capitalization and the stock price of the company to analyze the cheapness of a company relative to income statement measures such as price/cash flow, price/earnings, or price/revenues or balance sheet measures such as price/book value of equity. Finally, stock prices can move substantially based on the concept of earnings or revenue surprise, where the stock price is expected to rise following a positive surprise where reported earnings or revenues exceeds the earnings or revenue expectation of the sell-side analysts.

Traditional data has been readily available and relatively straight forward to work with. Quantitative income statement and balance sheet data can be downloaded as panel datasets, with the reports of thousands of companies held in a structured database holding quarterly reports over historical time periods. Stock prices and analyst revenue and earnings estimates can be easily added, as they have been previously organized or tagged with the stock ticker by the data vendor.

Once the raw data is available, quantitative analysts can run backtests of this data, testing the historical response of stock prices to some combination of value and growth factors, as well as earnings or revenue surprises. The goal, of course, is to build a model where the combination of factors is predictive of future stock prices.

The academic literature on return anomalies documents a number of traditional quantitative factors that have predicted excess returns over long periods of time. Note that while these factors may earn excess returns relative to the market over long periods of time, they may also lag the market for significant periods of time. For example, while value stocks have outperformed growth stocks from 1980 to 1994, value underperformed from 1995 to 1999. More recently, value stocks beat growth stocks from 2000 to 2016, while growth triumphed from 2017 to 2020. Perhaps, value stocks outperform in weak stock markets when risk aversion is high and growth stocks outperform in strong stock markets when investors are risk seeking.

Fama and French pioneered studies on the returns to small cap and value stocks [Fama and French, 1992]. Value stocks track companies where the stock price is low relative to some measure of earnings or value, such as price-to-earnings, price-to-book, price-to-sales or price-to-cash-flow. For each dollar of stock price, value investors are buying more earnings, book value, revenues or cash flow than can be obtained by buying growth stocks. However, value stocks tend to have lower expected growth in these metrics compared to growth stocks.

Similarly, stocks with smaller market capitalizations have historically outperformed stocks with larger capitalizations over long periods of time. In the US, small stocks outperformed large stocks from 1981 to 1997 with a brief benefit for growth in 1990, while large stocks benefited from outperformance in 1998-1999. Small stocks once again outperformed from 2000 to 2018, while larger stocks had higher returns in 2019–2020.

Excess returns to stocks with price momentum were documented by Carhart [Carhart, 1997] as well as Jagadeesh and Titman [Jegadeesh and Titman, 1993]. Stocks that have outperformed the market over the trailing six to twelve months are likely to outperform over the subsequent six to twelve months. Conversely, stocks that have underperformed over a similar period are likely to continue to underperform. Note that the effect is the opposite in the short run, as stocks that have outperformed over the last month tend to underperform over the next few weeks.

Lawrence Brown pioneered the research [Brown, 1997] on earnings momentum and earnings surprise. Due to positive momentum in a company's business, positive news regarding earnings often lasts longer than one quarter. If markets were highly efficient, a positive earnings surprise would lead to an immediate and complete rise in the stock price to reflect this new information. However, a post-earnings-announcement drift has been documented, where investors could buy stocks after the earnings surprise is announced and still be able to outperform the market. Earnings surprises may be predicted by noting that analysts are increasing their earnings estimates during a calendar quarter.

Research from Richard Sloan has focused on quality stocks and accounting accruals [Sloan, 1996]. Investors focusing on quality of earnings seek to buy stocks where net income and earnings are highly correlated. That is, when companies report high quality profits that are received mostly in cash, there is a lower probability that earnings will subsequently be restated. When accounting accruals are high, net income is elevated relative to the cash flow received by companies with higher earnings quality. Companies that report higher net income by deferring expenses and accelerating revenues tend to find it more difficult to sustain the level of reported earnings growth.

There is also a strong signal in the change in shares outstanding of a company. Stocks that are reducing shares outstanding through buyback programs tend to outperform, as their purchase of stock increases the demand for the stock and reduces the supply. Earnings per share is defined as net income divided by shares outstanding. As the number of shares outstanding is reduced, earnings per share rises, even when net income is unchanged. Stocks that are increasing the number of shares outstanding tend to underperform, whether that issuance comes through employee stock options, secondary sales or stock swap mergers. Investors may also track the purchases and sales of stock by insiders, such as the CEO, CFO and board members. When insiders are buying shares, the company may outperform. Conversely, when a large number of insiders are selling a large portion of their shares, the company may underperform. Note that insider trading can be a signal for the entire stock market, where a sharp increase in insider sales across the entire market may predict a decline in the stock market index.

Modeling these stock market anomalies is relatively straight forward, as the data was available in a structured manner, mapped by dates, stock tickers and columns of fundamental data for each stock. Backtesting engines and various datasets are readily available from commercial vendors. There were two downsides, though, to this use of traditional data. First, because the data was readily available and highly structured, large numbers of investors had access to the same data and often built trading strategies using highly correlated signals. That is, many investors derived the same models based on the limited number of factors and the rich academic literature just explained. Second, the data was not very granular, as most of the new information came only when each company in the database reported their earnings, income statement and balance sheet on a quarterly basis. While stock prices and analyst estimates could change on a daily basis, most of the data disclosed by companies was available only four times each year. Because of the small number of data points on each company, models using traditional data are most useful for time periods of one to six months or longer. That is, the main value added is at the time of the next quarterly report, as many stocks have the majority of their annual idiosyncratic price movements during the one week of each calendar quarter when earnings are reported.

While users of this traditional data generally waited until earnings announcements to earn their excess returns, investors deploying alternative data may be able to profit over a shorter time period. While the holding period of traditional models may be denominated in months, models deploying alternative data may have holding periods of days to weeks. That is, while traditional models may profit from holding a stock after an earnings announcement, alternative data models may profit by predicting that earnings announcement.

1.4 ALTERNATIVE DATA

In contrast to the highly structured world of most traditional data sources, alternative data is unstructured, more messy, high volume and high frequency. Alternative data does not come from a single source already organized by stock ticker, as was the case of traditional data such as income statements, balance sheets, earnings estimates or stock prices.

The goal of alternative data is to build a forecast of corporate revenues and earnings more quickly than can be accomplished using the quarterly data disclosed by publicly traded companies. Alternative data comes in many forms, some of which are available in real time. While traditional data is available for all companies and is available in easily accessible database formats, many of the alternative data sources focus on a more narrow sector of companies. That is, every public company reports income statement and balance sheet data, while only healthcare companies report drug reactions to regulators and motor vehicle bureaus only report on the sales of automobile manufacturers.

A very promising avenue of exploration for alternative data is to derive the revenue of privately held companies. In the US, there are 3,000 public companies with revenues over 100 million dollars, while there are over 30,000 private companies with similar revenues. Due to regulatory requirements, public companies offer regular transparency regarding revenues and profits. Private companies, however, are considered more difficult to analyze, as they do not have the required disclosures of public companies. Using alternative data to estimate the revenues and profits of private companies may be a lucrative endeavor, not only due to the larger universe of private companies but also due to the potentially less efficient pricing of private companies. While public companies are priced in a transparent and competitive market, owners of private companies may not readily know the valuation of their company. Private equity investors may be able to invest in a private company at a lower valuation multiple than a similar public company, especially if that private company sells all or part of its equity without a competitive bidding process.

1.4.1 Alternative data vs. Big data

The definition of big data varies by user, but it is clear that data is big if it can't fit on a single desktop computer. Big data may be defined by its volume and its velocity. Investors using tick data, or the second by second pricing of every stock or option in a market, are likely using big data, as are investors who seek to download a substantial portion of the global news or social media flow in real time. However, some users may be able to process alternative data using a structured database on a single computer, especially if the data has already been cleaned and structured by an alternative data vendor. Rather than processing all of the credit card transactions that may predict a single company's revenue, an investor may simply choose to purchase weekly revenue estimates for each stock from a vendor, who has already organized the data by time and ticker.

There are tradeoffs to using alternative data vendors who clean the data and process the signals. Highly processed data may be more expensive and more widely distributed than the unfiltered raw data, which makes it easier to use. Buyers of aggregated data or buy vs. sell signals experience higher data costs, but lower costs to build the team and infrastructure necessary to clean and organize the data and build their own proprietary signals. While it may be expensive to build this team and infrastructure, investors that incur this expense may be able to find more unique trading signals that may maintain their value for longer periods of time. Data that is widely distributed and easier to use may be deployed by a larger number of investors who may more quickly arbitrage away the value those signals provide.

Let's explore some of the categories of alternative data.

1.4.2 Social Media Postings and Natural Language Processing

A key difference between traditional data and alternative data is that traditional data is typically quantitative in nature. Much of the alternative data is available in text form and must be analyzed using methods such as natural language processing (NLP). Analyzing text can be very messy, as the author of each document or online post has different writing skills, writing styles and some documents may be full of misspellings, acronyms, or even emojis. In 2021, analysts reviewing posts on Reddit needed to determine the meaning of emojis such as money bags, rocket ships, diamonds and hands.

Consider both the value and the difficulty of determining public sentiment regarding a specific company by monitoring social media posts. Application programming interfaces (APIs) are available to read massive numbers of postings to social media sites such as Twitter and Facebook. This data can be valuable in determining sentiment of investors regarding specific companies on a very short term basis, such as hours to weeks. However, the volume of this data can be overwhelming and the comments aren't always mapped to companies. The process for sorting this data is to first determine which posts are related to specific companies and then determine whether the content is expressing a positive or negative sentiment relative to each specific company. While some posters on Twitter make ticker tagging easier by including \$TSLA or \$FB in a post, other social media platforms to not have posting conventions that are this easy to follow. Once attributing a specific post to a specific company, the analyst next needs to determine the sentiment of that post. The most simple way to determine sentiment is to count the number of positive and negative words. However, this method destroys the context, so more complex NLP methods seek to use chains of words to retain the meaning of phrases. Other challenges of NLP are that different industries have different vocabularies that may need to be learned in order to effectively analyze the text.

Evaluating standard news sources is another key usage of NLP. News may be easier to work with than social media data, as it may be more likely to be tagged or linked to a specific company and may be written in a more formal way that avoids misspellings, emojis and acronyms.

NLP methodology can also be used to interpret company filings with regulators and analysis of audio transcripts from company earnings conference calls. Public companies have obligations to file information with national securities regulators. These include structured data in the form of income statements and balance sheets and unstructured textual data, such as the management discussion and analysis section of an annual report. CEOs and CFOs of publicly traded companies also frequently meet with investors and equity analysts in regular conference calls to discuss quarterly results.

Algorithms can be used to transcribe the voice call into text as well as to mine the voice recording for clues regarding the emotion of the speakers. A key goal of NLP is to compare current filings and calls with those archived from past quarters, focusing on the change in the language, emotion and sentiment from one year to the next.

Both private companies and public companies may have filing requirements with governmental regulators. Governments in many countries sponsor agencies that compile complaints from customers and employees as well as safety issues with pharmaceuticals or manufactured goods. Analysts using NLP tools may be able to predict when the complaints are nearing the point of public disclosure or a mandatory recall of a firm's products. While these signals may be specific to a limited number of industries, such as banking, travel or health care, being able to predict when a regulator will announce potential sanctions on a company can lead to profits on short sales or avoided losses on long positions.

1.4.3 Online Activity

Besides the voluminous social media feeds, other Internet-based information can also be useful for informing investment decisions. Investors may be able to profit from analyzing trends in the behavior of the users of search engines. Services such as Google trends can show changes in the frequency of terms being searched on the Internet. Sharp increases in search activity for a company or one of their products may be indicative of future revenue growth.

Web crawling is used to search and archive data from specific web sites across the Internet. For example, investors may be able to profit from noticing an increase in job postings and hiring activity at a specific firm, which may point to future announcements of revenue growth. Other goals of web crawling may be to find new web pages either through a domain name system (DNS) registration service or for web pages that have been posted by companies that aren't yet fully public. Recently, investors purchased stock in Affirm after noticing a new web page with both Amazon and Affirm in the URL. Having this information just hours or days before an announcement of a partnership between Amazon and Affirm allowed investors to profit nearly 50% on a very short-term trade.

Alternative data may also be used in due diligence for investment managers, such as when a pension fund wishes to invest in a hedge fund. By reviewing information on Revelio Labs, an analyst may note the number of employees at a specific fund that have skills and experience in AI and machine learning. In recent years, hedge funds employing a significant number of skilled quantitative analysts have outperformed the hedge fund industry by an average of 2% to 5%.

1.4.4   Consumer Activity

For companies engaged in business-to-consumer (B2C) transactions, there are many interesting ways to predict revenue. Some alternative data providers access geolocation data from satellites, drones or smartphones. By estimating the foot traffic at retail outlets, investors hope to be able to estimate revenue growth at specific locations where a company's goods are sold.

On a broader scale, consumer spending can also be estimated through emails and credit card transactions. While some credit card processing companies sell aggregate transactions data to estimate GDP or income growth in a certain geography, others allow access to the revenues at specific companies or locations. When reading emails that are sold by email providers, investors are searching for invoices that can be used to estimate the revenue of specific companies.

The online advertising industry makes extensive use of cookies, which are files stored on the computer of each user that document the web sites visited by each user. Online advertisers tend to pay lower prices for general advertisements of small purchases, such as soda manufacturers or specific television shows, but will pay much higher prices when the cookies predict a large purchase is imminent, such as a new car or a piece of business equipment.

1.4.5   Weather and Satellite Data

Some investments may have profits that are correlated to changes in the weather. For example, ski resorts may have lower revenue and profits during times of lower snowfall and higher temperatures, while beach resorts have the opposite problem of greater precipitation and lower temperatures. Those who invest in commodity futures may use satellite photos to estimate crop yields, increasing investments in corn or wheat when photos show reduced crop yields through drought or flood, while reducing investments when it is clear that crop production exceeds market estimates. Farmers can increase their productivity through algorithmically guided tractors that use satellite imagery to optimize the amount of water, seed and fertilizer applied to each 0.1 hectares of land.

1.4.6   Government Statistics

Other sources of alternative data include reports and statistics that have been filed with local and national regulatory agencies. These datasets tend to be specific relative to a given industry. For example, analysts of automobile manufacturers may purchase data from the registration of newly sold vehicles, while investors in home building stocks may be interested in the number of building permits issued for new homes.

1.4.7 Data Ethics

Note that these last two areas of online activity and consumer activity may bring legal risk to the vendor or the user of alternative data. In the case of web scraping, users and vendors of alternative data must understand the policies of the web sites from which the data is being sourced. If the web site specifically prevents harvesting of its data, the user or the vendor may have some legal liability. Data from emails, credit cards and geolocation must carefully deal with personally identifiable information (PII). It is the responsibility of the data vendor to make sure that the data has been anonymized before being sold to the user. Users of alternative data should refuse to accept any dataset where it is clear that the data is non anonymous or the data can somehow be recreated.

1.5 GROWTH OF ALTERNATIVE DATA

As discussed in Section 1.2, in the recent past, new forms and sources of data have emerged in abundance. Companies are using such data to make financial decisions, analyse market behavior, forecast cashflow, identify trends and gain competitive advantage. In comparison with the traditional data, alternative data is generated from broader set of novel sources and provides unique insights into investment opportunities. Demand and usage of alternative data are increasing exponentially. Due to the economy of scale, alternative data becomes more accessible for end users. This chapter provides an overview of growth history and potential of alternative data market, growth drivers and their positive impact.

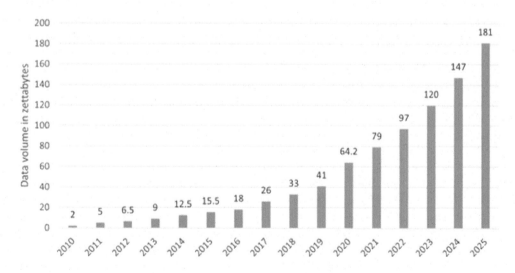

Figure 1.1   Volume of data created and replicated worldwide *(source: IDC)*

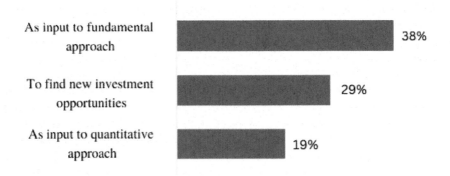

Figure 1.2 Private Equity uses of Alternative Data

In financial market use of alternative data has grown substantially. Applications range from predictive analysis of stock market direction, frequently combined with sentiment analysis, to identification and prevention of financial fraudulent activities. These multiple applications have increased dependence on alternative data. According to survey by Oxylabs and Censuswide, "63% of respondents have started using alternative data to improve their decision making". According to Statista, buy-side companies spent 1.088 billion US dollars on alternative data in 2019. In 2020, 1.708 billion US dollar has been spent. The survey report generated by Global Alternative Data Market in 2020, "the global alternative data market size will grow at a 44% compound annual growth rate (CAGR). By 2026, it will have reached $11.1 billion." According to the report by Grand View Research, in 2020 Alternative Data Market was expected to grow at CAGR of 58.5%, reaching $69.36 billion by 2028.

According to CISCO, 5.3 billion people will use internet services and 29.5 billion networked devices will exist on this planet by 2022. There will be 3.6 global devices and connections per capita [Cisco, 2021]. This increasing number of network devices and internet users clearly predicts the increase in growth of alternative data volume, impacting financial market. According to TabbForum, 38% of data is used as input to fundamental analysis, 29% of it is utilized to find new investment opportunities and 19% is used as input to quantitative approach.

A report published by F. Norrestad at Statista on September 20, 2021 suggested that approximately 50% of hedge funds managers known as Alternative data market leaders utilized more than seven Alternative datasets globally whereas only 8% rest of the market used at least seven alternative datasets [Norrestad, 2020]. This report highlighted the difference between these two classified groups. It further stated: "Using two or more alternative datasets was the most popular approach across both groups with 85% of market leaders and 77% of the rest of the market doing this".

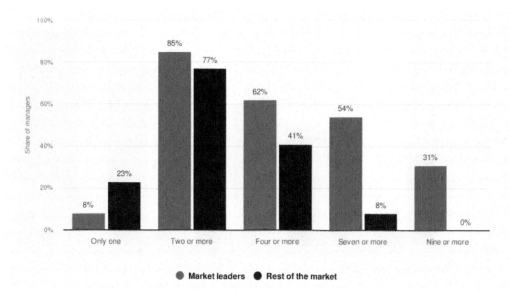

Figure 1.3 Number of alternative datasets used by hedge fund managers worldwide in 2020, by alternative data experience level

Deloitte Center for financial services stated: "Those firms that do not update their investment processes within that time frame could face strategic risks and might very well be outmaneuvered by competitors that effectively incorporate alternative data into their securities valuation and trading signal processes". In today's era, most of the firms rely on Alternative data besides traditional datasets to make their financial decisions more effective.

ALTERNATIVE DATA GROWTH DRIVERS

Increased Number of Smart Cities

Alternative market is expected to grow because of rapid development of smart cities across the globe. Many developed countries are establishing smart cities to manage a city in a systematic manner via data acquisition and analysis. Various alternative data governance schemes are adopted to make the digital life more successful. Smart cities projects are more focusing on empowering cities and developing technologically advanced systems build in this datafied world [Mercille, 2021]. Alternative data market is expected to grow in investment in smart cities. According to Statista, Technology spending on smart city initiatives worldwide will reach 189.5 billion US dollars in 2023.

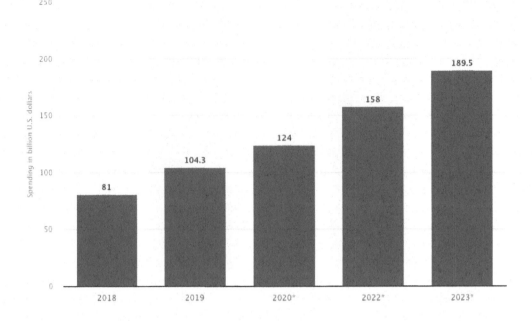

Figure 1.4 Technology spending on smart city initiatives worldwide from 2018 to 2023 (in billion U.S. dollars)

Increase in Internet Penetration

Internet technology and digitalization is increasing day by day; hence, companies are investing more and more in alternative data. The rapid growth of internet penetration paved the path to numerous applications and increases the data acquired from its users. This will ultimately boost the demands of alternative data in the near future.

Increasing Adoption of 5G Networks

5G internet network is one of the hot topics these days, and people are demanding faster internet services. This increasing adoption of 5G network will serve as a driving force for growth of alternative data demands. Better network technologies will accelerate the adoption of alternative data by making high speed data communication and efficient networking. 5G internet technology offers increased bandwidth which will allow low latency rate and more data transmission in a short period of time. According to Statista annual report and prediction, 5G subscription will reach 3 billion by 2025.

Favorable Government Initiatives

Alternative data market growth is significantly impacted by government initiatives to promote technological advancements in various countries across the

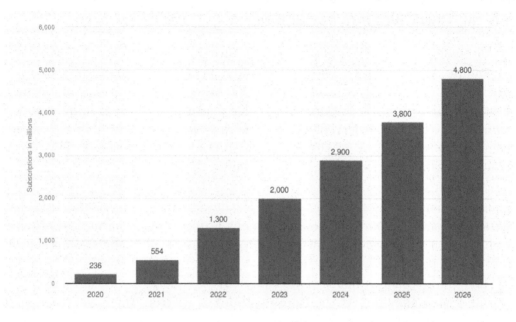

Figure 1.5 Forecast number of mobile 5G subscriptions worldwide from 2019 to 2026

globe. Such projects operate with collaboration with tech firms that focus on use of internet, data security, protection of personal data, risk aversion of data privacy violation and advancement in data sciences. Such initiatives are expected to prove as beneficial impact on alternative data market growth.

Growing Interest in Stock Trading

Alternative data demand is increasing on a greater pace due to rising interest in stock trading. Nowadays, numerous stock trading companies are adopting the use of alternative data approaches for better prediction and outcomes. These novel interests have positive impact on non-traditional and unconventional data acquisition and strategies. These strategies include various floating rate bond funds and bottom-up corporate bonds.

In Conclusion, alternative data is one of the growing domains for research. The use of alternative data in addition to traditional data is used increasingly to make decisions; examples are business expansion, product launch, investment and so on. The major drivers of alternative data include increased number of smart cities, growing interest in stock trading, favorable government initiatives, increased adoption of 5G networks and rising trend of internet penetration. In contrast to conventional data, alternative data is created by means of different sources and gives exceptional knowledge of market trends, customer demands and financial information from different perspectives. More and more companies are using alternative data to make better financial decisions.

1.6 ALTERNATIVE DATA PROVIDERS

As discussed, alternative data is various, unstructured and is extremely large in volume. Despite big potential values that people can acquire by exploiting alternative data, using it is impossible for most individual users or business users who allocate little resource for it. The reason is that the volume, velocity, value, veracity and variety of alternative data make it unavailable to model using, so it needs a lot of effort to be collected and structured. This is the place where alternative data vendors come in to play.

Alternative data vendors or providers are those who solve the above problems and monetize from it. They create the data from their operation or collect data from several sources, then implement a bunch of preparatory organization work to transform the raw data into a form that is ready to be analyzed and pulled insights from.

Like many other industries, alternative data vendors range significantly in size and what they do. At the time this book is written, there are three main kinds of alternative data providers in the market. Firstly, they can be well-known existing market data companies such as Bloomberg. They sell their own alternative datasets and simultaneously create data markets that connect their customers to third-party data vendors. The second and also the most common kind of alternative data vendor is start-ups. They are usually established with a core technical team that collects data from many sources, cleans and transforms data into datasets that can be sold directly to customers or transferred to the customers' system via API technology. Some of them even aggregate many datasets together to construct a trading strategy or produce signals and sell them to investing and trading funds. The last kind of alternative data vendor is big companies which are not traditionally associated with this area but possess exhaust data. Data exhaust refers to the data generated as trails or information byproducts created from all digital or online activities, such as storable choices, actions and preferences such as log files, cookies, temporary files and even information that is generated for every process or transaction done digitally. This data can be very revealing about an individual, so it is very valuable to researchers, marketers and business entities. These companies therefore may sell the data directly to data users or cooperate with other alternative data vendors or a consultancy to help them. One example of this kind of data vendor is Mastercard, which sells its customer transaction data.

The data vendors use their expertise in alternative data processing to monetize these datasets, which includes structuring the data, creating data products, marketing and selling data to users and so on.

The world of alternative data is extremely fragmented, so it is hard for one company to provide all kinds of data. Usually, each data vendor specializes in one or some certain types of alternative data. Below we present some renowned data vendors and their specialties in the market.

Bloomberg is the global leader in business and financial data, news and insight [Bloomberg, 2022]. Using the power of technology, they connect the world's decision makers to accurate information on the financial markets—and help them make faster, smarter decisions.

RavenPack is one of the leading data analytics providers in financial services [RavenPack, 2022]. They help organizations extract value and insights from large amounts of information such as internal documents, emails, instant messages, support tickets and other textual data. Using proprietary NLP technology, RavenPack solves the problem of unstructured and hard-to-process textual data and helps users to extract valuable insights quickly and easily by transforming their unstructured text into structured data. Their process is astoundingly quick and easy.

Yewno deals with creating intelligence data that deals with a knowledge graph [Yewno, 2022]. It is currently being applied across financial services, education and government to help deliver products that make better business decisions. Yewno primarily offers company, research and pricing data.

Tickdata's core product is clean, research—ready, global historical intraday data [TickData, 2022]. Their offering includes institutional-grade quote and trade history from the world's top financial markets, from the Americas to Europe to Africa to Asia to Australia. They cover Equities, Futures, Options, Forex and Cash Indices.

Stocktwits is a large social network for investors and traders, where millions of investors and traders are saying in real time about the stocks, crypto, futures and forex market [StockTwits, 2008]. From this activity, they offer huge tweet databases that can be transformed into sentiment score of these assets, which are then used to create trading rules, perform simulation or back testing, create neural network prediction models and so on.

MarketPsych produces the global standard in financial sentiment and ESG data derived from thousands of news and social media outlets [MarketPsych, 2022]. They develop alpha generation and risk management models based on behavioral principles and also integrate personality testing and practice management tools to help advisors better serve their clients.

Refinitiv is an American-British global provider of financial market data and infrastructure [Refinitiv, 2018]. They provide financial software and risk solutions—delivering news, information and analytics, enabling transactions and connecting the global users. Refinitiv has comprehensive and trusted data covering investment banking, funds, bonds, earnings, macroeconomic indicators, FX, commodities and more, gathering from historical to real-time insights. Their data is delivered via a portfolio of market data products and services, so all users—from traders and investors to wealth and asset managers, as well as risk, compliance, strategy and advisory managers—get the content they need, in their required format.

Alternative data vendors often offer the products as aggregated datasets or as a straight data feed, through APIs. Aggregated data, the less expensive option, is structured, and, therefore, is easier to work with and slot directly into an investment model. But these advantages also make these datasets more widespread to the market, which at the same time is the culprit of their less alpha potential. When there are more people make decision of long or short assets in the market based on the same information, the assets' value will quickly reflex that information, so the alpha added by the information is diminished more quickly.

According to Gene Ekster, CEO of Alternative Data Group and an alternative-data professor at New York University [Gossett, 2021], the aggregated datasets usually lack depth since users will lose the "ability to really dig and mine the data in unique ways". They may also be not truly representative, so users could be swayed by selection bias. Most of the data providers' techniques and methodologies are black-box systems, which are also unavailable for customer inspection, thereby exacerbating aggregate errors due to lack of transparency.

Several years ago, some sportswear retailers inserted an asterisk between the two Lus in their reports: Lu * lulemon, instead of Lululemon. The analyzers didn't have Lu*lu keyword, which made it seem like sales volume had dropped significantly [Gossett, 2021]. This aggregating conclusion led to some short bets, which incurred calamitous loss because Lululemon indeed reported a great quarter. If these people had the raw data, they should have seen the fact and not traded against that.

For these reasons, a raw data feed is considered much more beneficial than aggregated one. But a purely unaltered dataset, with no transformation applied, is essentially just data exhaust. "Any hopes it would provide value would have to be weighed against the considerably heavy clean-up lift" [Gossett, 2021].

The best solution is a live API feed with automatic conversion and structure as much as possible. Under this approach, entity mapping and ticker tagging is a big challenge. This means we have to assign a company reference or brand alias back to its unique stock symbol and correct name. For example, "Facebook" needs to map back to FB and Meta Platforms Inc; "Amazon" needs to map back to AMZN and Amazon.com Inc. And not all references are so direct. Maybe a Twitter user sarcastically references Facebook's or Amazon's slogan while including a typo—"that's powerfull". A hedge fund might want that sentiment included in its investment analysis, but it would need highly sophisticated AI to detect the reference merely. In many cases, it doesn't stop at ticker symbols.

Some alternative data vendors have been focusing on tackling the tagging and mapping challenge. They call their product referential data, which contains all different ways if referencing a given entity, company, or security, mapped back in a way that facilitates the data analysis.

The appearance of alternative data vendors has significantly eased the process of applying alternative data of financial practitioners and open a new genre for alternative data in finance. Financial analysts hinge on alternative data to better their predictions on stocks return. Quality predictions can give an investor a more significant edge over the competition and drive higher profits. Fund managers use alternative data to create signals about the market and entities they are investing on, which help them to enhance investment profit. Hedge funds and investment banks also include alternative data in their risk management process. Some assign ESG values for their assets based on alternative data, from which they can understand fundamental values and assess related risks more precisely. Some use alternative data to produce signals that warn them of potential collapse situations.

Alternative data vendors need to explore and build datasets that are valuable to and recognized among their target customers, so that their data can be monetized.

However, there are differences in the quality and price of data. Some data intermediaries even go for a quantity-over-quality approach while aggregating datasets with high ticker coverage but not necessarily insightful ticker coverage. Therefore, not every data is valuable to users. One might put a lot of effort and resources into exploiting and analyzing a dataset but gain no value from it, or the profit it creates cannot cover the huge cost of utilizing it. Any ability to diminish the turnaround time between acquisition and analysis is valuable.

People start using alternative data may keep questioning how they know if a dataset is going to be valuable. It could take six months of R&D, and they have to buy it first. They will not know how much additional profit the data is going to create until much later.

Neuravest is one of the companies focused on cracking that enigma and may be considered an intermediary of the intermediaries. They associate with 42 other alternative-data vendors and work to validate datasets before integrating them into machine-learning investment models.

Raw data is fed into the system to generate a data qualification report as what the company calls it. In this system, the data is measured along 12 checkpoints before being reasonably qualified to be incorporated into a model. These 12 checkpoints include a time indicator which measure the time length before a signal drops value, a price action distribution following a given event, such as a news announcement that stimulates social-media conversation.

After validation, the data is cleaned, tagged and normalized before building the model to create testable investment arguments. By aggregating uncorrelated data sets, the models aim to identify constituent stocks and assets that are about to experience unusual volatility relative to similar stocks.

With the growing market size for alternative data, more and more players are entering, providing alternative data for investors and hedge fund managers. As we are living in an extremely competitive buying and selling environment, it is important for everyone to capture whatever data they can get their hands on to gain a competitive advantage.

1.7   CHALLENGES OF USING ALTERNATIVE DATA

Users of traditional, simpler and structured data usually end up building models with correlated signals and investing in trades with eroding alpha. The more users of a dataset and the more structured the datasets are, the harder it is to get a return advantage because so many other investors have access to the same data. A crowded trade also has risk management implications on exit. The quest for differentiated signals and a longer-lasting return advantage is making contemporary investors gravitate towards alternative data. The potential rewards of deploying alternative data are often accompanied by challenges, many of which can be significant.

The sheer volume and richness of alternative data mean that investment managers have a likely opportunity to come up with unique signals and potentially more sustainable return advantages. However, the challenges of using alternative data are also directly related to the volume and complexity of the data. Noise and missing values also stymie their facile, instant use. It takes substantial effort to clean the data, map the data to specific securities and test the data within a firm or portfolio's existing frameworks (sets of data and models). Unfortunately, not every dataset will be additive to the models currently employed; so, factor in wasted effort to clean, load, compare and test datasets that may never be deployed. Also, users of alternative data will need to perform a cost-benefit analysis before employing the data. The costs are not only to purchase the data, but also to build a team to test, clean and load the data. The benefits are not the performance of the data on a stand-alone basis, but the value when added to an investor's existing set of frameworks. As more models, factors, and data are included in a trading system, the more challenging it is to find the next framework that is value-additive. Generating innovative insights by mashing up multiple datasets is one of the key challenges for any alt data expert, yet the results can add tremendous amount of value to their investment debate.

Some examples come to mind. On a Monday in late April 2019, a boutique research firm—using publicly available corporate aircraft and flight data—reported that a Gulfstream V jet belonging to Occidental Petroleum had been sighted the day before in Omaha, Nebraska, the hometown of famed investor Warren Buffett. A day later, Buffett's Berkshire Hathaway announced its intention to invest $10 billion in a preferred stake in Houston-based Occidental, a move that ultimately proved critical in helping Occidental trump Chevron in a takeover battle for Anadarko Petroleum. More immediately, however, Berkshire's official announcement of its involvement amplified price movements and trading volume in Occidental and Chevron shares relative to the prior day—when information about the jet data was known to just a handful of eagle-eyed investment sleuths.

Aircraft and flight logs are examples of "alternative data"—non-reportable information that some investors increasingly are embracing to gain an edge. Other examples include private opinion polls before a key political referendum; satellite and drone imagery of ports and mall parking lots; credit-card transactions; job postings; patent and trademark filings; and social-media sentiment, including ratings and reviews of products and services. Unlike financial and other data reportable to market regulators, which is readily available and widely disseminated, alternative data is often proprietary, affording its collectors or buyers information that other investors do not have.

Yet, the leap from today's alternative data to a future investment consensus is neither immediate nor guaranteed. Data vintage, authority and provenance matter. Any changes in the data generation or collection process also can cause problems for those relying on the information to make investment decisions.

Just ask Bloomberg, which constructed and published a Tesla Model 3 production volume estimator, based on Vehicle Identification Numbers (VINs) from official U.S. government sources, social media reports and input from owners of Tesla's electric cars. The Model 3 Tracker launched in February 2018. For most of 2018, as the market focused on Tesla's production ramp-up for the Model 3, Bloomberg's alternative-data-based estimate proved a good predictor of the company's quarterly output. Model 3 production numbers were a key driver of Tesla's stock price in late 2018 as well as in 2019's first quarter.

In early 2019, however, a rise in Tesla exports led to changes in how many extra VINs the company registered pre-production to keep on hand a buffer of the numbers. These data—generation and collection—changes caused the Bloomberg tracker to veer off course, overestimating production numbers for the first quarter of 2019 by 26%. Tesla stock sold off early April of that year, when the actual production and delivery numbers were announced, catching many investors relying on the Bloomberg tracker off guard.

To cite another example, the 2016 U.S. Presidential Election was a significant black eye for election forecasters. Relying on alternative data, generated via opinion polls, almost all the pundits had predicted Hillary Clinton to be the victor. Similarly, the Brexit vote during June of 2016 and the election of Scott Morrison as the Australian Prime Minister were additional instances where alternative data-informed methods led analysts astray. Investors across many asset classes—ranging from currencies to equities—were stunned by the volatility resulting from the surprise election results. Many firms and pundits have recently talked about backing off from election forecasting.

New strategic surprises also expose new data streams. As the recent pandemic unleashed its fury, new sources of data (e.g., the number of people going through the airport TSA checkpoints daily in the US) emerged. Investors started to focus on the robustness of the reopening across different sectors of the economy and were looking for multiple, new clues.

AI has recently emerged as a premium refiner of alternative data, especially to deal with surprises arising from missing, incomplete and uncertain values; and, to normalize for transient effects, such as traffic-flow numbers affected by inclement weather, health warnings or construction.

However, maximizing the utility of alternative data likely requires a hybrid analysis, combining humans and machines. Computers are able to process large amounts of data rapidly, while humans are able to apply judgment and empathy in ambiguous situations, often guided by past experiences, some from unrelated domains. Just as a pilot must interpret conflicting readings or flight-deck alerts in the cockpit, an analyst should use alternative data in concert with other information to make better choices or mitigate risk. Fund managers making significant use of AI and alternative data have shown the potential to increase returns relative to managers who are not yet making use of these tools. They have done so, by typically using more unique and often unstructured datasets, and by employing AI to scale human expertise. Prudent use of AI can help overcome human fatigue and short attention spans, helping an investor build a lasting advantage using carefully curated alternative data.

There are a few significant caveats to keep in mind while employing alternative data. The first is that the data needs to imply a relatively direct connection to asset prices—a potential takeover usually implies a higher price for the company being acquired, an uptick in implied product demand points to increasing revenues for the manufacturing company in the near term; and, a candidate's improved election forecast may point to imminent implementation of previously telegraphed policy changes. The second caveat is that often there is a significant delay between what is hinted by the data and a majority of the market participants acting on or realizing it. Consensus view or stock sentiment may not change overnight, and an asset can continue to remain mispriced. If the adjustment to pricing of a security takes a long time to materialize, it erodes the internal rate of return (IRR), turning a potentially good investment decision into a mediocre one.

Finally, as previously stated, there is also the challenge of dealing with the data in an ethical manner. Ethical harvesting of data is performed when all of the data is sourced legally from sources that do not prohibit the use of the data. Users of alternative data are expected to comply with Global Data Protection Regulation (GDPR) and other rules regarding data usage and storage. It is imperative that users of alternative data work with completely anonymized data and work with vendors who are compliant with GDPR and other regulations; and are highly respectful of protecting PIII.

Caveat (alternative data) emptor!

1.8   SUMMARY DISCUSSION AND CONCLUSION

A few decades ago, a financial modeler and an analyst depended mainly on small number of traditional data sources to analyze and make decisions. In recent times, this fast-evolving field of alternative data as used in financial analytics is emerging as a major disruptor of traditional applications of financial modelling. Thus all instances such as decision-making in asset allocation, predicting the market, risk management and credit scoring are one way or another affected. In this Handbook we have focused mainly on the above areas of financial applications. We have highlighted that many financial (expert) practitioners have already used alternative data to create trading strategies, investment strategies and risk management applications. We believe this field will grow rapidly and contents of this Handbook provides the readers an early entry into this exciting field.

REFERENCES

Aldridge, I. (2015). Trends: all finance will soon be big data finance.

Barigozzi, M. and Hallin, M. (2017). A network analysis of the volatility of high dimensional financial series. *Journal of the Royal Statistical Society: Series C (Applied Statistics)*, 66(3):581–605.

Berg, T., Burg, V., Gombović, A., and Puri, M. (2020). On the rise of fintechs: Credit scoring using digital footprints. *The Review of Financial Studies*, 33(7):2845–2897.

Bertolotti, A. (2020). Effectively managing risks in an esg portfolio. *Journal of Risk Management in Financial Institutions*, 13(3):202–211.

Bloomberg (2022). Bloomberg. https://www.bloomberg.com/. Accessed: 2022-06-23.

Brown, L. D. (1997). Analyst forecasting errors: Additional evidence. *Financial Analysts Journal*, 53(6):81–88.

Carhart, M. M. (1997). On persistence in mutual fund performance. *The Journal of finance*, 52(1):57–82.

Cheney, J. S. (2008). Alternative data and its use in credit scoring thin-and no-file consumers. *FRB of Philadelphia-Payment Cards Center Discussion Paper* (08-01).

Cisco, U. (2021). Cisco annual internet report (2018–2023) white paper. 2020. *Acessado em*, 10(01).

Cong, L. W., Li, B., and Zhang, Q. T. (2021). Alternative data in fintech and business intelligence. In *The Palgrave Handbook of FinTech and Blockchain*, pages 217–242. Springer.

De Spiegeleer, J., Höcht, S., Jakubowski, D., Reyners, S., and Schoutens, W. (2021). Esg: a new dimension in portfolio allocation. *Journal of Sustainable Finance & Investment*, pages 1–41.

Djeundje, V. B., Crook, J., Calabrese, R., and Hamid, M. (2021). Enhancing credit scoring with alternative data. *Expert Systems with Applications*, 163.

Donoho, D. L. et al. (2000). High-dimensional data analysis: The curses and blessings of dimensionality. *AMS Math Challenges Lecture*, 1(2000).

Dourish, P. and Gomez Cruz, E. (2018). Datafication and data fiction: narrating data and narrating with data. *Big Data & Society*, 5(2).

Engle, R. F. (2000). The econometrics of ultra-high-frequency data. *Econometrica*, 68(1):1–22.

Fama, E. F. and French, K. R. (1992). The cross-section of expected stock returns. *The Journal of Finance*, 47(2):427–465.

Fan, J., Lv, J., and Qi, L. (2011). Sparse high dimensional models in economics. *Annual Review of Economics*, 3:291.

Fang, B. and Zhang, P. (2016). Big data in finance. In *Big Data Concepts, Theories, and Applications*, pages 391–412. Springer.

Goldfarb, A. and Tucker, C. (2019). Digital economics. *Journal of Economic Literature*, 57(1):3–43.

Goldstein, I., Spatt, C. S., and Ye, M. (2021). Big data in finance. *The Review of Financial Studies*, 34(7):3213–3225.

Gossett, S. (2021). What is alternative data and why is it changing finance? *Built In.*

Henriksson, R., Livnat, J., Pfeifer, P., and Stumpp, M. (2019). Integrating esg in portfolio construction. *The Journal of Portfolio Management*, 45(4):67–81.

In, S. Y., Rook, D., and Monk, A. (2019). Integrating alternative data (also known as esg data) in investment decision making. *Global Economic Review*, 48(3):237–260.

Jame, R., Johnston, R., Markov, S., and Wolfe, M. C. (2016). The value of crowdsourced earnings forecasts. *Journal of Accounting Research*, 54(4):1077–1110.

Jegadeesh, N. and Titman, S. (1993). Returns to buying winners and selling losers: implications for stock market efficiency. *The Journal of Finance*, 48(1):65–91.

Kolanovic, M. and Krishnamachari, R. T. (2017). Big data and AI strategies: machine learning and alternative data approach to investing. *JP Morgan Global Quantitative & Derivatives Strategy Report.*

Lewis, M. (2014). The wolf hunters of wall street: an adaptation from'flash boys: A wall street revolt,'by michael lewis. *The New York Times Magazine March*, 3.

MarketPsych (2022). Marketpsych: financial markets sentiment and esg data. https://www.marketpsych.com/. Accessed: 2022-06-23.

Mercille, J. (2021). Inclusive smart cities: beyond voluntary corporate data sharing. *Sustainability*, 13(15):8135.

Mitra, G., Baigurmanova, D., and Jhunjhunwala, S. (2020). The new landscape of financial markets: Participants and asset classes. *Available at SSRN 4104690.*

Monk, A., Prins, M., and Rook, D. (2019). Rethinking alternative data in institutional investment. *The Journal of Financial Data Science*, 1(1):14–31.

Norrestad, F. (2020). Alternative data: Hedge fund manager use globally 2020. https://www.statista.com/statistics/1169968/alternative-data-hedge-fund-managers-global/.

O'Leary, D. E. (2013). Big data', the 'internet of things'and the 'internet of signs. *Intelligent Systems in Accounting, Finance and Management*, 20(1):53–65.

Quinlan, B., Kwan, Y., and Cheng, H. (2017). Alternative alpha: unlocking hidden value in the everyday.

RavenPack (2022). Ravenpack: technology and insights for data-driven companies. https://www.ravenpack.com/. Accessed: 2022-06-23.

Refinitiv (2018). Refinitiv financial solutions – financial technology experts. https://www.refinitiv.com/. Accessed: 2022-06-23.

Sloan, R. G. (1996). Do stock prices fully reflect information in accruals and cash flows about future earnings? *Accounting Review*, pages 289–315.

Stock, J. H. and Watson, M. (2009). Forecasting in dynamic factor models subject to structural instability. *The Methodology and Practice of Econometrics. A Festschrift in Honour of David F. Hendry*, 173:205.

StockTwits (2008). Stocktwits – the largest community for investors and traders. https://stocktwits.com/. Accessed: 2022-06-23.

Syed, A., Gillela, K., and Venugopal, C. (2013). The future revolution on big data. *Future*, 2(6):2446–2451.

The Organization for Economic Cooperation and Development (2013). *Exploring Data-Driven Innovation as a New Source of Growth: Mapping the Policy Issues Raised by "Big Data"*. OECD Digital Economy Papers.

TickData (2022). Tick data. https://www.tickdata.com/. Accessed: 2022-06-23.

Wang, D., Liu, X., and Chen, R. (2019). Factor models for matrix-valued high-dimensional time series. *Journal of Econometrics*, 208(1):231–248.

Wang, P. (2008). *Financial Econometrics*. Routledge.

Xiao, C. and Chen, W. (2018). Trading the Twitter sentiment with reinforcement learning. *arXiv preprint arXiv:1801.02243*.

Yewno (2022). Yewno – transforming information into knowledge. https://www.yewno.com/. Accessed: 2022-06-23.

$$(\mathrm{I})$$

ALTERNATIVE DATA: PROCESSING AND IMPACT

Contemplation and Reflection on Using Alternative Data for Trading and Fund Management

David Jessop

Columbia Threadneedle Investments

CONTENTS

2.1 INTRODUCTION

In this chapter, we consider how we use alternative data in the investment process, asking the question of how we go about implementing this usage—the journey from the initial investment question to a reliable, regularly updated signal. Perhaps mixing our metaphors, this journey involves at least three sets of players who, as with the fairies, lovers and mechanicals in Shakespeare's A Midsummer Night's Dream, tend to interact little, if at all—and when they do chaos and misunderstanding often ensues. The players in our drama include investment professionals, data scientists (or quants), and various parts of a firm's IT department. Ideally, they will all act together as one team; but this requires effort and understanding.

DOI: 10.1201/9781003293644-2

We will use the phrase "data scientist" to describe someone working with fundamental PMs and "quant" as someone who is investing using a quantitative / systematic process. There is a significant overlap in both the skills and experience needed for both.

Why do we want to use alternative data? There are at least two separate contrasting approaches practised by finance analysts these we label as (i) quantitative(quant) analysis and (ii) fundamental analysis. The first uses the data directly in the investment process as a source of return: the second would be using the data as another source of information to inform our (more fundamentally based) investment decisions. A top-down example of the second could be "nowcasting" the state of the economy (or more recently return-to-the-office) - is the economy growing today faster or slower than the market thinks?

What is nowcasting?

[Banbura et al., 2013] define now-casting as "the prediction of the present, the very near future and the very recent past. The term is a contraction for now and forecasting ...". They go on to say "now-casting is relevant in economics because key statistics on the present state of the economy are available with a significant delay. This is particularly true for those collected on a quarterly basis, with Gross Domestic Product (GDP) being a prominent example."

The basic idea of nowcasting is to use data which is published either earlier or at a higher frequency than the official numbers in order to obtain an "early estimate" of the measure in question. For example, in nowcasting US GDP we could use the weekly non-farm payroll number. The New York Federal Reserve bank used to publish a regular nowcast based on 34 macroeconomic variables (they suspended it in September 2021 due to "the uncertainty around the pandemic and the consequent volatility in the data have posed a number of challenges to the Nowcast model"). Authors such as [Babii et al., 2022] have added in both financial data and textual data (using a machine learning based model) to improve the estimates. There are a number of papers which now use alternative data as inputs into these types of models.

2.2 WHERE SHOULD OUR STORY BEGIN?

Where should our story begin? The answer is "With a question". This is true for both of our use-cases (although the type of question obviously differs). Writing at the start of 2022, as at least some of the world seems to be coming out of the stresses imposed by the COVID 19 pandemic, an obvious question could be "How many people are returning to work"?, or perhaps the opposite— "How many people have moved to permanently work from home?" Another example could be "I think that companies with happy employees should be more profitable in the future, how can we measure this"? A final example would be "Can we forecast the future carbon intensity of the companies in this portfolio"?

For quantitative investors the obvious question is "Can I make money from this dataset"? but even here we need to consider questions of trading horizon, portfolio construction, how to cope with a short data history. Perhaps the question should be "Does this dataset add a new, orthogonal, source of return to my existing model"?

Why is it important to start with a question? With the question comes a set of beliefs (which can be encoded in a Bayesian framework if necessary) which can guide the data scientist or quant in their choice of research direction and techniques. Too often the author has seen the process begin with "We've got this great new dataset - how can we use it?" but that, at least in this author's opinion, shouldn't be the impetus. The starting point should come from someone with a real-world need.

Let us suppose we have our question—what should we then do? The first thing the data scientist needs to do is pin down exactly what is being measured or forecast. Examples of this could be

- What do we want the output to be?

- At what horizon?

- What beliefs do we have about how the world works?

- How often should we update the model?

- What will it tell us?

- Are we forecasting or just "nowcasting"?

- What will we do with the output?

To give a concrete example we will use the final question above – "Can we forecast the future carbon intensity of the companies in this portfolio"? Questions that need answering include

- Do we want to forecast carbon, or carbon intensity (carbon / sales), or financed carbon (carbon / enterprise value)? And using Scope 1, or 1 & 2, or 1, 2 & 3 emissions? *[What's the output?]*

- Are we forecasting next year or a few years in the future? *[What horizon?]*

- The raw carbon data is only annual and tends to start in 2010 *[How often should we update the model?]*

- Engagement or divestment? *[What will we do with the output?]*

Understanding the source and meaning of the data is important. Scope 3 measures all the emissions the company is indirectly responsible for both up and down its supply chain. This measure has many problems: it isn't known with any measure of accuracy and it tends to double count emissions are but two. Returning to Scope 1 and Scope 2 of emissions we observe that these two measures of a company's carbon output capture very different aspects of a company's processes. Understanding these differences will almost certainly lead to differences in modelling approaches. Scope 1 emissions measures the amount of green-house gases (GHGs) the company makes directly. Scope 2 is primarily driven by the company's electricity usage.

Hence modelling Scope 1 emissions is very much a company specific problem—what is the company doing to improve its efficiency? Here textual analysis of a company's announcements around climate change could be of value.

Scope 2 is different—a company can aim to be more efficient but the method of electricity production (which is unlikely to be under a company's control) is likely to make a bigger difference. For example, in the UK in 1990, 72.0% of electricity generated came from coal; this had fallen to 1.8% by 2020. Similarly, the percentage of UK electricity from low carbon sources rose from 2.7% in 2000 to 43.1% in 2020 [Authority, 2021]. These changes were very much driven by government policy; but they have the effect of reducing UK companies' Scope 2 emissions significantly without any direct action from the companies.

Getting this understanding of the real-world behaviour and drivers of the variables we wish to forecast is important. It leads to better models which can be more readily understood by those who wish to use the outputs.

2.3   WHAT DATA SHOULD WE USE?

The next question we need to contemplate is that of the data sources we choose to use. There are many things that can influence this (budget being one of them) but one basic question is should we buy a "ready-made" dataset, or should we attempt to acquire it ourselves? In our view, this depends on the use for the data. For a return signal then data we acquire ourselves is likely to be more valuable for longer—a commercially available dataset will be analysed by many people and any signal within it is going to be exploited (and hence arbitraged away) quite quickly. For an "additional input" type of analysis then this signal arbitrage aspect is less important.

If we wish to acquire our own dataset there are questions around intellectual property—for example, does the website from which you scrape the data allow this usage? This could be the topic of a whole chapter in itself—we will just highlight it here.

2.4   WHAT MODELS SHOULD WE BUILD?

We have our question; we have our aims, and we have our data: we need to start the process of building our model. There is an instinctive desire today just to throw data into some form of machine learning model, obtain the output and then present the results. But what are we trying to measure? If we consider the "return-to-office" style of question then there is no history against which we can compare, so any form of supervised learning isn't available to us.

The top-level choice here can be expressed as the choice between a structural model, i.e., one where we impose our belief of the relationship between the variables (a linear regression being the most commonly used of these) and a machine learning model where we allow the ML algorithm find the form of the relationship.

In the first case, we claim to know the form of f in $y = f(x)$ and are trying to directly find the parameters of f which minimise some error term. If we use machine learning, then we don't care about the functional form of f - we just care about the output.

If we choose to use a structural model, then how do we decide upon the form of f? Ideally, we would approach this from a theoretical point of view—we have a mental model (explicit or implicit) of how the world works. A simple example: what is the relationship of asset returns to inflation?

If we were to approach this theoretically, we might start from a discounted cash flow model. For fixed income assets the numerator of our model—the coupons from the bond—are fixed and if inflation rises it seems likely the discount rate would rise, so the price of a bond would be expected to fall. For equities the situation is less obvious—earnings, and hence dividends, are nominal numbers so could be expected to rise in the future with higher inflation, but the discount rate would also rise. Which of these effects is going to overweigh the other is not obvious.

Plotting the two time series against each other gives the chart shown below.

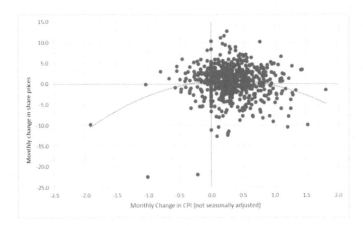

Figure 2.1   Stock returns against Inflation
Source: St Louis Federal Reserve Economic Data (https://fred.stlouisfed.org).
The chart shows the monthly percentage change in the US CPI All Items
against the monthly percentage change in Total Share Prices for All Shares
for the United States. Both series are not seasonally adjusted.

From the picture above, we see that there appears to be a sweet spot.[1] If
inflation is too high or negative then stocks do worse. We will return to this
topic below but this leads to a subtle problem which is that of multiple testing.
Given we have looked at the data we know (roughly) what to expect and hence
when we find the parameters are significant would not be surprising.

Be careful in your choice of error function

The choice of error function is often not considered. In some cases, for
example linear regression, it is just a part of the algorithm. However, this
acceptance of the default option can lead to problems.

[Patton, 2011] discusses the situation if the variable you are trying to
forecast is latent – the example in the paper is volatility. Other such latent
variables would include economic growth. We cannot observe volatility –
all we can do is create an estimate of volatility from the prices we observe.
Patton points out we have to be very careful with our choice of error
function otherwise this "can lead to incorrect inferences and the selection
of inferior forecasts over better forecasts."

---

[1]The maximum of the fitted line occurs at a monthly inflation rate of 0.26% - approximately 3% per annum. The analysis here is simply suggestive – the inflation and stock
returns are contemporaneous; the CPI could have been revised.

Another example of this came when the author was trying to build a model to forecast carbon intensity of companies. The obvious choice of error function was the percentage error:

$$\frac{forecast - actual}{actual} \tag{2.1}$$

as the model was being built on a pooled dataset. Although the model was in general relatively successful there were some outliers with very large percentage errors. The answer was obvious – we want companies to reduce their carbon emissions to zero. If a company is close to achieving this aim the denominator of our fraction approaches zero and the error tends to infinity.

One particular problem with ML models, however, is that they are not at all transparent—you can't explain why they produce the output they do. This lack of insight into the relationship between the inputs and the outputs can be a significant drawback in explaining a model to a fund manager. They want to see if their intuitive beliefs are true. There has been an enormous effort to overcome this issue, and there are now some good approaches that can be used to gain at least some insight into what's going on within the model. See for example [Lundberg and Lee, 2017] or [Ribeiro et al., 2016].

There are many other problems in building ML models. One is that our intuition fails in higher dimensions. As [Domingos, 2012] says "... our intuitions, which come from a three-dimensional world, often do not apply in high-dimensional ones. In high dimensions, most of the mass of a multivariate Gaussian distribution is not near the mean, but in an increasingly distant 'shell' around it; and most of the volume of a high dimensional orange is in the skin, not the pulp. If a constant number of examples is distributed uniformly in a high-dimensional hypercube, beyond some dimensionality most examples are closer to a face of the hypercube than to their nearest neighbor. And if we approximate a hypersphere by inscribing it in a hypercube, in high dimensions almost all the volume of the hypercube is outside the hypersphere. This is bad news for machine learning, where shapes of one type are often approximated by shapes of another."

Another thing that could drive your decision is the amount of data available. Most companies have only reported their carbon emissions for the past 10 years (at best). If we have to set aside at least one value for training and perhaps one for testing, then we have potentially reduced our dataset by 25%. Hence, perhaps a structural model where we define the relationship between carbon and time would have some benefits. We could also go ahead and fit the model using a Bayesian approach which would have the advantage of quantifying the uncertainty around our forecasts.

2.4.1 The problems of multiple testing

The issue of multiple testing is one that has been known about for statisticians for many years, but it has only relatively recently been acknowledged within the field of finance (for example see [Harvey, 2017] and many of Marcos Lopez de Prado's papers).

What do we mean by multiple testing? At a simple definition it is the problem of looking at the dataset more than once—the second time we run analysis on a dataset we know something: the first approach we took didn't work. As [Romano et al., 2010] say "this scenario is quite common in much of empirical research in economics. Some examples include: (i) one fits a multiple regression model and wishes to decide which coefficients are different from zero; (ii) one compares several forecasting strategies to a benchmark and wishes to decide which strategies are outperforming the benchmark; (iii) one evaluates a program with respect to multiple outcomes and wishes to decide for which outcomes the program yields significant effects."

One example of the multiple testing problem is fitting parameters to a trading strategy. Let's suppose we want to use a moving average cross over strategy for trading a few commodity futures. The obvious question is what two look back periods should we use for our moving averages? Given there is no theoretical underpinning to this strategy then we have no prior view as to what to use, so we use a simple search strategy, running a back test of our trading strategy a number of times with different values of the look back periods for the long- and short-term averages. There is no reason to believe that the in-sample "optimal" pair will work out of sample.

[Bailey et al., 2014] show that if we sample performance statistics S_k about a trading strategy which follow a Gaussian distribution with mean μ and standard deviation σ, then the expected value of the maximum of these statistics is

$$E(\max[S_k]) \approx \mu + \sigma\left((1-\gamma)Z^{-1}\left[1 - \frac{1}{N}\right] + \gamma Z^{-1}\left[1 - \frac{1}{N}e^{-1}\right]\right) \quad (2.2)$$

where γ is the Euler-Mascheroni constant (~ 0.5772) and $Z^{-1}[.]$ is the inverse of the standard Gaussian cumulative distribution function.

If we take the mean as zero and the standard deviation as one and S_k as the Sharpe ratio, then in the figure we see that even if we sample only 10 strategies the best one is expected to have a Sharpe of 1.57; 100 samples give us an expected maximum of around 2.5; 1000 samples gives 3.25. We think we have a Sharpe ratio of something quite spectacular but the out of sample expectation is zero.

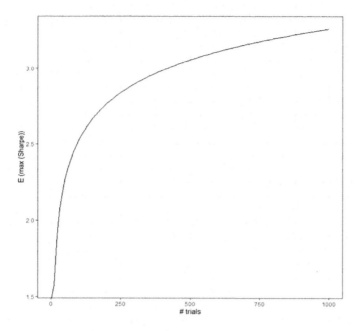

Figure 2.2   E(max(Sharpe)) function

2.4.2   Does this answer the question?

Having done all this experimentation and exploration, we return to the fund manager who asked the question in the first place. We then have to explain our results. The author has seen too many times the data scientist / quant / statistician trying to explain the results using their own jargon: *t-stats*, *significance* etc or even worse, *equations*; which tends to have the outcome of confusing the fund manager rather than enlightening them.

The secret here is to tell a story. What did you find? What can it tell the analyst? Part of this could be placed earlier in this chapter—the question of what one is being asked to deliver. Are we trying to forecast something or just find a story from within the data.

2.5   MOVING OUR CODE INTO PRODUCTION

Assuming the analysis we have just undertaken is not a one-off request, then there will be the need to move the code we used to run our analysis from the data scientist's development environment into production. This brings the third set of players into our story: the IT department. The interaction between the data scientist (or quant) and the professionals within IT leads to its own set of problems. Some of these are similar to those with the data scientist and the fund manager—the language used by data scientists can be somewhat alien to IT people. And similarly, the jargon used by IT professionals can confuse as well. There are problems of priorities: you might want to get this up and running tomorrow but the IT team are working in three-week sprints[2] and so whatever you have can't be released into production until the end of at least this sprint.

The other major issue is that the IT team do not tend to know exactly what analysis the data scientist has done, nor the techniques they use, nor why; and they don't have an intuitive feel for the data.[3] This lack of understanding of the process (and the underlying problem) is generally not an issue on a day-to-day basis—but it becomes a very big issue if the code goes wrong or the data being processed has errors in it. If the IT team now own the code how do they know what to debug or how to check the output after debugging?

If the data scientist is anything like the author, then much of the above process of experimentation will happen using an R (or python) script, or possibly a jupyter notebook. In some cases, there will be commented out code or code that doesn't go anywhere as it was realised that an approach was a dead end or could be improved upon. This is very much the norm and to be expected when doing "scientific" computing[4] but obviously doesn't work in a production environment. It tends not to work when you come to update the output in a month's time!

The code is potentially all in one (rather long) file—the author has come across examples with over 1,000 lines of code—and is hard to understand. A particular example the author inherited was code with variable names in Italian. These made sense to the (unsurprisingly Italian) analyst who wrote the code, but they were less than enlightening to the author, who's knowledge of Italian is, to say the least, limited.

Perhaps the data scientist has gone further and developed a simply Shiny app (see shiny.rstudio.com) where the fund managers have a dashboard available on the internal network where the results of the analysis are visualized and interactive.

Passing this skein of code with UI development (if it's a shiny app), dead ends, comments (or the lack thereof) over to the IT department and expecting them to understand it and convert it into a robust, reproducible process which can run reliably is going to lead to disappointment all round.

Another source of friction can simply be the choice of programming language: not many IT developers know R; more probably know python. However, it is likely that they will be more comfortable in other languages.

---

[2]"What's a sprint?" I hear you ask. It's a term that has come out of agile development techniques to mean a relatively short period of time (a week to a month) in which development is carried out. "But what's agile development?" As you can see the terminology or jargon from one world can confuse people outside of it!

[3]For example if one sees a one day stock return of very close to -50% the author's instinct would be a missed 2:1 stock split, not the value of the company halving.

[4]By scientific computing in this sense, I mean using a computer to try to research and solve a particular problem—and in particular a problem where you're not certain how you're going to solve it until you've played around with the data somewhat.

How can one make this process easier and less prone to error? Most of the work has to fall with the data scientist, perhaps sad to say. The first step is to refactor the code. Take out any dead ends; ensure that the comments are helpful and accurate; ensure that the variable names make sense.[5]

The ideal thing is to put most of the code into a package or a library, and as part of that migration develop test code which can be automatically executed when building the package.

Testing code

What do we mean by code tests? As an example in R we have the following code:

```
add <- function (x, y) return (x + y)
test (add) <- function ()
\{
checkEquals (2, add (1, 1))
checkException (add (1, "Fred"))
\}
```

The first line obviously creates a function called *add* which adds two numbers together . The second block of code creates a test function which checks that $1 + 1 = 2$ and that you get an exception when the integer 1 is added to the string "Fred". When we run the testing code it calls *add* twice and checks the output.

This example is obviously very simplistic but say the calculation was the risk of a portfolio from a linear factor model. The inputs to all the calculations are

— a vector of portfolio weights, w

— a matrix of factor sensitivities, B

— a matrix of factor covariances, F

— a vector (or diagonal matrix) of asset specific risks, D

and the calculation (using vector notation) is $w^T BFB^T w + w^T Dw$. The values of the first half and second half of this calculation will also be an output. In order to test this calculation, one could create a case with, say, three assets and two factors which can be easily run and compared to the known output.

The ideal is that you have enough tests that every (significant) line of code is tested for both things working and things going wrong. This can be a large volume of test code, but in the end it will be worth it.

An extension to this idea is called Test Driven Development where we actually write the tests before we write the code to pass them.

---

[5]An ex-colleague of the author tends to use either foo and bar or names of fruit as temporary variables when initially developing functions.

If the IT team want to tidy your code (or even move it to another language), then having automated testing can help to avoid errors. If the code has been tidied or translated it is, in our view, a good idea to retain the original code and develop it in parallel (perhaps in a different environment) and compare the outputs. This is particularly valuable in the case where the language has been changed.

One example the author developed had to be rewritten from R into C/C++ for reasons of computational speed. The original code was refactored into functions with sensible names and parameters, and then the C code was written using functions with exactly the same names and parameters. The author developed a number of R wrappers for some of the key C routines so the output of the C code and the original R code could be compared.

This duplication of code and functionality does lead to the situation of supporting two code bases: but in this case the benefit of developing in R (with the benefits of easy array access and graphics) and then bringing along the C code to match and being able to compare the outputs outweighed any costs of keeping two sets of code up to date,

The discussion so far assumes that the data is available. The next step is to think about the process that downloads the data—there are various clever tools out there that can automate downloading data and putting it into a database (of some form) and these should be used rather than the hack that the data scientist used as a one off. There are many reasons to use these tools: they handle errors well; they can trigger the processing routines when the data has (finally) arrived; they can be automated and also they can produce helpful error messages.

What happens if this downloading process doesn't work, or if the data has changed format? Who is responsible for chasing down data that hasn't downloaded or is incomplete? How do they know there's a problem in the first place? Answer is that the data scientist (probably) has to write code to check the data. A simple example (although not always true): do portfolio weights add up to 100? Have we downloaded roughly the same amount of data as yesterday? When we parse the data (if it's a web page say) do the results look sensible?

One of the biggest problems that no one seems to have really automated successfully is that of mapping—taking our raw data and associating it with companies and then to the individual bonds or equities in which we are interested. When a company has a corporate action (share splits, new bond issues or takeovers and mergers), then the identifier of the equity or bond tends to change, or a new one is introduced. How do we ensure that firstly we pick up this change and secondly track the data back to the previous identifier?

2.6   ENVIRONMENTS

In the IT world, there are at least two, if not three, environments in which we can work. There's a development (DEV) one, perhaps a user acceptance testing (UAT) one and then finally a production (PROD) one. The code in the PROD environment cannot just be changed—it has to go through a release process and be signed off by various interested parties. In most firms, there is a separation of duties. Developers are not allowed read / write access to the production environment; other support people have to do the release.

There are two big issues the author has come across when working across environments. The first is library versioning—when developing the analysis one uses the latest version of a library, but the production environment has an earlier version which either works differently or doesn't have the brand-new bit of functionality that was needed. Python's virtual environments are a partial solution to this, although there remains the problem of getting the three environments to align.

The second is ensuring that the data in any development environment database is identical to that in the production environment. An obvious solution to this, assuming one is only reading from the database, is for the development code to look at the production database—but that tends not to be allowed (for reasons that the author sometimes find rather hard to understand). Taken to an extreme, this can mean that all three environments have to do the data downloading, mapping and loading separately—and someone has to check that three processes have run rather than just one. Synchronising the data between the environments is another idea—every day or week take the data from the production environment and load this into the other two. If this isn't done, then it is almost invariable that the development code will be running on data which is subtly (or even significantly) different from that in the production (or UAT) environment—and hence if there are differences in output between our development code and the final product, then tracking down the source of the difference becomes much harder.

2.7   IS THIS THE END OF OUR STORY?

This chapter has been written, by the necessity of the printed word, as a linear exposition, but the real world doesn't work in such a way. As the data scientist, we develop something, the fund manager decides that although the results are useful what they actually wanted to obtain was the answer to a different question[6] and so the data exploration process and model building has to be

---

[6]A useful skill to learn is figuring out what's the real question!

revisited. The code is written and then moved into a production environment—and then the web site from where the data is obtained changes its format and so the data download process has to be (rather rapidly) rewritten and tested.

Or some shock comes along and the world changes and the relationship between our input data and the output we require which our long search has found stops working (see the comment above from the New York Fed about their nowcasting model). Hence, we have to first make assumptions about the "new world order" and then attempt with a minimal amount of data to build a new model.

All of these examples, and many more, turn the subject of this chapter—the process of taking alternative data and turning it reliably and regularly into investment insights is a never-ending iterative process of refinement.

REFERENCES

Authority, U. S. (2021). UK Energy in Brief. http://www.gov.uk/government/statistics/uk-energy-in-brief-2021.

Babii, A., Ghysels, E., and Striaukas, J. (2022). Machine learning time series regressions with an application to nowcasting. *Journal of Business & Economic Statistics*, 40(3):1094–1106.

Bailey, D. H., Borwein, J. M., de Prado, M. L., and Zhu, Q. J. (2014). Pseudomathematics and financial charlatanism: the effects of backtest over fitting on out-of-sample performance. *Notices of the AMS*, 61(5):458–471.

Banbura, M., Giannone, D., Modugno, M., and Reichlin, L. (2013). Nowcasting and the real-time data flow. In *Handbook of Economic Forecasting*, volume 2, pages 195–237. Elsevier.

Domingos, P. (2012). A few useful things to know about machine learning. *Communications of the ACM*, 55(10):78–87.

Harvey, C. R. (2017). Presidential address: the scientific outlook in financial economics. *The Journal of Finance*, 72(4):1399–1440.

Lundberg, S. M. and Lee, S.-I. (2017). A Unified Approach to Interpreting Model Predictions, NIPS'17: Proceedings of the 31st International Conference on Neural Information Processing Systems. *Advances in Neural Information Processing Systems*, 30:4768–4777.

Patton, A. J. (2011). Volatility forecast comparison using imperfect volatility proxies. *Journal of Econometrics*, 160(1):246–256.

Ribeiro, M. T., Singh, S., and Guestrin, C. (2016). " why should i trust you?" Explaining the predictions of any classifier. In *Proceedings of the 22nd ACM SIGKDD international conference on knowledge discovery and data mining*, pages 1135–1144.

Romano, J. P., Shaikh, A. M., Wolf, M., et al. (2010). Multiple testing. *The New Palgrave Dictionary of Economics (Online Edition)*.

CHAPTER 3

Global Economy and Markets Sentiment Model

Jacob Gelfand

Northwestern Mutual, Investment Risk Management

Kamilla Kasymova

Northwestern Mutual, Investment Risk Management

Seamus M. O'Shea

Northwestern Mutual, Managed Investments

Weijie Tan

Northwestern Mutual, Investment Risk Management

CONTENTS

DOI: 10.1201/9781003293644-3

3.1 INTRODUCTION

3.1.1 In Pursuit of Alpha

Investment research involves collecting diverse information from various sources, and piecing these together to form a mosaic; this mosaic is then used to obtain a view of the risks and opportunities in the market or within an asset class. The task is laborious, prone to bias and human error and naturally limited in its scope. In our implementation of the Global Economy and Markets Sentiment (GEMS) model, we have considered news sentiment analysis and have addressed these shortcomings.

Fortunately, globalization, the internet and the proliferation of cheap and powerful computing technologies have made information accessible in volumes and at speeds unimaginable twenty-five years ago. As a result, even the most oppressive and controlling regimes must now go to great lengths to guard state secrets and prevent the voices of their citizenry from being heard. Yet while these advances liberalized information transmission, they also introduced new challenges for the investment management discipline. Twenty-five years ago, possessing information itself often conferred an advantage, or "edge" as its typically known by in financial markets. Today, however, it is insufficient for an analyst to simply amass data. Instead, it is the uniqueness of the information, and in particular the speed with which an analyst can process it that may afford him or her an edge. Said differently, markets are highly efficient. By the time news has reached the front pages or broadcasts of global media outlets, asset prices have already moved.

Practically speaking, then, to consistently generate positive benchmark-relative returns, or "alpha", an investor must not only stay abreast of news flow, but ideally ahead of it.

3.1.2 Sentiment Analysis

Investors aim to continually stay ahead of news flow. Part of this process entails assessing the credibility of news, ignoring the noise and distilling what is left into salient and comprehensible data points. This is non-trivial. Media manipulation by opposing political and economic interests, explicit or implicit controls governing flow and information asymmetries more broadly create formidable obstacles to a researcher's ability to form objective viewpoints of current events. What's more, these challenges do not just afflict a single asset class, geographical region, or political paradigm.

One can reasonably assume, however, that the challenges become more pronounced in a sovereign risk setting given disparate political, social and economic forces, and the influence of the state apparatus in countries with limited tradition of democratic governance, rule of law and public transparency. This can be especially acute in many emerging markets countries. Nevertheless, a tightly controlled or even distorted information transmission mechanism does not necessarily imply an informational vacuum. Even under the oppressive Communist regime of the Soviet Union and its satellite states, careful observers (aka, "Kremlinologists") were able to glean critical political or economic insights from local news sources. For example, Khrushchev's denunciation of Stalin (1956) and Gorbachev's reform agenda (circa 1985) were correctly registered by diligent western analysts and investors simply by their reading between the lines of official communiques and evaluating their credibility.

There are recent examples, too. Kazakhstan's abandonment of the Tenge's peg to the US Dollar in 2015, and Russia's 2018 pension reform, to cite just two, both caught financial markets off guard,despite their long having been a subject of speculation in local media and internet blogs. Simply put, it is often through local media and rumor mills that potentially market-moving economic or political changes are first socialized. In many cases, this is well before any such rumors become reality, or even worthy of attention by the likes of Reuters, Bloomberg, or The Wall Street Journal.Systemically leveraging such information sources falls in the domain of what is commonly referred to as sentiment analysis.

At the Northwestern Mutual Investment Management Company ("NMIMC"), analysts and portfolio managers conduct sentiment analysis through the company's proprietary Global Economy and Markets Sentiment (GEMS) model. The model, which leverages the publicly available GDELT data set, empowers analysts by tapping into previously neglected information sources. It employs the power of AI, machine learning and power computing to manage scale, while at the same time mitigating human bias and error. In its current implementation, GEMS focuses on developing countries, where the NMIMC team often perceives information asymmetries to be most pronounced(for the reasons cited earlier). More specifically, the company's Emerging Markets portfolio managers and analysts use GEMS, and its associated frameworks, to cull insights from large volumes of global and local news originating in over 65 different languages. The principal objective is to reach better and more informed investment decisions, and to do so before the market does.

The rest of this chapter is organized in the following way. In Section 3.2 we introduce the novel data source, namely, the Global Database of Events, Location and Tone (GDELT). In Section 3.2.1, we consider its development and distribution under a joint initiative of Google, Georgetown University and a number of other partners. GDELT is an open source data initiative. In Section 3.2.2 Marching Ahead—"Boots On The ground" or "Boots In The Cloud"? we discuss the interaction of this data with our proposed model and the availability of this data in the clouds. In Section 3.3 GEMS Framework and Vulnerability Sentiment Indices, we introduce vulnerability sentiment indices (VSI); the VSI may also be interpreted as sentiment indicating the source of "risk". In Section 3.3.1, Country Vulnerability Sentiment Indices and in Section 3.3.2 Regional Vulnerability Sentiment Indices are discussed and elaborated. In Section 3.4 Investment Strategies, we discuss two types of strategies. Section 3.4.1 Systematic Strategies and Section 3.4.2 Discretionary Strategies highlight their contrasting styles. We summarize our findings in Section 3.5 Conclusions.

3.2 GDELT

3.2.1 GDELT

GDELT, or the Global Database of Events, Location and Tone, monitors the world's broadcast, print and web-based news from nearly every corner of the globe and in over 65 different languages. It identifies the people, locations, organizations, themes, sources, emotions, images and events driving our global society every moment of the day. It is the largest, most comprehensive and highest resolution open database of human society ever created. Kalev Leetaru, a former adjunct faculty member at the Georgetown University and a Senior Fellow at George Washington University's Center for Cyber and Homeland Security, founded the GDELT project with a vision to codify the entire planet into a structured and searchable format so users could better understand the world.Here we would like to explain to the readers that the GDELT project is supported by Google Ideas,Google Cloud, Google and Google News, the Yahoo! Fellowship at Georgetown University, BBC Monitoring, the National Academies Keck Futures Program, Reed Elsevier's Lexis Nexis Group, JSTOR, DTIC and the Internet Archive.

GDELT uses some of the world's most sophisticated natural language processing (NLP) and data mining algorithms to compile a real-time immutable record of global society that can be visualized, analyzed and modeled. Given the highly structured and accessible nature of the data,its range of applications makes it useful to different disciplines and, in particular, to Wall Street trading desks. For example, the ability for a user to query a topic of interest and quickly analyzes a global and comprehensive set of viewpoints, regardless of their originally published language, offers the prospect of better-informed decision making.

Three proprietary data streams populate GDELT. The first codifies events around the world into over three hundred categories. A second records the people, places, organizations, themes and emotions characterizing those events; it is updated every fifteen minutes. The third data stream catalogues the visual

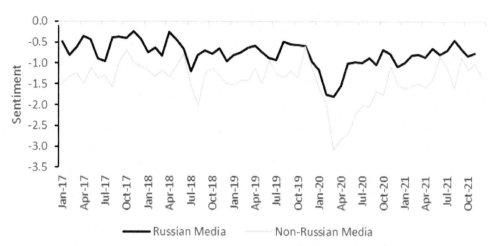

Figure 3.1 News sentiment of the Russian economy as expressed in domestic and foreign media

narratives of the world's media in real time. GDELT's Global Content Analysis Measures system (GCAM) scores each content's "tone" on a scale ranging from −100 (extremely negative) to +100 (extremely positive). Common values fall between −10 and +10; a score of zero corresponds to a neutral tone.

By leveraging the GDELT dataset, the GEMS model can see beyond the reach of western media, enabling a more holistic perspective of events, and a glimpse into a world void of language barriers and biases. This is important because stark differences often exist between the narratives, perspectives and interpretations of events put forth by global-oriented news outlets and those put forth by localized news sources. In other words, there exists media bias. Consider Figure 3.1, for example, which compares the sentiment of the Russian economy as expressed in both Russian and non-Russian media.

The sentiment conveyed in the local Russian news consistently outranks that of the non-Russian news. Nothing about this phenomenon, in and of itself, jumps out as remarkable. After all, fans of Europe's Champions League tournament can expect Manchester United to receive more favorable press in England than it does in Spain, just as Real Madrid will enjoy better treatment in Spain than it does in England. Which media source, then, is the impartial one? It's unclear, but the answer may well be none of the above. Since either can be manipulated, propagandized, or used to advance an agenda, blindly trusting one over the other risks imprudence. Yet many western-based analysts routinely favor content from well-known global media sources, sell-side research shops and independent global macro research providers, despite its obvious vulnerabilities to subtle, and in many cases not so subtle, biases. These phenomena and predispositions are discussed further in the section "Marching ahead - 'Boots on the ground' or 'Boots in the cloud'?".

In the interest of extending these research analysts, the benefit of the doubt, one might argue that news originating in English travels faster than foreign language originated news. The global media conglomerates—predominately English-language oriented—have correspondents in every major city, affiliate sharing agreements in place and lightning-fast networks—features that should confer a significant speed advantage. Or perhaps simply fewer people, computer programs and globally oriented news outlets mine foreign language news to the extent they do English language news. Consider the following two examples drawn from GDELT data.

On or about October 17, 2019 global media began directing attention to protests in Chile. The protests, a near immediate response to public transportation fare hikes, quickly escalated in the Chilean capital, turning violent and compelling the government to instate curfews and declare a state of emergency. Unsurprisingly, Chilean risk assets fell sharply in response. The peso lost 8.5% of its value against the US Dollar, domestic equity markets sold off by over 10% and local and hard currency bond spreads widened. The coverage volume intensity, a measure of the global news coverage of these events, increased dramatically. Confining the dataset to Chilean media alone, however, reveals that the same measure had started moving notably higher almost a week earlier (Figure 3.2).

More recently, protests originating in the gas-producing western region of Kazakhstan, sparked by the imposition of higher fuel prices on January 1, 2022, spread across the country, evolving into a widescale revolt against corruption, poverty and inequality. Protesters aimed their discontent against the incumbent administration, President Tokayev and de facto ruler Nursultan Nazarbayev, the latter of whom had presided over Kazakhstan since the Soviet era. Figure 3.3 depicts the rapid deterioration in sentiment towards Nazarbayev in Kazakh media, nearly two weeks prior to the outbreak of protests. By contrast, the decline in sentiment reflected in global (that is, non-Kazah) media became evident only after protests had erupted.

The above examples, along with many others, invalidate the notion that English-language originated news travels faster than that originating in non-English languages. Further, they suggest that unaltered translations of local news may offer the prospect of differentiated information, uncolored by foreign biases and often retrievable before it is picked up by the global news outlets. The GEMS model targets this very information, thereby allowing users an opportunity to augment and improve their investment processes.

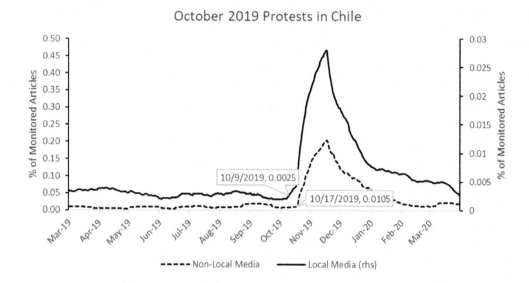

Figure 3.2 New coverage of October 2019 Chilean protests in local and non-local media

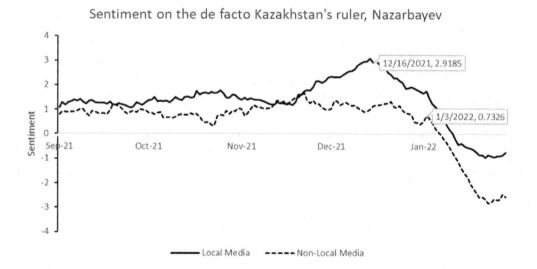

Figure 3.3 Sentiment toward de facto Kazakhstan's ruler Nazarbayev, in local and non-local media

3.2.2 Marching Ahead—"Boots On The ground" or "Boots In The Cloud"?

Although the Efficient Market Hypothesis (EMH) is a widely accepted doctrine within the investment discipline, academics and investors typically view it through the lens of equity markets. Hence, many of the largest *fixed income* investment managers employ dozens of research analysts who closely monitor earnings calls, press releases and credit rating agency actions, attend investor relations meetings and industry conferences and conduct onsite visits to stay abreast of news flow relevant to their area of coverage.

By and large, however, this appears to be a mostly futile exercise, since the US Investment Grade ("IG") corporate bond market, for one, is highly efficient as well. Corporate bond analysts are privy to the same data that their equity analyst counterparts are and both groups compete with a public corporation's executives, board members and employees—each of whom subtly influences stock and bond prices through acquisition and disposition, ESPP ("Employee Stock Purchase Plan") activity and through the exercising of stock options, decisions that in turn are influenced by their informational edge, however, small. As a result, there is limited opportunity for the IG corporate bond analyst to identify and capture alpha from heavily followed, and therefore well understood, corporations. Most serious studies support this: information ratios of widely used corporate bond indexes are typically no higher than about 0.10.

Bonds issued by non-US domiciled corporations in the developed world may offer improved room for alpha given geographic and linguistic barriers, as well as differing reporting regulations. Even then, however, the generally lower spread premiums and higher transaction costs (owing to their lower liquidity) accorded these bonds may render any informational edge or insight immaterial. By contrast, tradeable debt issued by Emerging Markets sovereign and corporate entities may offer better opportunities for alpha. Language, geographic, political and regulatory barriers are arguably higher, as is the variability in content quality and timeliness of issuer reporting. What's more, institutional credibility, competing policymaker objectives and widely varied sources of GDP—among many other factors—lower the propensity for investor groupthink and introduce greater information asymmetries.

To address such challenges, US and global investment banks, which arrange financing for Emerging Markets borrowers, regularly sponsor investor trips that allow their Sales and Trading clients to meet with local officials, visit the host country's state institutions and develop an improved sense of familiarity. In addition, many of the largest clients themselves may even have local offices, where they employ analysts fluent in the language and culture (the local presence, of course, also helps them attract and retain assets). Broadly speaking, one can think of this approach to information gathering and analysis as having "boots on the ground".

For an investment management firm that sends their research analyst(s) on investor trips, or considers employing one locally, the presumption is the same: a boots on the ground presence enriches the investment analysis, leading to superior decision making and in turn higher returns. There is some validity to this. Following Russia's annexation of Crimea in 2014, for example, many otherwise well-informed and pragmatic analysts leaned heavily on the collective viewpoints of western media, becoming virtually certain that a sanctions-induced economic collapse was imminent. By contrast, investors who traveled to Russia and spoke firsthand with statesmen and laymen alike quickly saw the folly in these views.

Unfortunately, a boots on the ground research approach, however, sound in theory, raises the risk that common cognitive biases may cloud the investment decision making process. Investor trips, for example, are often carefully choreographed by both the underwriters and the borrowing institutions and in some cases by the host governments. They select the timing and location of the meetings, control the messaging, pick the wine and cheese served at the cocktail hours and arrange for the hotels, transportation and leisure opportunities curated to impart cultural education and awareness. In short, investors risk returning home with a *perceived* understanding of where they have just been, the well-documented *familiarity bias*. Secondly, the organizers of conferences and investor trips often maintain lucrative business relationships with their guests that span across several different business areas. As such, their willingness to challenge political, economic, or philosophical suppositions held by the client can only stretch so far. Instead, at some point the client will hear a message better tailored to her pre-existing beliefs, introducing *confirmation bias*. Finally, a prestigious, highly sought after and invitation-only conference or investor trip may prove fertile ground for *group think* or a *bandwagon effect bias*.

Admittedly, a more permanent boots on the ground presence in the form of a local office with fluent and resident research analysts can help mitigate such biases. Yet such a setup can be prohibitively expensive for many investors. Nor is there any guarantee they will eliminate biases, or improve investment decisions, or yield higher returns.

The GEMS model, in essence a "boots in the cloud" approach, can either complement a boots on the ground effort or serve as an alternative altogether. It can be tailored for both qualitative and quantitative uses. Importantly, it allows for repeatable, sterilized and objective analysis.

3.3 GEMS FRAMEWORK AND VULNERABILITY SENTIMENT INDICES

Broadly speaking, quantitative finance offers two approaches to the construction of investment models and strategies. The first uses mathematical constructs, sometimes with limited economic intuition, to validate empirical observations. Examples include statistical arbitrage and factor-based strategies. The second approach deploys mathematical apparatus to replicate frameworks used by traditional fundamental or "global macro" analysts. The use of natural language processing and AI to analyze corporate earnings calls and financial

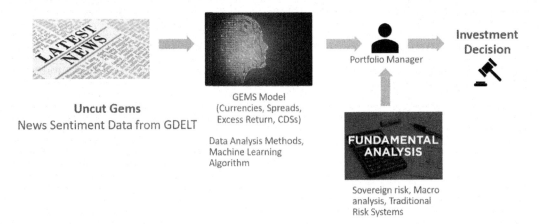

Uncut Gems
News Sentiment Data from GDELT

GEMS Model
(Currencies, Spreads,
Excess Return, CDSs)

Data Analysis Methods,
Machine Learning
Algorithm

Portfolio Manager

**Investment
Decision**

FUNDAMENTAL
ANALYSIS

Sovereign risk, Macro
analysis, Traditional
Risk Systems

Figure 3.4 GEMS (Global Economy and Markets Sentiment) Framework

statements, typically the domain of fundamental analysts, belong to this latter approach.

The GEMS model marries elements of both in a so-called quantamental approach. As Figure 3.4 illustrates, the GEMS framework processes macroeconomic, fiscal and political news flow and sentiment ("uncut gems") from the GDELT database. It then refines this raw data via a series of economic, mathematical and statistical transformations to produce a set of measures called *Vulnerability Sentiment Indexes* (VSIs) [Casanova et al., 2017]. Having single, blended sentiment measures allows for a number of interesting adaptations involving statistical analysis and algorithm design. One of the model's common use cases, in fact, entails plotting VSIs against observable financial market indicators such as foreign exchange rates, credit spreads, or swaps prices. This, in turn, allows a user to lean on technical analysis rubrics to generate risk and investment signals. Other use cases involve the development of rules-driven, adaptive and self-learning systematic trading strategies.

3.3.1 Country Vulnerability Sentiment Indices

Since the user specifies the sources, topics, or themes of the uncut gems and also shapes the refining process, VSIs can be finely tailored to their needs. In its current implementation, for example, GEMS relies on country specific VSIs. Each country VSI, in turn, consists of an *External* Vulnerability Sentiment Index (EVSI) and an *Internal* Vulnerability Sentiment Index (IVSI).

The EVSI measures the tone surrounding a country's relations with its neighbors and trading partners and its role in foreign affairs. The number of effective trading partners is determined by the inverse of the Herfindahl-Hirschman Index, a common measure of market concentration, which is calculated using the total trade volume between a country and its key counterparties. In their raw or "uncut" form, relevant datapoints are queried from three distinct sources: the country's local (i.e., domestic) media, the trading partner's local media, all other media sources. Consider Russia, which has vast hydrocarbon resources and supplies about 30% of Europe's current energy needs. A codependency exists: the price of natural gas impacts Russia's fiscal accounts, while German households rely on Russian gas to heat their homes. To compute Russia's EVSI, the GEMS model scans Russia's domestic media, scoring the sentiment of news concerning natural gas and Germany. It does the same with Germany's domestic media, evaluating any mentions of natural gas and Russia, then repeats the exercise using all other foreign media, searching for news involving natural gas and Germany and Russia. The model considers each of Russia's primary trading partners, exports, political alliances and political enmities. The scores of each unique combination are normalized and weighted to arrive at the EVSI. Similarly, the Internal Vulnerability Sentiment Index (IVSI) reflects sentiment concerning a country's internal affairs—for example, its political and social conditions, governing institutions and key drivers of its economy. Four component subindices make up the IVSI:

1. *Economic* index: reflects sentiment around metrics such as GDP growth and unemployment.

2. *Fiscal* index: reflects sentiment of a country's fiscal deficit or surplus, tax regime, welfare and social security programs, etc.

3. *Monetary* index: reflects sentiment around monetary factors like inflation and interest rates.

4. *Institutional* index: captures sentiment of a country's institutions, government, rule of law, ease of doing business, etc.

A country's overall VSI is a weighted sum of its constituent EVSI and IVSI measures, where the weights vary across countries. Countries with a greater number of significant trading goods and a wider dispersion of trade partners are considered less susceptible to external pressures than countries with fewer—as such, the EVSI will contribute less to their overall VSI. Indonesia, whose trade is well-diversified across the Asia-Pacific region and beyond, will have a lower EVSI weighting than does Mexico, whose trade is heavily tilted towards the United States and Canada.

3.3.2 Regional Vulnerability Sentiment Indices

In addition to country sentiment, investors may also care about regional sentiment. To accommodate this, GEMS offers VSIs for regions such as Central

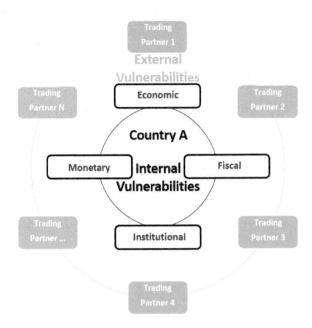

Figure 3.5   Country Vulnerability Sentiment Index

America and the Caribbean (CAC), Latin America and the Commonwealth of Independent States (CIS). For simplicity, and to maintain consistency with country VSIs, the regional VSIs are constructed by weighting constituent country VSIs. The weights of each country depend on five attributes: GDP, trade, investor benchmark exposure, volume of news and volatility of news (see Figure 3.6). Naturally, a country with a higher GDP receives a greater weight in the regional VSI. *Trade*, proxied by the gross volume of imports and exports, captures the economic interdependence between two countries. Countries with large trade volumes are deemed of greater importance to the geographical region; as such, they receive higher weights in the regional VSI. The presumption is that sentiment surrounding countries with large trade volumes impacts not only themselves, but also their regional trading partners.

The other three attributes are more dynamic. *Index* takes into account the degree to which a country's financial risk assets factor into widely followed market benchmarks. For example, a country whose hard currency debt obligations comprise 20% of a popular bond market index will receive greater weight, all else equal, than a country whose bonds comprise just 5% of that index. *News volume* tallies significant events—both the frequency of such events and the level of significance therein. If the volume of country-specific news spikes well above its long-term average, it may signal the presence or onset of an extraordinary and potentially market-moving event that can impact the region to which the country belongs. Finally, *news volatility* measures the frequency and amplitude of material changes in sentiment and is used in part to

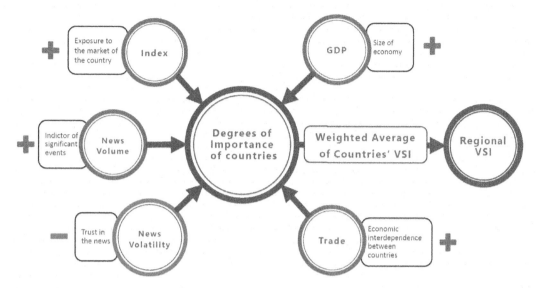

Figure 3.6   Region Vulnerability Sentiment Index

assess credibility. This makes intuitive sense: if country sentiment fluctuates wildly, users may well discredit it, or at least discount its significance. Financial markets do the same. Traders will learn not to reprice a country's risk assets as hurriedly as they may another's if related headlines frequently contradict one another. The CRITIC method [Diakoulaki et al., 1995], a widely used method to determine objective weights in multiple criteria weight setting problems, is used to weight the five inputs.

3.4   INVESTMENT STRATEGIES

Understanding the GEMS framework and its potential financial markets applications allows a user to develop and test different investment strategies. One set of test strategies involves the analysis of foreign exchange markets. As a single measure, exchange rates are complex reflections of a country's relative economic health. Of course, they also reflect macroeconomic activity and technical forces that drive decision-making by financial markets participants. Since FX markets are liquid, highly efficient and typically the first to reprice in response to market news, they serve well as a proxy for risk within the GEMS model, in both systematic and discretionary investment strategies.

In systematic strategies, trading and investment decisions follow precise and formulated rules. Discretionary strategies, by contrast, have no predefined rules. Instead, the investor forms decisions based on observations, fundamental research and ultimately their own professional judgement. To start, two different systematic strategies are discussed: (1) a trend-based FX trading strategy

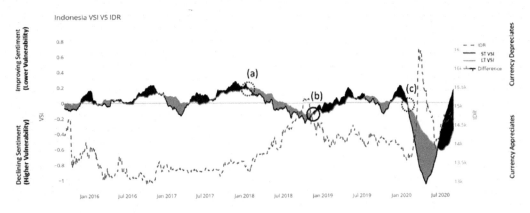

Figure 3.7 Indonesia's Vulnerability Sentiment Index (VSI) Crossovers: Short- and Long-Term Sentiment functions

involving short-term and long-term VSI functions and (2) a portfolio strategy involving active positions in benchmark sovereign risk.

3.4.1 Systematic Strategies

3.4.1.1 Crossover VSI strategies

Consider a simple hypothesis: meaningful improvements in a country's economic, fiscal, institutional, monetary and external trade conditions lead to local currency appreciation. In the GEMS environment, this hypothesis translates to "an increasing country VSI leads to a higher exchange rate for the domestic currency".

The Crossover VSI strategy uses short-term and long-term VSI functions to identify trading signals. Borrowing a convention from Technical Analysis [Peterson, 2016], instances where the short-term sentiment function (ST VSI) crosses above the long-term sentiment function (LT VSI) imply improving sentiment and thereby indicate a possible buy signal (that is, expected local currency appreciation). Conversely, when the ST VSI falls below the LT VSI, suggesting deteriorating sentiment, the result is a sell (currency depreciation) signal.

Figure 3.7 provides an example using Indonesia. On the chart, the gray and black lines represent the short-term and long-term Indonesia VSI functions; they refer to the left vertical axis. The dashed line is the US Dollar / Indonesian Rupiah spot rate. Circles denote three distinct trading signals.

The data presented in Figure 3.7 indicates that the ST and LT VSI functions may help a user identify inflection points in the value of a country's domestic currency. Consider the three highlighted trading signals:

Signal (a)—sell / currency depreciation; January 29, 2018

Three distinct events took place around this time. First, trade tensions between China and the US began intensifying. Heightened US-China trade tensions, and certainly the imposition of tariffs, would negatively impact other Asia-Pacific countries like Indonesia due to their ties with China's supply chain (nearly half of China's exports to the United States involve other countries through intermediate inputs and materials). Second, global markets were becoming increasingly concerned that monetary policy tightening in the US and Europe might trigger capital outflows from emerging markets countries. Third, Indonesia's Finance Minister announced that he expected the economy's growth rate to fall short of the target previously outlined in the 2017 state budget. Following the trade signal, the Rupiah started to depreciate against the Dollar.

Signal (b)—buy / currency appreciation; November 12, 2018

News flow around this time included mentions of higher-than-expected economic growth and renewed talks between Chinese President Xi and US President Trump to reach a trade deal at the upcoming G20 summit. Following this crossover, IDR began to appreciate.

Signal (c)—sell / currency depreciation; February 3, 2020

Like most of the global economy, Indonesia faced increasingly dire headwinds from COViD-19. Risk assets all over the world started selling off, IDR included.

In all three of these examples, a trading strategy designed to tilt risk posture in accordance with the signal would have captured much of the subsequent exchange rate move.

A variation of this strategy uses a proprietary trend function in conjunction with its first derivative. Specifically, instances where the trend function (Trend) crosses up through the signal function (Signal) denote a possible buy (currency appreciation) signal while those where Trend crosses down through Signal indicate a sell (currency depreciation) signal. In Figure 3.8, a buy signal occurs on May 4, 2020, likely a response to the news on April 7 that the Bank of Indonesia had secured a $60 billion repurchase facility with the US Federal Reserve to help shore up their liquidity amid a dollar shortage triggered by the COViD-19 pandemic. While the crossover took four weeks to materialize, the slope coefficient of the Trend function flipped from negative to (strongly) positive almost immediately following the announcement, hinting at a different way to define the strategy.

Finally, Indonesia boasts diversified trade across many key trading partners. Its Internal VSI (IVSI) component therefore makes up 70% of its VSI. By

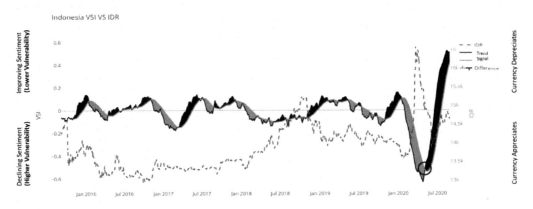

Figure 3.8 Indonesia's Vulnerability Sentiment Index (VSI) Crossovers: Trend and Signal Sentiment functions

Figure 3.9 Indonesia's Institutional Vulnerability Sentiment Index (IVSI) Crossovers: Trend and Signal Sentiment functions

drilling down into the individual components of the IVSI, one can see (Figure 3.9) that the Institutional VSI provides an earlier currency appreciation signal than does the overall VSI. This squares with intuition since the procurement of a repo line from the US Fed represents a vote of confidence in the country's institutions. Interestingly, it also hints at yet another way to parameterize the trade study.

Figure 3.10 provides a regional-level perspective. The Asia and Asia-Pacific region VSI includes China, Philippines, Indonesia, Malaysia, South Korea and India VSIs. Currency depreciation signals on January 29, 2018 and June 26, 2019 stemmed from an uptick in trade tensions between the US and China. Note that a depreciation signal in this case refers to an expected depreciation of a basket of currencies, where the weights of the constituent currencies in the basket correspond to the country weights used to construct the regional VSI. Since China's VSI has the largest weight in the Asia and Asia-Pacific region

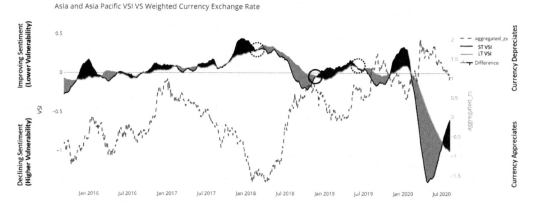

Figure 3.10 Asia and Asia-Pacific Region Vulnerability Sentiment Index (VSI) Crossovers: Short- and Long-Term Sentiment functions

VSI, any meaningful shift in sentiment regarding China would be apparent in the regional VSI. And, as alluded to before, heightened trade tensions or the imposition of additional tariffs between the US and China should negatively impact the economies of other countries in the region. Accordingly, intuition suggests that this ought to drive currency depreciation in those countries, too. On November 26, 2018—in between the two sell signals—there is a buy signal, driven by news of renewed talks between President Trump and President Xi. Following this, the Indonesian Rupiah begins to appreciate against the Dollar. To summarize, analyzing regional VSIs may be helpful in predicting spillover effects in other regional economies.

3.4.1.2 Portfolio Strategy

A second systematic strategy tests a slightly different hypothesis: countries with increasing VSIs offer higher forward returns on their sovereign bonds. The strategy overweights and underweights country exposures relative to a sovereign benchmark index based on country VSI levels. Early tests used the Barclays EM USD Sovereign index, with various time lags considered to account for possible delays between VSI changes and subsequent market movements. The index is modified to include only countries in the GEMS model. In addition, a scaling factor controls the magnitude of each tilt, with higher values corresponding to longer time lags [RavenPack, 2021].

Figure 3.11 shows a basic schematic of the trading strategy involving a two-month lag. For the sake of clarity, a two-month lag means that at the start of the third month the portfolio composition reflects benchmark weights in the third month and VSI values from the first month. During this third month, the portfolio and benchmark prices move in accordance with the market. At the beginning of the fourth month, active country exposures are reweighted

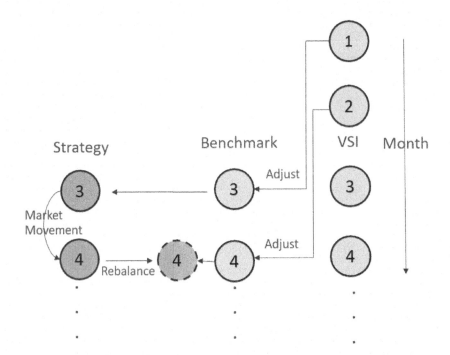

Figure 3.11 Portfolio Strategy

based on the new benchmark and the second month's VSI. Subsequent months work in the same manner.

Figure 3.12 and Figure 3.13 below display benchmark-relative returns and information ratios given a range of scalars and lag periods. Strategies that use two-to-six-month VSI lags outperform their benchmark, with the highest performance achieved with a four-month lag. Figure 3.14 plots the cumulative excess returns of the benchmark and portfolio under the four-month lagged strategy.

3.4.2 Discretionary Strategies

As detailed above, preliminary work suggests that the GEMS framework may prove useful in the design of algorithmic, high frequency, or, collectively, "fast money" trading strategies. In a long-only, largely bottom-up oriented investment process, the framework can also be accretive, complementing traditional research methods and helping an analyst or portfolio manager to strengthen or temper their investment conviction. A few examples serve to illustrate this.

In mid-January of 2019, Saudi Arabia's dollar-denominated bonds were lagging a fierce rally in the investment grade-rated segment of the Emerging Markets fixed income space. In fact, Saudi risk assets more broadly had underperformed since the prior October, when journalist Jamal Kashoggi was

Average Active Return

Scalar	0	1	2	3	4	5	6	7	8
0.5	-0.04	-0.01	0.01	0.01	0.01	0.01	0.00	-0.02	-0.01
1	-0.07	-0.02	0.01	0.02	0.03	0.03	0.00	-0.04	-0.02
1.5	-0.11	-0.03	0.01	0.03	0.05	0.04	0.00	-0.05	-0.02
2	-0.15	-0.05	0.01	0.03	0.07	0.06	0.00	-0.06	-0.03
2.5	-0.18	-0.07	0.00	0.03	0.09	0.08	0.01	-0.07	-0.03
3	-0.21	-0.10	-0.02	0.04	0.11	0.10	0.01	-0.08	-0.03

Figure 3.12 Strategy performance—Average Active Return [percent / month] as a function of scaling factor and time lag (months). Considers the September 2015 to August 2020 time period.

Information Ratio

Scalar	0	1	2	3	4	5	6	7	8
0.5	-0.25	-0.06	0.08	0.10	0.13	0.12	-0.04	-0.23	-0.12
1	-0.26	-0.07	0.06	0.10	0.15	0.14	-0.02	-0.23	-0.11
1.5	-0.26	-0.07	0.04	0.09	0.17	0.16	-0.01	-0.22	-0.10
2	-0.26	-0.08	0.02	0.09	0.19	0.18	0.01	-0.21	-0.09
2.5	-0.26	-0.10	-0.01	0.08	0.20	0.19	0.02	-0.20	-0.08
3	-0.26	-0.12	-0.03	0.08	0.21	0.20	0.03	-0.19	-0.07

Figure 3.13 Strategy performance –Information Ratio as a function of scaling factor and time lag (months). Considers the September 2015 to August 2020 time period.

killed at the Saudi embassy in Istanbul, Turkey. Yet in January 2019, the macroclimate turned favorable following the US Federal Reserve's "pivot" in its monetary policy stance. Interest rates had started falling and global oil prices began to stabilize following a precipitous fall during the prior quarter. What's more, Saudi Arabia had strong incentive to improve its image in advance of its widely expected IPO of state-owned Saudi Aramco later in the

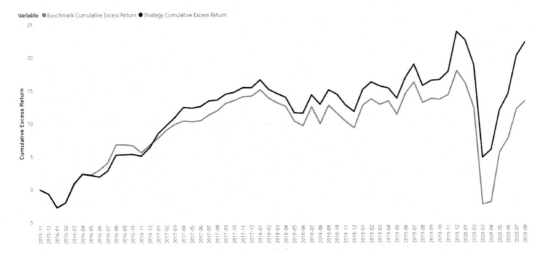

Figure 3.14 Cumulative portfolio and benchmark performance (%), four month lagged VSI strategy.

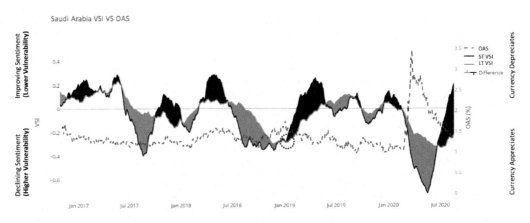

Figure 3.15 Saudi Arabia's Vulnerability Sentiment Index (VSI): Short- and Long-Term Sentiment functions

spring. Finally, the bonds still held A-ratings from the major credit ratings agencies. In short, the malaise around the country's investment prospects felt overdone. Since the underperformance of the bonds in the aftermath of the Kashoggi incident was largely sentiment driven (as opposed to, for example, a surprise deterioration in the country's sovereign credit metrics), it made sense to consider whether sentiment—as measured by Saudi Arabia's VSI—had begun to turn more favorable. The GEMS model output, depicted in Figure 3.15, suggested that, in fact, sentiment had already begun to turn positive. In Figure 3.15, Saudi Arabia's weighted average USD-denominated bond spread (OAS) is used in place of FX, given the Saudi Riyal's peg to the US Dollar. This observation helped strengthen the view that the bonds could rally.

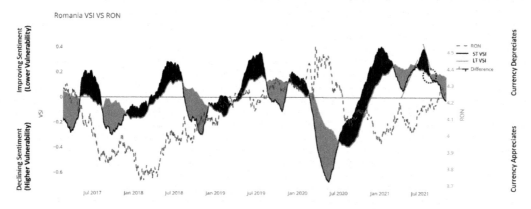

Figure 3.16 Romania's Vulnerability Sentiment Index (VSI) Crossovers: Short- and Long-Term Sentiment functions

A second example pertains to broad sentiment around Romania during 2021. The slope of the short-term sentiment function (see Figure 3.16) turned positive in mid-April 2021. At the time, spreads on Romania's hard currency bonds had been on a tightening trend. While they would tighten a further fifteen basis points following this inflection point, investors remained cautious on the credit, citing still meaningful political risks. In mid-July, however, the slope of the short-term sentiment function (ST VSI) flipped from positive to negative, prompting a renewed discussion of the country's political environment. When layered on top of other fundamental concerns, as well as spread levels that, while off their May tights, were still well inside of their 1Q2021 wides, the case for moving to a more defensive stance began to take shape. Just a few weeks later, right around the time when the long-term sentiment function (LT VSI) and the short-term sentiment function (ST VSI) crossed (negatively), spreads on Romanian bonds began to widen dramatically.

A final example merits attention. In late January 2020, nearly every country and regional VSI turned sharply negative. The speed with which negative sentiment in the Asia and Asia-Pacific region took hold, as measured by the negative slope coefficient of the short-term sentiment function, was especially pronounced (see Figure 3.17). If investors, politicians, or policymakers harbored any doubts about the credibility or durability of the international medical community's warnings regarding COViD-19, the GEMS VSIs would have helped to dispel them. Two weeks after the negative VSI signal, the S&P 500 began falling precipitously. At its trough on March 23, it had fallen 33%. Over this same time period the yield on the ten-year US Treasury note had been cut in half—by about 0.85%.

While the VSI construct in GEMS, as currently defined and implemented, is by no means perfect, it has proved capable of imparting quick and unbiased

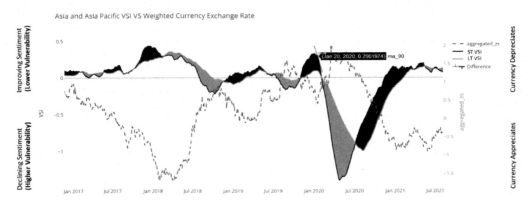

Figure 3.17 Asian Region Vulnerability Sentiment Index (VSI) Crossovers: Short- and Long-Term Sentiment functions

information that can help predict performance of certain financial risk assets. In the case of Saudi Arabia's dollar bonds in January 2019, the crossover of the short and long-term moving averages strengthened an investment team's conviction in their evolving thesis that the bonds were primed for a rally. In the same team's continual assessment of Romania in 2021, it was the change in slope of the short-term moving average, rather than a crossover of the long-term and short-term averages, that drew attention and hastened further investment dialog. And finally, in early 2020, the deterioration in regional and country-level VSI measures presaged the subsequent fall in global asset prices. In summary, the GEMS model can be used in different ways, at different times and with respect to different investment opportunities to help better inform decision-making.

3.5 CONCLUSION

In its application to modern finance, sentiment analysis can serve as a differentiated lens through which to identify and evaluate investment opportunities. The GEMS model, using proprietary techniques, dissects large quantities of data to help analyze economic, political, social and market sentiment across the globe, in the process uncovering previously neglected sources of information and risk. In its simplest use case, the model output complements existing investment processes, affording better decision making by portfolio and risk managers. It can also help inform the design and refinement of systematic trading strategies. The GEMS application may also reduce the need for certain market participants to conduct expensive and time consuming "boots on the ground" research. As an alternative "boots in the cloud" solution, it may help mitigate information asymmetry challenges that local research efforts cannot, while at the same time offering an unbiased and replicable analytical framework.

In addition to its financial markets applications, the GEMS framework promises compelling use cases for a variety of other disciplines including academia, government and central bank policymaking (particularly in emerging markets countries), and management consulting. And it may offer practical solutions for multinational corporations as well. Its highly customizable features allow these different users to construct their own sentiment measures, defined not just according to geographies, but by economic, cultural and linguistic themes, topics of interest and myriad other concerns. It follows naturally from this that the GEMS model serves as fertile ground for AI-driven refinement processes, leaving it uniquely well-positioned to benefit from, and capitalize on, continued advances in data and information technologies.

REFERENCES

Casanova, C., Ortiz, A., Rodrigo, T., Xia, L., Iglesias, J., et al. (2017). Tracking Chinese vulnerability in real time using Big Data. https://www.bbvaresearch.com/wp-content/uploads/2017/03/20170328_CVSI2.pdf.

Diakoulaki, D., Mavrotas, G., and Papayannakis, L. (1995). Determining objective weights in multiple criteria problems: the critic method. *Computers & Operations Research*, 22(7):763–770.

Peterson, R. L. (2016). *Trading on Sentiment: The Power of Minds Over Markets*. John Wiley & Sons.

RavenPack (2021). Developed markets sovereign bonds investing-enhancing style with sentiment. *Wilmott*, 2021(116):10–13.

(II)

COUPLING MODELS WITH ALTERNATIVE DATA FOR FINANCIAL ANALYTICS

Enhanced Corporate Bond Yield Modeling Incorporating Macroeconomic News Sentiment

Zhixin Cai

OptiRisk Systems Ltd, London, United Kingdom

Christina Erlwein-Sayer

Financial Mathematics, University of Applied Sciences HTW Berlin, Berlin, Germany

Gautam Mitra

CEO, OptiRisk Systems Ltd and UCL Department of Computer Science, London, United Kingdom

CONTENTS

DOI: 10.1201/9781003293644-4

4.1 INTRODUCTION

4.1.1 Background: Evolving Trading Venues of Fixed Income Products

Since electronic trading has been increasingly adopted in all sectors of financial markets, it is essential to highlight the trend of changing the fixed income trading situation and trade venues in the market. Fixed-income markets traditionally involve substantially higher amount of Over-The-Counter trades then equity markets, see e.g. [Bessembinder et al., 2020] for a study on Fixed Income trading in recent years. But trading patterns also in Fixed Income markets expand: trading and clearing are more frequently based on electronic trading. In general, electronic trading is the trading of financial instruments through an electronic system where two counterparties are matched and engage in negotiation and execute their orders. Electronic trading platforms have several forms of trading protocols:

- Request for quote (RFQ)—RFQ is a common trading protocol where users request prices on an order of a specific size by contacting the platform market makers. RFQ systems vary massively in terms of several aspects: disclosure of the participant identities; revelation of the sign of the orders (buy or sell) and execution details of the orders (executable quote or indicative quote). What usually happens in the fixed-income RFQ trading protocols is that the requests from clients are delivered only to dealers and only in restricted numbers. These systems are generally used in the markets with securities which have a wide range of numbers and varieties and which are traded without sufficient large dealers. Tradeweb is an example of using RFQ systems.

- Central limit order book (CLOB)—CLOB is a trading protocol where the outstanding bids and offers are stored in a queue and these orders must be executed according to a priority rule. Different from RFQ, the quotes of CLOBs are often transparent to the users before the trades and CLOB systems are mostly used for strategies that require a high-speed trade environment. Examples for CLOB protocols are BrokeTec and eSpeed.

- All-to-all: All-to-all trading protocol is an emerging channel of electronic bond trading and allows multiple parties to directly access one another, creating a highly transparent trading background. Examples for the All-to-all protocols are Tradeweb and MarketAxess.

Electronification has evolved to a more advanced level in the largest, highly liquid and more standardized markets such as the highly liquid sovereign bond market and fixed income futures. According to [Nagel, 2016], around 90% of the transaction of fixed income futures were traded electronically in 2015, even higher than cash equities, credit default swap (CDS) index and spot foreign exchange (FX). A little less electronic asset classes are US treasuries, European government bonds and standardized interest rate swap (IRS), whose electronification rate is nearly 70%. As for the corporate bond market, electronic trading is less prevalent because corporate bonds are traded less frequently due to the diversity of the needs of market participants. However, corporate bond trading via electronic protocols has risen in recent years. MarketAxess [MarketAxess, 2016] stated that around 13% of all the US investment-grade and high-yield corporate bonds were transacted on their electronic trading platforms. A recent study by [O'Hara and Zhou, 2021] finds that electronic bond trading is growing but still limited.

Varieties of innovations in electronic platforms have emerged in the corporate bond markets. One of the innovations that should be highlighted is the all-to-all trading platform. It is estimated that around 5% of investment grade and high-yield bonds which are electronically traded are transacted via all-to-all protocols [Nagel, 2016]. Since all-to-all platforms allow end clients to participate equally and directly with each other, the transaction costs are lower and liquidity is more accessible.

4.1.2 News Sentiment in Fixed Income Market

News sentiment and its use in modeling risk and credit spread stems from a well-known research area on investor sentiment in financial markets. We outline some research on the effect of investor sentiment and macroeconomic fundamentals on fixed income markets that have an effect of today's use of news sentiment and alternative data. The investor sentiment referred to in this section is a broad concept including all the sentiment generated from market, media, economic fundamentals and so on. Researchers have used several indicators or indexes to represent investor sentiment in their studies. Baker and Wurgler [Baker and Wurgler, 2006] proposed using six separate proxies (the average first-day return and number of initial public offerings, the average of the closed-end fund discount, NYSE stock turnover, the newly issued equity share and the dividend premium) to construct a sentiment index. Based on these proxy variables as sentiment measures, they found that investor sentiment has explanatory power on the cross section future stock returns.

Laborda and Olmo [Laborda and Olmo, 2014] adopted these sentiment proxies as well. However, they applied sentiment analysis in the fixed income market instead and found that market sentiment possesses the predictive power on U.S. treasury bond excess return and this effect is extraordinary even in the recession period.

Fernandes, Gama and Vieira [Fernandes et al., 2016] investigated the predictability of sovereign bonds by exploiting the irrational sentiment in Portugal and the Euro area. In their study, they regressed the Economic Sentiment Indicator from the European Commission on macroeconomic fundamentals and obtained the residuals as the irrational sentiment. Their research indicated that the investor sentiment in Portugal and the Euro area negatively impacted future sovereign bond yield spreads and this effect is more obvious in the bailout period.

Although scarce, several researchers have conducted research about the effect of sentiment on the corporate bond market. Nayak [Nayak, 2010] is inspired by the sentiment measure index created by Baker and Wurlger [Baker and Wurgler, 2006] and applies this sentiment measure to the study of the sentiment effect in the US corporate bond yield spreads. They found that US corporate bonds appear to be undervalued when the sentiment is pessimistic and overvalued when the sentiment is optimistic. This result indicates that corporate bond yield spreads are likely to be positively correlated with the sentiment in the market. In addition, the authors suggests that high-yield bonds are more vulnerable to the effect of sentiment than the investment grade bonds.

As plenty of empirical evidence demonstrates the valuable role of sentiment on the valuation of assets, the extent to which news sentiment affects the credit quality of corporate bond merits investigation. In our research, we investigate the impact of news sentiment on European corporate bond yield spreads. The news sentiment we discuss in the report is sentiment arising from machine learning textual analysis of news articles. RavenPack is the provider of the news sentiment data, its news analytic tool, RavenPack News Analytics (RPNA), provides structured sentiment data processed and transformed from reputable sources including Dow Jones Newswires, the Wall Street Journal and over 19,000 other news media sites. A big advantage of RPNA is that all the events in news articles and social media are assigned an entity-specific relevance and sentiment score. With these a data analytics, we can measure the news sentiment, identify the relevant entity and evaluate the extent to which the news is relevant to the entity.

News sentiment applied in credit risk assessment is a growing research area which merits further investigation. diBartolomeo [diBartolomeo, 2016] illustrates the role of news and sentiment for credit risk assessment of corporate debt and proposed that news flows help to improve the calibration of a contingent claims model. A more efficient and transparent estimate of key parameters and probability of default can be obtained. Apergis [Apergis, 2015] explored the predictive power of the newswire messages to forecast CDS spreads in the European market during the financial distress period. CDS spread is widely known as an important indicator of the quality of the associated bonds. Apergis (2015) argued that CDS spreads can also be influenced by the sentiment conveyed through news articles and social media. By comparing the out-of-sample forecasting results of ARIMA (without news) and ARIMAX (with news), it was found that news sentiment provides superiority in the forecasting of CDS spreads especially in the financial distress period. Inspired by the methodology of Apergis [Apergis, 2015], our research will examine the individual effect of positive and negative sentiment in forecasting corporate bond yield spreads.

4.1.3 Outline of the paper

The structure of this report in the following sections is as follows: Section 4.2 outlines the data preparation approach including both the bond and sentiment data. In Sections 4.3 and 4.4, several model set-ups are explained and techniques used to test the accuracy of models are provided. Section 4.5 illustrates the results of fitting and forecasting after using macroeconomic and firm-specific news sentiment as the explanatory factors. The concluding Section 4.6 summaries the findings and proposes some suggestions for the research direction in the future

4.2 DATA

4.2.1 Corporate Bond Data

The data source of our corporate bonds data set is Thomson Reuters DataScope. We investigated the corporate bonds issued by seven companies from three main countries in Europe, these companies are all listed in the Euro Stoxx 50 index. We analyze the stocks Adidas AG, Deutsche Bank AG and Munich Re Group from Germany; Banco Santander, S.A. and BBVA S.A. from Spain; Enel and Eni S.p.A. from Italy. All the corporate bonds we investigate are those issued after 1 January 2007. The time series of bond yields and news sentiment analysed cover the period from 1 January 2007 to 15 May 2017.

4.2.2 News Sentiment Data

The sentiment data used for the research is based on three news sentiment databases provided by RavenPack. The three databases cover the news sentiment in Germany, Spain and Italy and each of the three datasets is

TABLE 4.1 Bond database

Bond	Company	Country	Bond Description	In-Sample Period	Out-of-Sample Period
ADIDAS	Adidas AG	Germany	0.250 06/14/19 CVT PUT	2012/03/20-2014/10/15	2014/10/16-2017/05/15
DBKG1	Deutsche Bank AG	Germany	2.500 03/31/09 MATd	2008/01/21-2008/08/19	2008/08/20-2009/03/25
DBKG2	Deutsche Bank AG	Germany	4.537 06/20/14 MATd	2009/10/14-2012/02/14	2012/02/15-2014/06/17
MUVG	Munich Re Group	Germany	7.625 06/21/28 18 FRN	2007/01/02-2012/05/16	2012/05/17-2017/05/15
SAN1	Santander	Spain	4.250 05/06/13 MATd	2007/04/30-2010/05/06	2010/05/07-2013/04/30
SAN2	Santander	Spain	4.000 08/02/13 MATd	2011/08/05-2012/07/31	2012/08/01-2013/07/30
BBVA1	BBVA S.A.	Spain	6.200 07/04/23	2008/07/04-2011/09/13	2011/09/14-2014/11/04
BBVA2	BBVA S.A.	Spain	0.750 01/20/22	2015/01/15-2016/03/15	2016/03/16-2017/05/15
ENEL	Enel	Italy	5.250 01/14/15 MATd	2009/01/19-2012/01/18	2012/01/19-2015/01/12
ENI	Eni	Italy	1.064 06/29/15 FRN MATd	2009/09/28-2012/08/14	2012/08/12-2015/06/25

comprised of news sentiment data of several entities. RavenPack categorizes all the entities involved in the news by COMP (company), ORGA (organization), PEOP (people) and PLCE (place). For convenience, we name the three main databases as the Germany Sentiment Database, Spain Sentiment Database and Italy Sentiment Database separately. As RavenPack has identified the entity involved in news, it would be convenient to extract firm-specific news sentiment data from the corresponding sub database. In addition to the firm-specific news sentiment, we can also extract macroeconomic news sentiment from the main database. Since most of the macroeconomic announcements are made by government organizations, this paper also examines the impact of news sentiment with different government ministries as the related entities.

Each sub-database of news sentiment data includes but is not limited to the following information:

Year, date and hour: these three figures represent the time and date when the specific news was received by RavenPack.

Relevance: This is a relevance score which ranges from 0 to 100, indicating the extent to which the entity is related to the news item. The higher the value, the stronger the relevance. In our analysis, we set the threshold to be 60, i.e.

only the news sentiment with relevance score greater than 60 is considered as significantly relevant.

ESS-Event Sentiment Score: This score ranges from -1 to 1 in which a negative value represents the sentiment from a negative news story and a positive value represents positive sentiment. The greater the absolute value of ESS, the stronger the sentiment conveyed through the news story. When ESS is zero, it means the news sentiment is neutral.

What should be noted when using the raw data is that all bond trading date is presented according to UTC (coordinated universal time) standard; however, the date listed in the news sentiment database is based on the local time of the country for which the news is listed. Therefore, it is necessary to adjust the date and time of news sentiment databases to a universal time zone. In addition, all the news originally occurring during holiday and weekend needs to be shifted to the next working day. One important assumption in our research is that the opening time of the bond market is from 8.00 am to 18.30 pm (UTC) in each working day. We assume, first, that the news breaks during holidays and weekends to be effective at 8.00 am of the next working day and, second, that a piece of news that pops up after 18.30 pm impacts the bond on the next working day.

The following table summarizes the databases regarding both macroeconomic sentiment and firm-specific sentiment databases we may use in the models.

4.3   METHODOLOGY

4.3.1   News Sentiment Data Aggregation

Sentiment scores are categorized as positive and negative. This research is aimed at analyzing effects of positive and negative news sentiment both separately and aggregated. We therefore transform sentiment scores into the following figures:

Case 1: sum of positive and negative sentiment scores separately per day:

$$\text{possum} = \sum \text{ESS}^+ \text{ and negsum} = \sum \text{ESS}^- \tag{4.1}$$

where ESS^+ and ESS^- are positive and negative event sentiment scores in the analyzed sentiment database.

Case 2: average of positive/ negative sentiment scores separately per day:

$$\text{posmean} = \frac{1}{N^+} \sum \text{ESS}^+ \text{ and negmean} = \frac{1}{N^-} \sum \text{ESS}^- \tag{4.2}$$

TABLE 4.2 Database

Country code: Germany (DE), Spain (ES), Italy (IT)		
Germany News Sentiment Database		
Entity	Entity Type	No. of data relevance ≥ 60
Germany	PLCE	165155
Government of Germany	ORGA	62439
Parliament of Germany	ORGA	857
Central Bank of Germany	ORGA	343
Adidas AG	COMP	1531
Deutsche Bank	COMP	2401
Munich Re Group	COMP	578
Spain News Sentiment Database		
Entity	Entity Type	No. of data relevance ≥ 60
Spain	PLCE	97605
Government of Spain	ORGA	37480
Central Bank of Spain	ORGA	702
Banco Santander	COMP	7081
BBVA S.A.	COMP	5428
Italy News Sentiment Database		
Entity	Entity Type	No. of data relevance ≥ 60
Italy	PLCE	94163
Government of Italy	ORGA	42984
Central Bank of Italy	ORGA	610
ENEL	ORGA	2480
ENI	ORGA	6836

where N^+ and N^- are the number of positive ESS and negative ESS per day. If, for any day, we do not observe positive or negative ESS, we consider the ESS as 0 (neutral ESS).

Case 3: sum of all sentiment scores per day:

$$\text{aggregate} = \text{possum} + \text{negsum}. \tag{4.3}$$

This figure aggregates positive and negative sentiment scores.

Case 4: impact score, which includes exponential decay effect of news sentiment for daily aggregated positive or negative sentiment:

$$\text{Impact score} = \text{ESS} * e^{-\lambda(T-t)} \tag{4.4}$$

where T represents the closing time of the bond market and t is the time news is detected. Therefore, $(T - t)$ measures the difference between time the news break and market closing time.

The impact score with decay was firstly introduced by Yu [Yu, 2014].
The calculation method in our model is based on the IS score in Yu
and Mitra [Yu and Mitra, 2016]. With exponential decay, news story
only have half of the initial impact left after a specific time span. In
our models, we specify this time span as 90 minutes, the parameter λ
is determined through $1 * exp^{-90\lambda} = \frac{1}{2}$, therefore $\lambda = 0.0077$.
The daily impact score is again determined separately for positive and
negative news through

$$posimpact = \sum ESS^{+} * exp-0.0077 * (T - t) \tag{4.5}$$

and

$$negimpact = \sum ESS^{+} * exp-0.0077 * (T - t) \tag{4.6}$$

Case 5: average of positive and negative impact scores (IS) per day:

$$posimpactmean = \frac{posimpact}{N^{+}} and posimpactmean = \frac{posimpact}{N^{-}} \tag{4.7}$$

Case 6: Aggregate positive impact score and negative impact score:

$$sumimpact = posimpact + negimpact \tag{4.8}$$

4.3.2 Bond Yield Spread Calculation Method

In finance, a yield spread measures the difference between the yields of two
assets with same maturity. In our study, the ECB yield rates published by
European Central Bank are used as the representative of benchmark bond
yield. The Svensson parametric method (Svensson [Svensson, 1994]) is adopted
to calculate the yield spreads of the corporate bonds. Before obtaining the
spread, the Svensson function is used to obtain the Svensson yield (spot rate
of ECB AAA-rated bond with the same time-to-maturity as the corporate
bonds):

$$SvenssonYield_t(\tau) = \beta_0 + \beta_1 \frac{1 - \exp(-\frac{\tau}{\lambda_1})}{\frac{\tau}{\lambda_1}} + \beta_2 \left(\frac{1 - \exp(-\frac{\tau}{\lambda_1})}{\frac{\tau}{\lambda_1}} - \exp(-\frac{\tau}{\lambda_1}) \right)$$
$$+ \beta_3 \left(\frac{1 - \exp(\frac{\tau}{\lambda_2})}{\frac{\tau}{\lambda_2}} - \exp(-\frac{\tau}{\lambda_2}) \right)$$
$$\tag{4.9}$$

where six parameters $(\beta_0, \beta_1, \beta_2, \beta_3, \lambda_1, \lambda_2)$, which are quoted by ECB, are
utilized with the time-to-maturity τ. Then the yield spread is obtained by
subtracting the spot rate of ECB AAA-rated bond (Svensson Yield) from the
current corporate bond yield.

4.4 ARIMA MODELS

4.4.1 ARIMA

A classical ARIMA time series model is utilized in our study of sentiment effects. We use a difference-staionary ARIMA-model with order of difference d and ARIMA parameters p and q.

The procedure to fit an ARIMA model is the Box-Jenkins method (Box and Jenkins [Box and Jenkins, 1976]), which contains three stages: model identification, parameter estimation and diagnostic test. Using a similar procedure to determine the model parameters, we use the Akaike information criterion (AIC) as the key criteria for selection of parameter orders. According to the Akaike [Akaike, 1974], AIC is determined as

$$AIC = -2\log(L) + 2(p + q + k + 1) \qquad (4.10)$$

where L is the likelihood of data, p and q are the orders of AR and MA and k is the number of parameters in the model (the number of explanatory variables plus one constant intercept). To avoid overfitting of the models, we select the candidate models with AR parameter p ranges from 0 to 6 and MA parameter q from 0 to 3. In addition, we determine the parameter of differencing d by increasing it from 0 and use the smallest possible d that makes the time series stationary. After deciding on d, we choose the candidate models with the smallest AIC. Since the purpose of this report is to examine whether a simple ARIMA model can be enhanced by adding news sentiment as external regressor (ARIMAX model), the parameter set of ARIMA is chosen as close as possible to that of the corresponding ARIMAX models.

4.4.2 ARIMAX

An ARIMAX model refers to the autoregressive integrated moving average model with explanatory variable. The explanatory variables include macroeconomic sentiment and firm-specific sentiment. To examine the forecasting accuracy of ARIMAX with one external variable and ARIMAX with multiple external variables, we design the following experiments:

- Examine the forecasting accuracy with single external variables in Table 3-1-1; these models are specified as one-variable ARIMAX models with

$$\phi(L)(1 - L)^d \text{Spread}_t = \theta(L) \text{Sentiment}_t + \theta(L)\epsilon_t \qquad (4.11)$$

θ is the coefficient for the external variable *sentiment*, which is chosen from the sentiment variables given in Cases 1 to 6 above.

- Expand the univariate ARIMAX model to a model with two external variables. In the first experiment of univariate ARIMAX model, positive and negative news sentiment are separately included as external explanatory variable of the univariate ARIMAX model. The expression for the multivariate ARIMAX model with two explanatory variables can be expressed as:

$$\phi(L)(1-L)^d \text{Spread}_t = \theta_1(L)\text{Sentiment}_{1t} + \theta_2(L)\text{Sentiment}_{2t} + \theta(L)\epsilon_t$$
(4.12)

As one of the key condition for using ARIMA and ARIMAX model is that the time series data should be stationary, we apply an augmented Dickey–Fuller test (ADF) (Fuller [Fuller, 1976]) to detect the stationarity of the data. The null hypothesis of the augmented Dickey-Fuller test is that a time series exhibits the feature of a unit root process. The alternative hypothesis is that the time series sample is stationary or trend stationary, depending on the specific test used. In this study, we used the ADF test to see whether the time series sample is stationary.

4.4.3 Model evaluation

For model evaluation, we consider mean absolute error (MAE), root mean square error (RMSE), mean absolute percent error (MAPE) and mean absolute scaled error (MASE) as the main measures of prediction accuracy of the models. MAE and RMSE are scale-dependent error measures and measure the prediction error of the models on average.
MAPE is associated with the percentage error

4.5 PREDICTIONS OF CORPORATE BOND DATA

4.5.1 Time Series Stationarity Test

The following table shows the ADF test results of 10 corporate bonds time series. All the original time series data of the yield spread exhibit non-stationarity; therefore, they cannot be applied directly to the models. However, they are all stationary when a difference of one lag is taken. As discussed in Section 3, ARIMA and ARIMAX model includes a differencing parameter representing the maximum times of difference to make time series stationary. Based on the ADF results from Table 4.3, the differencing order d takes a value of 1 for all ARIMA and ARIMAX models.

TABLE 4.3 Augmented Dickey–Fuller test

Bond	p-value (original time series)	p-value (difference=1)
ADIDAS	0.9900	0.01
DBKG1	0.8803	0.01
DBKG2	0.9194	0.01
MUVG	0.3220	0.01
SAN1	0.8753	0.01
SAN2	0.4893	0.01
BBVA1	0.6679	0.01
BBVA2	0.1426	0.01
ENEL	0.8401	0.01
ENI	0.7109	0.01

TABLE 4.4 ARIMA In-sample Result

Result number	Bond	RMSE	MAE	MAPE
1	ADIDAS	0.107808	0.7840896	2.621932
2	DBKG1	0.04426089	0.02974929	2.900349
3	DBKG2	0.03307575	0.02292463	1.758099
4	MUVG	0.1401733	0.05545005	2.240148
5	SAN1	0.05301667	0.03537848	4.517674
6	SAN2	0.08413574	0.05915032	1.772193
7	BBVA1	0.03801894	0.0270464	1.850272
8	BBVA2	0.02565777	0.01912145	4.491516
9	ENEL	0.08353927	0.0504008	3.389179
10	ENI	0.09560577	0.05754326	22.44376

4.5.2 ARIMA Results

The following table 4.4 shows the in-sample calibration accuracy results of the ARIMA models for all the bonds: After training the ten ARIMA models with the sample time series data, we can forecast the future values of corporate bond spreads based on these fitted models.

In Table 4.5, it shows the out-of-sample forecast accuracy of the ARIMA models.

4.5.3 ARIMAX Results

4.5.3.1 ARIMAX with Country News Sentiment

First, we compare the model accuracy results of the ARIMAX models with the country macroeconomic news sentiment as external variables. Starting from the corporate bonds in Germany, the associated country macroeconomic news regard the topics: balance-of-payments, business-activity, consumption,

TABLE 4.5   ARIMA Out-of-sample Result

Result number	Bond	RMSE	MAE	MAPE
1	ADIDAS	0.3300	0.2159	2.8460
2	DBKG1	0.1815	0.0744	2.7813
3	DBKG2	0.0665	0.0305	6.7680
4	MUVG	0.0569	0.0377	3.7316
5	SAN1	0.0945	0.0547	3.8240
6	SAN2	0.0563	0.0367	2.5495
7	BBVA1	0.1057	0.0264	0.7900
8	BBVA2	0.0264	0.0190	3.4746
9	ENEL	0.0604	0.0336	2.4654
10	ENI	0.0499	0.0251	4.6916

TABLE 4.6   ADIDAS BOND ARIMAX with country news sentiment (in-sample)

External Variable(s)	RMSE	Diff	MAE	diff	MAPE	diff
G1possum	0.10680	−0.001013	0.07805	−0.70604	2.6083	−0.0136
G1negsum	0.10683	−0.000980	0.07805	−0.70604	2.6288	0.0069
G1posmean	0.10690	−0.00091	0.07793	−0.70616	2.6220	0.0001
G1negmean	0.10688	−0.000924	0.07791	−0.70618	2.6216	−0.0003
G1aggregate	0.10691	−0.000898	0.07797	−0.70611	2.6248	0.0029
G1posimpact	0.10678	−0.001025	0.07787	−0.70622	2.6166	−0.0053
G1negimpact	0.10684	−0.000964	0.07798	−0.7061	2.6285	0.007
G1sumimpact	0.10691	−0.000893	0.07801	−0.7061	2.6268	0.0049
G1pos/negsum	0.10668	−0.001124	0.07781	−0.70628	2.6168	−0.0051
G1pos/negimpact	0.10674	−0.001064	0.07784	−0.70625	2.6190	−0.0029
G1posimpactmean	0.10699	−0.000821	0.07824	−0.70585	2.6220	0.000033
G1negimpactmean	0.10691	−0.000901	0.07806	−0.70603	2.6286	0.0066
G1pos/negimpactmean	0.10689	−0.000919	0.07803	−0.70606	2.6280	0.0061

credit, crime, domestic product, elections, employment, foreign-relations, government, housing and industrial accidents, among others.

In the above table, the first column represents the external variables in the ARIMAX models. What should be noted is that the columns with "G1pos/negsum", "G1pos/negimpact", "G1pos/negimpactmean" are the multivariate ARIMAX models with two external variables. "G1pos/negsum" represents two external variables in the multivariate ARIMAX model which are G1possum and G1negsum. "G1pos/negimpact" corresponds to two external variables: G1posimpact and G1 negimpact, "G1pos/negimpactmean" stands for G1posimpactmean and G1negimpactmean. "diff" is the abbreviation of difference, which measures the gap between the error measure of ARIMAX model and that of corresponding ARIMA model. For example, in Table 4.6, the RMSE of ARIMAX model with G1possum as the external variable for ADIDAS bond is 0.1067952. Subtracting the RMSE of simple ARIMA model for

TABLE 4.7 In sample model accuracy DBKG1 Bond ARIMAX models with Country News Sentiment

External Variable(s)	RMSE	diff	MAE	diff	MAPE	diff
G1possum	0.044229	−0.00003	0.029748	−0.00002	2.9005	0.000198
G1negsum	0.044201	−0.00006	0.029773	0.00002	2.9000	−0.00031
G1posmean	0.044260	−0.00000	0.029737	−0.00001	2.8992	−0.00115
G1negmean	0.043851	−0.00041	0.029875	0.000126	2.90936	0.009011
G1aggregate	0.044157	−0.000103	0.029781	0.00003	2.90113	0.000776
G1posimpact	0.043992	−0.00027	0.029482	−0.00027	2.87808	−0.02227
G1negimpact	0.044247	−0.00001	0.029838	0.00009	2.9083	0.007951
G1sumimpact	0.044081	−0.00018	0.029828	0.00008	2.90827	0.00792
G1pos/negsum	0.044155	−0.00011	0.02978	0.00003	2.90048	0.000131
G1pos/negimpact	0.043977	−0.00028	0.029580	−0.00017	2.88683	−0.01352
G1posimpactmean	0.044107	−0.000154	0.029645	−0.0001	2.88702	−0.01333
G1negimpactmean	0.044261	0.00000	0.029751	0.0000	2.900482	0.000133
G1pos/negimpactmean	0.044096	−0.00016	0.029745	0.0000	2.896088	−0.00426

ADIDAS bond yield spread (0.107808, as shown in Table 4.6) from 0.1067952 obtains the difference as -0.0010128. Therefore, when an ARIMAX with news sentiment improve the model accuracy, the value of diff is negative. From Table 4.6, it can be found that both positive country news sentiment and negative country news sentiment improve the in-sample model accuracy as they reduce RMSE, MAE and MASE of ARIMAX model. However, the improvement of in-sample model performance does not guarantee that forecast accuracy will be improved as well in the out-of-sample test. Out-of-sample model accuracy is a more straightforward measure when examining how much the prediction accuracy is enhanced by using ARIMAX models with news sentiment.

As observed in Table 4.7 and Table 4.8, country news sentiment scores and impact scores improve the model accuracy and it is not limited to the in-sample model performance. Positive average sentiment score (G1posmean), negative average sentiment score (G1negmean), positive impact score (G1posimpact), overall impact score (G1sumimpact) and average positive impact score (G1posimpactmean) improve three of four performance measures. Negative impact score (G1negimpact), average negative impact score (G1negimpactmean) and the pair of variables: G1pos/negimpactmean enhance all four model accuracy measures. Overall, these results prove that using negative country sentiment can improve the forecast of Deutsche Bank corporate bond yield spread.

Table 4.9 and Table 4.10 present the model accuracy of the univariate ARIMAX models with country news sentiment for another Deutsche Bank bond. It can be found that average positive sentiment score (G1posmean), aggregate sentiment score (G1 aggregate) and average positive impact (G1posimpactmean) improve the forecast of Deutsche Bank bond. It seems

TABLE 4.8 Out-of-sample model accuracy: DBKG1 Bond ARIMAX models with Country News Sentiment

External Varibale(s)	RMSE	diff	MAE	diff	MAPE	diff
G1possum	0.1818255	0.0003591	0.07445284	0.0000251	2.77875	−0.00254
G1posmean	0.1814307	−0.0000357	0.07441523	−0.000013	2.781334	0.000042
G1negmean	0.1808681	−0.0005983	0.07417954	−0.00025	2.788869	0.007577
G1aggregate	0.1821014	0.000635	0.0747986	0.000371	2.775641	−0.00565
G1posimpact	0.1826718	0.0012054	0.0743033	−0.00012	2.773576	−0.00772
G1negimpact	0.1813182	−0.0001482	0.07435342	−0.000074	2.775602	−0.00569
G1sumimpact	0.1816988	0.0002324	0.0740542	−0.00037	2.756752	−0.02454
G1pos/negsum	0.1820783	0.0006119	0.07489974	0.000472	2.77841	−0.00288
G1pos/negimpact	0.1825001	0.0010337	0.07410882	−0.00032	2.762585	−0.01871
G1posimpactmean	0.1815903	0.0001239	0.07397393	−0.00045	2.759834	−0.02146
G1negimpactmean	0.1814557	−0.0000107	0.07441574	−0.000012	2.78103	−0.00026
G1pos/ negimpactmean	0.181428	−0.0000384	0.07378516	−0.00064	2.753941	−0.02735

TABLE 4.9 In-sample model accuracy: DBKG2 Bond ARIMAX models with Country News Sentiment

External Variable(s)	RMSE	diff	MAE	diff	MAPE	diff
G1possum	0.033007	−0.000068	0.02299	0.000066	1.761683	0.003584
G1negsum	0.03304355	−0.0000322	0.023006	0.000081	1.766524	0.008425
G1posmean	0.033076	−0.00000003	0.022923	−0.000002	1.757887	−0.000212
G1negmean	0.033000	−0.00007599	0.022896	−0.000029	1.756951	−0.001148
G1aggregate	0.033076	0.0000001	0.022924	0.000000	1.758067	−0.000032
G1posimpact	0.032957	−0.000119	0.022936	0.000012	1.757815	−0.000284
G1negimpact	0.032928	−0.0001475	0.022968	0.000043	1.763744	0.005645
G1sumimpact	0.033044	−0.00003205	0.022947	0.000023	1.761202	0.003103
G1pos/negsum	0.032993	−0.00008256	0.023033	0.0001081	1.76641	0.008311
G1pos/negimpact	0.032866	−0.0002101	0.022937	0.000012	1.760048	0.001949
G1posimpactmean	0.033065	−0.00001049	0.022874	−0.000051	1.754528	−0.003571
G1negimpactmean	0.03295	−0.0001242	0.022873	−0.000052	1.75236	−0.005739
G1pos/ negimpactmean	0.032943	−0.0001327	0.022832	−0.000092	1.749506	−0.008593

that the results from DBKG1 bond and DBKG2 bond are opposite. However, these two results are acceptable when comparing the in-sample period and out-of-sample periods of these two bonds. As discussed in Section 2.1, the first Deutsche Bank corporate bond yield spread is modeled by using the data sample during 2008/01/21–2008/08/19, which is the early stage of recession period. It is reasonable that corporate bond spreads are sensitive to negative country news in a crisis period. Furthermore, Deutsche Bank belongs to the banking industry which is more vulnerable to the unfavorable news stories in

TABLE 4.10 Out-of-sample model accuracy: DBKG2 Bond ARIMAX models with Country News Sentiment

External Variable(s)	RMSE	diff	MAE	diff	MAPE	diff
G1posmean	0.066467	−0.00000346	0.030519	−0.00000238	6.767676	−0.000254
G1aggregate	0.066465	−0.00000497	0.030522	0.00000041	6.767901	−0.000029
G1posimpactmean	0.066432	−0.00003892	0.030529	0.00000752	6.766368	−0.001562

TABLE 4.11 In-sample model accuracy: MUVG Bond ARIMAX models with Country News Sentiment

External Variable(s)	RMSE	diff	MAE	diff	MAPE	Diff
G1possum	0.140167	−0.0000065	0.055477	0.0000265	2.241178	0.00103
G1negsum	0.140078	−0.0000953	0.055497	0.0000471	2.240834	0.000686
G1posmean	0.140078	−0.0000955	0.055583	0.000133	2.249474	0.009326
G1negmean	0.140141	−0.0000322	0.055439	−0.000012	2.242864	0.002716
G1aggregate	0.140125	−0.0000482	0.055443	−0.0000072	2.239434	−0.00071
G1posimpact	0.140164	−0.0000094	0.055411	−0.000039	2.237451	−0.0027
G1negimpact	0.140096	−0.0000777	0.05547	0.0000204	2.238471	−0.00168
G1sumimpact	0.140131	−0.0000419	0.055498	0.0000477	2.241905	0.001757
G1pos/negsum	0.140078	−0.0000954	0.055499	0.000049	2.240908	0.00076
G1pos/negimpact	0.140095	−0.0000786	0.055458	0.00000813	2.237684	−0.00246
G1posimpactmean	0.140115	−0.000058	0.05544	−0.00001	2.240552	0.000404
G1negimpactmean	0.140173	−0.0000003	0.055465	0.0000148	2.240981	0.000833
G1pos/negimpactmean	0.140019	−0.000154	0.055221	−0.00023	2.230686	−0.00946

TABLE 4.12 Out-of-sample model accuracy: MUVG Bond ARIMAX models with Country News Sentiment

External Variable(s)	RMSE	Diff	MAE	diff	MAPE	diff
G1possum	0.056849	−9.44E-05	0.037634	−5.90E-05	3.727977	−0.00362
G1posmean	0.056839	−0.000104	0.037835	0.000142	3.740062	0.008464
G1posimpact	0.056848	−9.56E-05	0.037643	−5.00E-05	3.72783	−0.00377
G1posimpactmean	0.056982	3.80E-05	0.037821	0.000129	3.730185	−0.00141
G1negimpactmean	0.056934	−9.68E-06	0.037682	−1.10E-05	3.730316	−0.00128
G1pos/negimpactmean	0.056784	−0.000159	0.037718	2.50E-05	3.704889	−0.02671

that time period. The second bond is modeled with the sample data range from 2009/10/14 to 2012/02/14, when the shadow of financial crisis is fading and economy is recovering. In this stage, positive news sentiment has a significant positive impact on corporate bond yield spread.

It is observed in Table 4.11 and Table 4.12 that both positive and negative country news sentiment can enhance the modelling performance. The most

TABLE 4.13 Out-of-sample model accuracy: SAN1 Bond ARIMAX models with Country News Sentiment

External Variable(s)	RMSE	diff	MAE	Diff	MAPE	Diff
S1negmean	0.0944791	−0.00003052	0.0546984	−0.0000162	3.823579	−0.00039
S1posimpactmean	0.0944708	−0.00003881	0.0548322	0.0001176	3.846613	0.022643
S1negimpactmean	0.0944459	−0.00006372	0.0546908	−0.0000238	3.822997	−0.00097
S1pos/ negimpactmean	0.0944175	−0.00009205	0.0548045	0.00008989	3.844933	0.020963

TABLE 4.14 Out-of-sample model accuracy: SAN2 Bond ARIMAX models with Country News Sentiment

External Variable(s)	RMSE	Diff	MAE	Diff	MAPE	Diff
S1possum	0.0555011	−0.00076693	0.0363472	−0.000396	2.550476	0.000958
S1negsum	0.0562744	0.00000642	0.0367332	−0.0000103	2.553928	0.00441
S1negmean	0.056314	0.00004602	0.0367059	−0.0000376	2.542578	−0.00694
S1aggregate	0.0560616	−0.0002064	0.036521	−0.000222	2.559233	0.009715
S1posimpact	0.0556724	−0.0005956	0.0367861	0.00004262	2.5571	0.007582
S1sumimpact	0.0561974	−0.00007055	0.0367901	0.00004666	2.560074	0.010556
S1pos/negsum	0.055159	−0.001109	0.0360762	−0.000667	2.530981	−0.018537
S1pos/negimpact	0.0555673	−0.00070071	0.0370319	0.0002885	2.575867	0.026349

significant improvement is observed when using multivariate ARIMAX model with average impact score (G1posimpactmean) and average negative impact score (G1negimpactmean) as two external variables. The out-of-sample performance result shows that RMSE is reduced by 0.000159, MAPE is reduced by 0.02671. The models for MUVG corporate bond yield spreads use the sample range from 2007/01/02 to 2012/05/16, which covers the financial crisis in 2008 and the economic recovery period. This result coincides with the results found from Deutsche Bank corporate bonds models.

The next four tables present the out-of-sample forecast performance of the ARIMAX models for the corporate bond yield spreads of four bonds in Spain.

In Spain, country news sentiment data can be used to enhance the forecast of corporate bond yield spread as well. In Table 4.13, average negative sentiment (S1negmean) and average negative impact score (S1negimpactmean) used as external variables in ARIMAX model increase all four model performance measures. Comparable to DBKG1 corporate bond yield spread, we used the sample data before the economic downturn when predicting corporate bond yield spread of SAN1 bond. A similar result suggests that negative news sentiment increases the forecast accuracy of the corporate bond yield spreads before the economic recession. The findings from Table 4.14 and Table 4.15 suggest that multivariate ARIMAX model with both positive and

TABLE 4.15 Out-of-sample model accuracy: BBVA1 Bond ARIMAX models with Country News Sentiment

External Variable(s)	RMSE	Diff	MAE	diff	MAPE	Diff
S1pos/ negimpactmean	0.1061405	−0.000232	0.0268598	−0.0004142	0.803647	−0.009524

TABLE 4.16 Out-of-sample model accuracy: BBVA2 Bond ARIMAX models with Country News Sentiment

External Variable(s)	RMSE	Diff	MAE	Diff	MAPE	Diff
S1possum	0.026417	−1.82E-05	0.0190084	−2.96E-05	3.468441	−0.006195
S1negsum	0.02627988	−0.00015534	0.0189607	−7.73E-05	3.464999	−0.009637
S1negmean	0.02643375	−1.47E-06	0.019046	7.98E-06	3.475935	0.001299
S1aggregate	0.02642596	−9.26E-06	0.0190193	−1.87E-05	3.47119	−0.003446
S1posimpact	0.02644177	6.55E-06	0.0190048	−3.33E-05	3.469971	−0.004665
S1negimpact	0.02633415	−0.00010107	0.0190084	−2.97E-05	3.475015	0.000379
S1pos/negsum	0.02625775	−0.00017747	0.0189378	−0.0001002	3.461608	−0.013028
S1posimpactmean	0.0263607	−7.45E-05	0.0187348	−0.0003033	3.417428	−0.057208
S1pos/ negimpactmean	0.02635634	−7.89E-05	0.0187269	−0.0003112	3.416036	−0.0586

negative country news sentiment data can enhance the prediction of SAN2 and BBVA1 corporate bond yield spreads. The common feature of the yield spreads of SAN2 and BBVA1 is that the spreads change dramatically from in-sample period to out-of-sample period: SAN2 has its yield spreads decreases dramatically and BBVA1 has its yield spreads increase greatly. In Table 4.16, the modelling results for BBVA2 also suggests using multivariate ARIMAX model in predicting corporate bond yield spreads and preferring negative country news sentiment.

4.5.3.2 ARIMAX with Government News Sentiment

Secondly, we compare the out-of-sample model performance of ARIMAX model with the government macroeconomic news sentiment as external variables.

In Table 4.17, DBKG1 ARIMAX model results suggest that both positive and negative Germany government news sentiment can be incorporated in the forecasting models to improve the model performance. They also suggest that positive government news sentiment, in general, is better than negative one in terms of increasing forecast accuracy as its RMSE, MAE, MAPE and MASE are much smaller. However, DBKG2 ARIMAX model results show that it is not necessary that positive Germany government news sentiment is a better

TABLE 4.17 Out-of-sample model accuracy: DBKG1 Bond ARIMAX models with government News Sentiment

External Variable(s)	RMSE	Diff	MAE	diff	MAPE	diff
G2possum	0.1813338	−0.0001326	0.07430372	−0.00012	2.781107	−0.00019
G2posmean	0.1814413	−0.0000251	0.07435303	−0.000075	2.776743	−0.00455
G2negmean	0.1808201	−0.0006463	0.07426214	−0.00017	2.782508	0.001216
G2posimpact	0.1801367	−0.0013297	0.07293078	−0.0015	2.734623	−0.04667
G2sumimpact	0.1812326	−0.0002338	0.07393925	−0.00049	2.764949	−0.01634
G2pos/negimpact	0.1802388	−0.0012276	0.07297465	−0.00145	2.735725	−0.04557
G2posimpactmean	0.1808013	−0.0006651	0.07392818	−0.0005	2.764062	−0.01723
G2negimpactmean	0.1813064	−0.00016	0.07428592	−0.00014	2.771233	−0.01006
G2pos/ negimpactmean	0.1807149	−0.0007515	0.07387129	−0.00056	2.758059	−0.02323

TABLE 4.18 Out-of-sample model accuracy: DBKG2 Bond ARIMAX models with government News Sentiment

External Variable(s)	RMSE	Diff	MAE	diff	MAPE	diff
G2posmean	0.066469	−0.00000143	0.030572	0.0000511	6.774728	0.006798
G2negmean	0.066463	−0.00000696	0.030529	0.00000819	6.768363	0.000433
G2posimpact	0.066466	−0.00000491	0.030516	−0.00000514	6.767209	−0.000721
G2posimpactmean	0.066458	−0.00001277	0.03053	0.00000894	6.76909	0.00116
G2negimpactmean	0.066478	0.00000761	0.030518	−0.00000274	6.767882	−0.000048
G2pos/ negimpactmean	0.066447	−0.00002346	0.030526	0.00000448	6.76952	0.00159

TABLE 4.19 Out-of-sample model accuracy: MUVG Bond ARIMAX models with government News Sentiment

External Variable(s)	RMSE	Diff	MAE	diff	MAPE	diff
G2posmean	0.056914	−0.000029	0.037652	−0.000041	3.726412	−0.00519
G2negmean	0.056844	−0.0001	0.037618	−0.000074	3.724966	−0.00663
G2negimpactmean	0.056942	−0.00000151	0.037721	0.0000287	3.738231	0.006633

input in the ARIMAX model than the negative news sentiment. RMSE can be reduced most by using both average positive impact score (G2posimpactmean) and average negative impact score (G2negimpactmean). However, to reduce MAE, MAPE and MASE, using positive government impact score is a better option than negative government impact score.

MUVG Bond results (Table 4.19) suggest that negative government news sentiment is a better external regressor for ARIAMX in forecasting yield spreads than positive government news sentiment.

TABLE 4.20 Out-of-sample model accuracy: SAN1 Bond ARIMAX models with government News Sentiment

External Variable(s)	RMSE	Diff	MAE	diff	MAPE	diff
S2negsum	0.0943485	−0.000161	0.0544954	−0.000219	3.818407	−0.00556
S2aggregate	0.0943434	−0.0001661	0.0545516	−0.000163	3.820006	−0.00396
S2posimpact	0.0945708	0.00006121	0.0547935	0.0000789	3.828757	0.004787
S2negimpact	0.094371	−0.0001386	0.0545636	−0.000151	3.829043	0.005073
S2sumimpact	0.0943548	−0.0001548	0.0545538	−0.000161	3.826772	0.002802
S2pos/negsum	0.0943458	−0.0001638	0.0546639	−0.000051	3.822513	−0.00146
S2pos/negimpact	0.0943952	−0.0001144	0.0545974	−0.000117	3.831777	0.007807

TABLE 4.21 Out-of-sample model accuracy: SAN2 Bond ARIMAX models with government News Sentiment

External Variable(s)	RMSE	Diff	MAE	diff	MAPE	diff
S2possum	0.056320	0.0000519	0.036619	−0.000124	2.54952	0.000002
S2negsum	0.056288	0.0000202	0.036730	−0.000014	2.549052	−0.000466
S2aggregate	0.056283	0.0000146	0.036739	−0.000004	2.549451	−0.000067
S2posimpact	0.056213	−0.000055	0.036762	0.0000182	2.553779	0.004261
S2negimpact	0.056286	0.0000182	0.036735	−0.000008	2.549152	−0.000366
S2sumimpact	0.056293	0.0000248	0.036734	−0.000010	2.548449	−0.001069
S2pos/negsum	0.056331	0.0000628	0.036608	−0.000135	2.548928	−0.00059
S2pos/negimpact	0.056233	−0.000036	0.036754	0.0000102	2.553443	0.003925

TABLE 4.22 Out-of-sample model accuracy: BBVA1 Bond ARIMAX models with government News Sentiment

External Variable(s)	RMSE	diff	MAE	diff	MAPE	diff
S2negsum	0.106365	−0.000008	0.027251	−0.000023	0.81263	−0.00055
S2posimpact	0.106130	−0.000242	0.027760	0.000486	0.82076	0.00759
S2sumimpact	0.106298	−0.000074	0.027219	−0.000055	0.81205	−0.00112

From Table 4.20 to Table 4.23, results from the modelling of SAN1 and BBVA2 yield spreads suggest using negative government sentiment in the ARIMAX model, however, the models for SAN2 and BBVA1 yield spreads suggest applying both positive and negative government sentiment as two external variables in the ARIMAX model to enhance prediction accuracy of corporate bond yield spread.

For the government news sentiment in Italy, the results show that using negative Italian government news sentiment data as external variables in the ARIMAX model can improve the forecast of ENEL and ENI corporate bond yield spread. In addition to the government news sentiment, we also investigate the parliament news sentiment in Germany. All the ARIMAX models with

TABLE 4.23 Out-of-sample model accuracy: BBVA2 Bond ARIMAX models with government News Sentiment

External Variable(s)	RMSE	diff	MAE	diff	MAPE	diff
S2possum	0.026312	−0.000123	0.01899	−0.000044	3.46435	−0.010285
S2negmean	0.026322	−0.000114	0.01884	−0.000203	3.44276	−0.03188
S2aggregate	0.026422	−0.000014	0.01901	−0.000025	3.47002	−0.004617
S2posimpact	0.026423	−0.000013	0.01903	−0.000006	3.47362	−0.001018
S2negimpact	0.026295	−0.000141	0.01895	−0.000084	3.46117	−0.013465
S2sumimpact	0.026412	−0.000023	0.01883	−0.000203	3.43628	−0.03836
S2pos/negsum	0.026324	−0.000111	0.01900	−0.000037	3.46580	−0.008846
S2pos/negimpact	0.026313	−0.000122	0.01896	−0.000076	3.46294	−0.011691
S2negimpactmean	0.026289	−0.000146	0.01889	−0.000149	3.44878	−0.025854
S2pos/ negimpactmean	0.026460	0.0000250	0.01896	−0.000078	3.46517	−0.00946

TABLE 4.24 Out-of-sample model accuracy: DBKG1 Bond ARIMAX models with Central Bank News Sentiment

External Variable(s)	RMSE	diff	MAE	diff	MAPE	diff
G10negsum	0.1813968	−0.0000696	0.07502644	0.000599	2.826585	0.045293
G10negmean	0.1813968	−0.0000696	0.07502644	0.000599	2.826585	0.045293
G10aggregate	0.1814474	−0.000019	0.07435474	−0.000073	2.77972	−0.00157
G10negimpact	0.1814663	−0.0000001	0.07442273	−0.000005	2.781328	0.000036
G10negimpactmean	0.1814663	−0.0000001	0.07442273	−0.000005	2.781328	0.000036

positive parliament news sentiment enhance the forecast of corporate bond yield spreads of Germany corporate bonds.

4.5.3.3 ARIMAX with Central Bank News Sentiment

Thirdly, we compare the out-of-sample model performance of ARIMAX model with the central bank macroeconomic news sentiment as external variables.

Tables 4.24, 4.25 and 4.26 suggest that both positive and negative German central bank news sentiment can enhance ARIMAX model accuracy, but negative news sentiment is more useful for predicting DBKG1 and Munich Rep Group corporate bond yield spread. Whereas positive news sentiment is more effective in predicting DBKG2 corporate bond yield spread.

As observed in Tables 4.27 to 4.30, both positive and negative Spain central bank news sentiment are helpful in enhancing the prediction model of corporate bond yield as well.

The evidence from Italian bond yield spreads ARIMAX models proves that Italian central bank news sentiment enhances the performance of

TABLE 4.25 Out-of-sample model accuracy: DBKG2 Bond ARIMAX models with Central Bank News Sentiment

External Variable(s)	RMSE	diff	MAE	diff	MAPE	diff
G10possum	0.066468	−0.000002	0.030519	−0.000002	6.76785	−0.00008
G10negsum	0.066460	−0.000010	0.030521	0.0000003	6.76860	0.00067
G10posmean	0.066257	−0.000214	0.03054	0.0000184	6.76885	0.00092
G10negmean	0.066462	−0.000009	0.030533	0.0000118	6.77390	0.00597
G10aggregate	0.066502	0.000032	0.030498	−0.000023	6.76692	−0.00101
G10posimpact	0.066468	−0.000002	0.030547	0.0000259	6.76632	−0.00161
G10sumimpact	0.066469	−0.000001	0.030555	0.0000343	6.76632	−0.00161
G10pos/negsum	0.066469	−0.000001	0.030517	−0.000004	6.76859	0.000657
G10pos/negimpact	0.066494	0.0000234	0.030571	0.0000498	6.76757	−0.00036
G10posimpactmean	0.066407	−0.000063	0.030503	−0.000018	6.76182	−0.00611
G10negimpactmean	0.066465	−0.000005	0.030506	−0.000015	6.76580	−0.00213
G10pos/negimpactmean	0.066403	−0.000067	0.030487	−0.000034	6.75928	−0.00865

TABLE 4.26 Out-of-sample model accuracy: MUVG Bond ARIMAX models with Central Bank News Sentiment

External Variable(s)	RMSE	diff	MAE	diff	MAPE	diff
G10negsum	0.056962	0.00001862	0.037684	−0.00000837	3.730235	−0.00136
G10negmean	0.057009	0.00006555	0.037687	−0.00000577	3.729357	−0.00224
G10aggregate	0.056978	0.00003465	0.037703	0.00001009	3.73134	−0.00026
G10pos/negsum	0.057107	0.00016307	0.037734	0.00004154	3.730559	−0.00104
G10negimpactmean	0.056847	−0.00009615	0.037735	0.00004215	3.736468	0.00487

TABLE 4.27 Out-of-sample model accuracy: SAN1 Bond ARIMAX models with Central Bank News Sentiment

External Variable(s)	RMSE	diff	MAE	diff	MAPE	diff
S4posimpact	0.0945064	−0.0000032	0.054708	−0.0000071	3.824281	0.00031
S4posimpactmean	0.0945094	−0.0000002	0.054713	−0.0000017	3.824104	0.00013

TABLE 4.28 Out-of-sample model accuracy: SAN2 Bond ARIMAX models with Central Bank News Sentiment

External Variable(s)	RMSE	diff	MAE	diff	MAPE	diff
S4negsum	0.056286	0.000018	0.03666	−0.000087	2.533493	−0.016025
S4negmean	0.056395	0.000127	0.03680	0.0000574	2.540976	−0.008542
S4negimpact	0.056308	0.000040	0.03674	−0.000005	2.548477	−0.001041
S4sumimpact	0.056251	−0.00002	0.03676	0.0000189	2.551535	0.002017
S4pos/negsum	0.056429	0.000161	0.03675	0.0000042	2.536715	−0.012803

ARIMAX model (see Tables 4.31 and 4.32). Positive news sentiment increases ARIMAX model performance when forecast the future yield spreads of ENEL bond. Both positive and negative central bank news sentiment boost the model performance for ENI corporate bond. Overall, central bank news sentiment is

TABLE 4.29 Out-of-sample model accuracy: BBVA1 Bond ARIMAX models with Central Bank News Sentiment

External Variable(s)	RMSE	diff	MAE	diff	MAPE	diff
S4possum	0.106358	−0.0000141	0.0272547	−0.00001935	0.812027	−0.001144

TABLE 4.30 Out-of-sample model accuracy: BBVA2 Bond ARIMAX models with Central Bank News Sentiment

External Variable(s)	RMSE	diff	MAE	diff	MAPE	diff
S4possum	0.026435	−0.00000025	0.019040	0.0000019	3.475069	0.000433
S4posmean	0.026434	−0.00000093	0.019039	0.0000014	3.474981	0.000345
S4aggregate	0.026279	−0.00015592	0.018770	−0.000268	3.422546	−0.05209
S4posimpact	0.026432	−0.00000285	0.019038	−0.0000001	3.474728	0.000092
S4pos/negsum	0.026244	−0.00019148	0.018726	−0.0003123	3.416012	−0.058624
S4pos/negimpact	0.026252	−0.00018363	0.018731	−0.0003075	3.417056	−0.05758
S4posimpactmean	0.026432	−0.00000301	0.019038	−0.0000002	3.474707	0.000071
S4pos/negimpactmean	0.026256	−0.0001797	0.0187367	−0.0003013	3.418033	−0.056603

TABLE 4.31 Out-of-sample model accuracy: ENEL Bond ARIMAX models with Central Bank News Sentiment

External Variable(s)	RMSE	diff	MAE	diff	MAPE	diff
I3possum	0.060421	0.00000186	0.033614	−0.00000046	2.465119	−0.00025
I3posmean	0.060428	0.00000953	0.033607	−0.0000077	2.464256	−0.00112
I3posimpactmean	0.06045	0.0000313	0.033617	0.00000195	2.464636	−0.00074

TABLE 4.32 Out-of-sample model accuracy: ENI Bond ARIMAX models with Central Bank News Sentiment

External Variable(s)	RMSE	diff	MAE	diff	MAPE	diff
I3possum	0.049783	−0.0000762	0.025041	−0.0000905	4.67624	−0.015343
I3posmean	0.049845	−0.0000137	0.025085	−0.0000464	4.68441	−0.007169
I3negmean	0.049871	0.0000119	0.025151	0.00001892	4.69431	0.002725
I3aggregate	0.049739	−0.00012	0.024971	−0.000161	4.66740	−0.024179
I3posimpact	0.049836	−0.0000232	0.025104	−0.0000272	4.68699	−0.004591
I3negimpact	0.049837	−0.0000222	0.025093	−0.0000387	4.68775	−0.003833
I3sumimpact	0.049825	−0.0000337	0.025089	−0.0000429	4.68554	−0.006044
I3pos/negimpact	0.049814	−0.000045	0.025078	−0.0000537	4.68445	−0.007136
I3posimpactmean	0.049845	−0.0000136	0.025104	−0.0000277	4.68798	−0.003598
I3negimpactmean	0.049632	−0.000227	0.02509	−0.0000418	4.665761	−0.02582
I3pos/negimpactmean	0.049852	−0.0000072	0.02515	0.00001857	4.692461	0.00088

TABLE 4.33 Out-of-sample model accuracy: DBKG1 Bond ARIMAX models with Company News Sentiment

External Variable(s)	RMSE	diff	MAE	diff	MAPE	diff
G12negsum	0.181511	0.000044	0.0741690	−0.00026	2.771208	−0.01008
G12posmean	0.181434	−0.000032	0.0742758	−0.00015	2.773017	−0.00828
G12negmean	0.181077	−0.000390	0.0744242	−0.000004	2.782636	0.001344
G12aggregate	0.181460	−0.000007	0.0744595	0.000032	2.781699	0.000407
G12posimpact	0.181415	−0.000052	0.0742727	−0.00016	2.771831	−0.00946
G12sumimpact	0.181427	−0.000039	0.0743202	−0.00011	2.775049	−0.00624
G12pos/negsum	0.181520	0.000054	0.0743603	−0.000067	2.782107	0.000815
G12pos/negimpact	0.181444	−0.000022	0.0743874	−0.00004	2.777852	−0.00344
G12posimpactmean	0.181426	−0.000040	0.0744033	−0.000024	2.780244	−0.00105
G12negimpactmean	0.181456	−0.000011	0.0743934	−0.000034	2.780484	−0.00081
G12pos/ negimpactmean	0.181426	−0.00004	0.07437876	−0.000049	2.779787	−0.00151

TABLE 4.34 Out-of-sample model accuracy: DBKG2 Bond ARIMAX models with Company News Sentiment

External Variable(s)	RMSE	diff	MAE	diff	MAPE	diff
G12possum	0.0665214	0.00005099	0.030552	0.0000309	6.767826	−0.000104
G12negsum	0.0663716	−0.00009884	0.030424	−0.0000968	6.737376	−0.030554
G12posmean	0.0664647	−0.00000572	0.030522	0.00000131	6.768079	0.000149
G12negmean	0.0664703	−0.00000011	0.030528	0.00000688	6.768584	0.000654
G12aggregate	0.0663893	−0.00008111	0.030463	−0.0000581	6.754191	−0.013739
G12posimpact	0.0664445	−0.00002594	0.030561	0.00003963	6.772564	0.004634
G12pos/negsum	0.0664075	−0.00006297	0.030439	−0.0000823	6.738256	−0.029674
G12posimpactmean	0.066458	−0.00001212	0.030597	0.00007638	6.778589	0.010659

of value to corporate bond yield spread prediction. Like country news sentiment, positive central bank news sentiment seems more helpful in the prediction of yield spreads during the economic recovering period and negative central bank news sentiment are more likely to be useful in enhancing the prediction when economy has recovered.

4.5.3.4 ARIMAX with Firm-Specific News Sentiment

From Table 4.37 to Table 4.40, we can see the performance accuracy of ARIMAX models with company specific news for the corporate bonds. It is suggested that both positive and negative company news sentiment reduces the forecast error when predicting Deutsche Bank and ENEL corporate bond yield spread, but the effect of negative company news sentiment is more evident

TABLE 4.35 Out-of-sample model accuracy: MUVG Bond ARIMAX models with Company News Sentiment

External Variable(s)	RMSE	diff	MAE	diff	MAPE	diff
G13negsum	0.056938	−0.00000539	0.037688	−0.00000455	3.731116	−0.00048
G13negmean	0.056916	−0.00002746	0.037661	−0.00003189	3.72822	−0.00338
G13negimpact	0.056861	−0.00008277	0.037634	−0.0000583	3.728563	−0.00304
G13posimpactmean	0.056952	0.00000862	0.037684	−0.00000816	3.731782	0.000184
G13negimpactmean	0.056919	−0.00002504	0.037677	−0.00001506	3.730797	−0.0008
G13pos/ negimpactmean	0.056926	−0.00001772	0.037669	−0.00002349	3.730974	−0.00062

TABLE 4.36 Out-of-sample model accuracy: SAN1 Bond ARIMAX models with Company News Sentiment

External Variable(s)	RMSE	diff	MAE	diff	MAPE	diff
S13possum	0.0944969	−0.00001265	0.0547115	−0.00000305	3.803129	−0.02084
S13negsum	0.0947227	0.00021308	0.0547509	0.00003632	3.80666	−0.01731
S13posmean	0.0944732	−0.00003635	0.0547456	0.000031	3.821869	−0.0021
S13negmean	0.0945783	0.00006867	0.0547682	0.00005365	3.815714	−0.00826
S13aggregate	0.0943814	−0.0001282	0.0546831	−0.0000314	3.829988	0.006018
S13pos/negsum	0.0946515	0.00014193	0.0547722	0.0000576	3.821905	−0.00206
S13pos/negimpact	0.094811	0.00030137	0.0552922	0.0005776	3.821388	−0.00258
S13posimpactmean	0.0944598	−0.00004974	0.0547195	0.00000493	3.825346	0.001376
S13negimpactmean	0.0945146	0.00000499	0.0546783	−0.0000363	3.818245	−0.00572
S13pos/ negimpactmean	0.0944691	−0.00004047	0.0546849	−0.0000296	3.819758	−0.00421

TABLE 4.37 Out-of-sample model accuracy: SAN2 Bond ARIMAX models with Company News Sentiment

External Variable(s)	RMSE	diff	MAE	diff	MAPE	diff
S13posimpactmean	0.0562537	−0.00001428	0.036718	−0.0000255	2.543799	−0.005719
S13negimpactmean	0.0562525	−0.00001547	0.0367324	−0.0000111	2.54957	0.000052
S13pos/ negimpactmean	0.0562333	−0.0000347	0.0367081	−0.0000353	2.544042	−0.005476

than that of positive news sentiment. MUVG corporate bond yield spread is affected by negative company news sentiment mostly. ENI bond, one BBVA bond and two Banco Santander corporate bonds are sensitive to both positive and negative sentiments.

TABLE 4.38 Out-of-sample model accuracy: BBVA2 Bond ARIMAX models with Company News Sentiment

External Variable(s)	RMSE	diff	MAE	diff	MAPE	diff
S14negsum	0.026427	−0.0000086	0.0190376	−0.0000005	3.474407	−0.000229
S14posmean	0.026439	0.0000039	0.0190816	0.0000435	3.482222	0.007586
S14negmean	0.026383	−0.0000522	0.0189483	−0.0000897	3.466734	−0.007902
S14aggregate	0.026549	0.0001142	0.0191477	0.0001096	3.495748	0.021112
S14posimpact	0.026425	−0.0000105	0.0190269	−0.0000112	3.472351	−0.002285
S14negimpact	0.026271	−0.0001641	0.0187497	−0.0002884	3.419178	−0.055458
S14sumimpact	0.026434	−0.0000017	0.0190239	−0.0000142	3.47185	−0.002786
S14pos/negimpact	0.026262	−0.0001736	0.0187486	−0.0002895	3.419399	−0.055237
S14posimpactmean	0.026412	−0.0000231	0.0190237	−0.0000143	3.472296	−0.00234
S14negimpactmean	0.026307	−0.0001279	0.0187672	−0.0002709	3.424921	−0.049715
S14pos/ negimpactmean	0.026425	−0.0000102	0.0190166	−0.0000214	3.471206	−0.00343

TABLE 4.39 Out-of-sample model accuracy: ENI Bond ARIMAX models with Company News Sentiment

External Variable(s)	RMSE	diff	MAE	diff	MAPE	diff
I11posmean	0.0498943	0.0000352	0.025121	−0.000011	4.692444	0.000863
I11negmean	0.0498766	0.0000175	0.025131	−0.000001	4.695559	0.003978
I11posimpact	0.0498381	−0.000021	0.025112	−0.000020	4.690841	−0.00074
I11negimpact	0.0498614	0.0000023	0.025113	−0.000018	4.693477	0.001896
I11sumimpact	0.0498485	−0.000011	0.025125	−0.000007	4.690457	−0.001124
I11pos/negimpact	0.0498432	−0.000016	0.025096	−0.000035	4.692916	0.001335
I11posimpactmean	0.0498521	−0.000007	0.025135	0.0000036	4.69611	0.004529
I11negimpactmean	0.0498288	−0.000030	0.025149	0.0000176	4.690447	−0.001134
I11pos/ negimpactmean	0.0498189	−0.000040	0.025159	0.0000269	4.695694	0.004113

TABLE 4.40 Out-of-sample model accuracy: ENEL Bond ARIMAX models with Company News Sentiment

External Variable(s)	RMSE	diff	MAE	diff	MAPE	diff
I10possum	0.060406	−0.000013	0.033604	−0.000011	2.472333	0.006962
I10negsum	0.060818	0.000399	0.033749	0.000134	2.463094	−0.00228
I10posmean	0.060418	−0.0000011	0.033614	−0.00000091	2.46538	0.000009
I10negmean	0.060611	0.000193	0.033541	−0.000074	2.44643	−0.01894
I10aggregate	0.06048	0.0000615	0.033614	−0.0000012	2.479014	0.013643

4.6 RESULTS AND DISCUSSION

This report examines the effect of macroeconomic news sentiment and firm-specific news sentiment on European corporate bonds. Using ARIMAX model with news sentiment as external variables improve the forecast performance compared to a simple ARIMA model. Positive country news sentiment is more effective in the economic recovery period, while negative country news sentiment predicts well the corporate bond yield spreads in a period of recession. Both positive and negative government news sentiment can enhance the one-step ahead forecast of spreads. Positive German parliament news sentiment enhances the prediction of German corporate bond but the effect of negative parliament news sentiment is limited. The effect of Central Bank news sentiment is mixed, but, overall, it also suggests that negative central bank news sentiment predicts the corporate bond yield spreads well in the recession period and positive central bank news sentiment does better in the recovery period. Compared to positive firm-specific news sentiment, in most cases, negative firm-specific news sentiment improves the prediction of corporate bond yields. For further research, we plan to investigate the combined effect of various categories of news sentiment in predicting corporate bond yield spread.

REFERENCES

Akaike, H. (1974). A new look at the statistical model identification. *IEEE Transactions on Automatic Control*, 19(6):716–723.

Apergis, N. (2015). Forecasting credit default swaps (cdss) spreads with newswire messages: evidence from European countries under financial distress. *Economics Letters*, 136:92–94.

Baker, M. and Wurgler, J. (2006). Investor sentiment and the cross-section of stock returns. *The Journal of Finance*, 61(4):1645–1680.

Bessembinder, H., Spatt, C., and Venkataraman, K. (2020). A survey of the microstructure of fixed-income markets. *Journal of Financial and Quantitative Analysis*, 55(1):1–45.

Box, G. E. and Jenkins, G. M. (1976). *Time Series Analysis: Forecasting and Control, revised ed.* Holden-Day.

diBartolomeo, D. (2016). Credit risk assessment of corporate debt using sentiment and news. In Mitra, G. and Yu, X., editors, *Handbook of Sentiment Analysis in Finance*. Optirisk Systems in collaboration with Albury Books.

Fernandes, C., Gama, P. M., and Vieira, E. (2016). Does local and euro area sentiment matter for sovereign debt markets? Evidence from a bailout country. *Applied Economics*, 48(9):816–834.

Fuller, W. A. (1976). *Introduction to Statistical Time Series*. John Wiley & Sons, New York.

Laborda, R. and Olmo, J. (2014). Investor sentiment and bond risk premia. *Journal of Financial Markets*, 18:206–233.

MarketAxess (2016). 4q 2015 fact sheet.

Nagel, J. (2016). Markets committee: electronic trading in fixed income markets. Technical report, Working paper, Tech. rep. 9789291974207, Bank for International Settlements.

Nayak, S. (2010). Investor sentiment and corporate bond yield spreads. *Review of Behavioral Finance*, 2(2):59–80.

O'Hara, M. and Zhou, X. A. (2021). The electronic evolution of corporate bond dealers. *Journal of Financial Economics*, 140(2):368–390.

Svensson, L. E. (1994). *Estimating and Interpreting Forward Interest Rates: Sweden 1992–1994 (No. w4871)*. National Bureau of Economic Research.

Yu, X. (2014). *Analysis of News Sentiment and Its Applications to Finance*. PhD thesis, Brunel University London, UK.

Yu, X. and Mitra, G. (2016). An impact measure for news: its use in (daily) trading strategies. In *The Handbook of Sentiment Analysis in Finance*, pages 288–309. Albury Books.

CHAPTER 5

Artificial Intelligence, Machine Learning and Quantitative Models

Gautam Mitra

OptiRisk Systems and UCL Department of Computer Science, London, United Kingdom

Yuanqi Chu

OptiRisk Systems

Arkaja Chakraverty

OptiRisk Systems

Zryan Sadik

OptiRisk Systems

CONTENTS

DOI: 10.1201/9781003293644-5

5.1 INTRODUCTION AND OVERVIEW

With increasing importance of mathematical probabilistic models and optimisation techniques in economic analysis, the neoclassical approach focused on mathematical elaboration of Homo economicus and added to Homo economicus its subjective character. Neoclassical finance embraces many pioneering theories including the mean-variance portfolio theory [Markowitz, 1952]; the capital structure irrelevance theorem [Miller and Modigliani, 1961]; the efficient market hypothesis (EMH) [Fama, 1995]; the capital asset pricing model (CAPM) [Sharpe, 1964]; [Lintner, 1969]; and the Black–Scholes option pricing model [Scholes and Black, 1973]. The main pillar of neoclassical finance, the efficient market hypothesis, has been the dominant paradigm in investing theory over the past three decades.

In the 1970s and 1980s, a new school of thought coined as behavioral finance took its birth. Behavioral finance emerged in response to the challenge faced by neoclassical economics in explaining market disruptions or anomalies phenomena. Risk can be quantified by probability calculation while uncertainty cannot be calculated and addressed exactly. Consequently, classical theories and mathematical computations are not sufficient to interpret and solve the paradox of irrationality. Limitations in classical and neoclassical theories supported the birth of a psychology-based explanation for asset prices. Defined as a multi-discipline underpinned by sociology, psychology, anthropology and other sciences, behavioral finance aims to understand and explain the factors that drive Homo economicus behaviors to deviate from rationality to irrationality when making investment decisions.

The backbone of behavioural finance is the prospect theory introduced by [Kahnemann, 1979]; [Kahneman and Tversky, 2013], which postulates human beings rank and weight losses and gains differently under risk and uncertainty conditions. This theory argues that financial decisions under risk are influenced by psychological biases and provides answers to key puzzles or anomalies arising from traditional finance theories. Expected utility theory [Neumann and Morgenstern, 1944] has been generally accepted as a normative model of rational choice under risk. However, the strong assumptions of expected utility theory, such as the absolute rationality and self-interests, make it too idealistic to portray the reality of economic agents' behavior patterns in decision making process. Prospect theory emerged as a leading descriptive alternative to expected utility theory, aiming to explain and predict how agents in economic settings "really behave" when making choices in various circumstances.

The development and progressive acceptance of behavioural finance have followed from the linking of the sentiment of the investors and other market participants. In the financial markets, it is commonly known that investors in the bullish regime are positive and optimistic, while in the bear regime, they are pessimistic and fearful of loss. AI&ML is used in mining various sources of text such as newswires, (micro)blogs and other sources of information. For a comprehensive discussion of the theory and multiple use cases, see [Mitra and Mitra, 2011]; [Mitra and Yu, 2016]. Whereas sentiment analysis is associated with asset price behaviors of equities and bonds (see [SenRisk, 2018]). Its connection to derivative products is noteworthy. [Smales, 2014] reports that the relationship between news sentiment and VIX is asymmetric and significantly negative, that is, negative news events correspond to large movements in VIX with a greater shift in prices induced by negative sentiment than positive sentiment. Furthermore, by examining the relationship annually, it is determined that news sentiment has an increasing strength of relation with fluctuations in VIX when periods of implied volatility increase. VIX is used as a consensus toward expected future stock market volatility where large values represent greater fear.

Regression analysis is one of the most frequently employed statistical techniques which aims at delineating the relationship between a response variable and one or more explanatory variables. Among the various regression models, logistic regression (LR) plays a vital role in classification analysis of binary and proportional responses [Hastie et al., 2009]. LR adds flexibility to traditional linear regression models through the employment of an appropriate link function, which allows for capacity to predict variables characterized as continuous or discrete, or any combination of both types. It is also preferred when the underlying normality assumption of the variables cannot be satisfied [Lee, 2004]. LR is a well-established approach in a wide range of applications and has been extensively investigated for its capabilities in predicting stock performance in financial markets [Gong and Sun, 2009]; [Luo et al., 2017]; [Ali et al., 2018]; [Zhou et al., 2019], among others. We refer to Section 5.3 for a more elaborate skeleton of the LR.

Before the ARCH (Autoregressive Conditional Heteroskedastic) process [Engle, 1982] was introduced for time series and econometric models, traditional statistical analysis of time series centred on conditional first moment (i.e., under the assumption of constant variance). The discovery that common measures of risk and uncertainty exhibited strong time-varying variation led to the development of modelling time variation in the second moment analysis. The stylized volatility clustering characteristics of a random variable, evolving through time, often refer to the momentum in conditional variance: the local variance is clustered in clumps of high or low values. Fluctuations with unusual volatility often occur for longer periods, while small fluctuations tend to be followed by small ones. Statistically, volatility clustering implies time-varying conditional variance. The ARCH model and its extension to the Generalized ARCH (GARCH) model [Bollerslev, 1986] have been proven to be extremely useful for capturing the serial correlation and the momentum in conditional variance, which are commonly observed in empirical financial time series. Compared with ARCH models, the GARCH processes are more parsimonious and provide solutions to the estimation issues raised by ARCH processes. As stated in [Brooks, 2014b], a GARCH (1,1) model is sufficient to capture the volatility clustering characteristics in real time financial series. Following the seminal work of [Engle, 1982] and [Bollerslev, 1986], studies on investigating financial market volatility with GARCH models and their extensions have been well documented (see, e.g., [Choudhry, 2000]; [Chiang and Doong, 2001]; [Alberg et al., 2008]; [Hung, 2011]; [Hu et al., 2018]; [Bouras et al., 2019]). A more detailed formulation of the GARCH processes is presented in Section 5.3.

In recent years, the rapid development of machine learning (ML) algorithms has unleashed remarkable innovation and improvements in quantitative research and practical investment in augmentation to traditional mainstream methods. ML mechanisms have proven capabilities to inherently capture the intrinsic non-linearity, non-stationarity and high complicacy in financial variables. Artificial neural networks (ANNs) are inspired by biological neural networks which constitute human brains [Brahme, 2014]. Artificial neurons are interconnected in linear or nonlinear forms to operate in parallel processing at multiple layers and are structured hierarchically in an ANN. Each connection serves to distribute signals from one neuron to another, which follows a chain fashion and the final output is retrieved after intermediate outcomes travel through the hidden layers. Nowadays ANNs have emerged as a powerful statistical modeling technique and have been popularly applied to financial problems such as stock index prediction and corporate bond classification. A comprehensive introduction and review of artificial neural networks can be referred to [Jain et al., 1996], [Priddy and Keller, 2005] and [Yegnanarayana, 2009]. ANNs have been demonstrated to be able to identify dynamic relationships between financial variables and predict future stock market valuations due to its learning, self-organizing and generalizing characteristics. Numerous articles have appeared recently that surveyed the application of ANNs for predicting the series of stock prices and trends. See, e.g., [Hu et al., 2018]; [Shastri et al., 2019]; [Thakkar and Chaudhari, 2020]; [Thakkar and Chaudhari, 2021]; [Anand, 2021]; [Chen et al., 2021]; [Zhang and Lou, 2021]; [Serrano, 2022] and among others.

Although ANNs have achieved remarkable achievements, the networks with high accuracy and satisfactory performance often tend to exhibit extremely complex internal structures, which decimate the interpretability of the model and make the neural networks as incomprehensible as a black box. In recent years, decision trees have become a prevalent ML technique owing to their simplicity and ease-to-interpret characteristics. Random forest (RF) developed by [Breiman, 2001] is an ensemble learning methodology for both classification and regression. RF integrates Breiman's bagging sampling technique [Breiman, 1996] and the idea of random feature selections introduced independently by [Ho, 1995]; [Ho, 1998] and [Amit and Geman, 1997]. Random forest applies a tree-like structure to generate predictions through a repetitive process of splitting based on logical statements. In the training phase, the decision trees are selected randomly and evaluated separately and the average of the outputs from the trees is produced as the best score. The random forest has received considerable attention from the research community in the effort to make effective decisions in stock market prediction [Tan et al., 2019]; [Lohrmann and Luukka, 2019]; [Nti et al., 2019]; [Ampomah et al., 2020]; [Sadorsky, 2021]; [Rakhra et al., 2021]; [Abraham et al., 2022]; [Zi et al., 2022]; [Srinu Vasarao and Chakkaravarthy, 2022], among others.

New variants of the random forest technique have also emerged in the finance arena to devote toward model performance enhancements in financial applications. To name a few exemplars, [Booth et al., 2014] explored an automated trading system based on performance-weighted ensembles of random forests. Their results demonstrated that recency-weighted ensembles of random forests generate superior results for both profitability and prediction accuracy compared with other ensemble techniques. [Lin et al., 2017] developed an ensemble random forest algorithm that employs the parallel computing capability and memory cache mechanism optimized by Spark. Their experiment results on the insurance business data from China Life Insurance displayed that the ensemble random forest algorithm outperformed other counterpart classification techniques in both performance and accuracy within the considered imbalanced data sets. [Xu et al., 2018] proposed a novel approach combining random forests and a signal processing method of the wavelet transform to forecast currency crises. The authors demonstrated that the enhanced methodology achieved a high level of accuracy in predicting currency crises. Another genre of models, namely, eXtreme Gradient Boosting (XGB) models use a method known as the 'boosting' technique. This technique refers to building a strong classifier from a number of weak classifiers. XGB minimizes a regularized objective function that combines a convex loss function and penalizes the model for complexity. It is called gradient boosting because it uses a gradient descent algorithm to minimize the loss when adding new models. It is quite a popular model in finance and is widely used for its better relative performance.

Ensemble learning (EL) is an umbrella term for techniques that combine multiple learning strategies to improve predictions or decrease variance and bias. Different ensemble learners, such as bagging [Breiman, 1996], boosting [Freund, 1995] and stacking/blending [Wolpert, 1992], have been extensively studied in the literature. The primary preponderance of EL is that by combining multiple algorithms, the errors of a single model will likely be compensated by other models, and therefore, the ensemble classifiers or regressors will yield a more precise overall prediction performance than the individual classifiers or regressors. A proliferation of studies has been dedicated to the employment of EL in stock market prediction [Weng et al., 2018]; [Gyamerah et al., 2019]; [Nti et al., 2020]; [Xu et al., 2020]; [Jiang et al., 2020], among others. A comprehensive review of the ensemble approaches and their challenges is provided in [Sagi and Rokach, 2018]. For further elucidation on the ensemble methods, we refer to Section 5.4.

All above models are classified under the following Taxonomy:

- *Descriptive Models* as defined by a set of mathematical relations simply describe and thereby in some sense predict how a physical, industrial, or social system may behave.

- *Normative Models* constitute the basis for (quantitative) decision-making by a superhuman following an entirely rational, that is, logically scrupulous set of arguments. Hence, quantitative decision problems and idealised decision-makers are postulated in order to define these models. An example is a two-person zero-sum game.

- *Prescriptive Models* involve a systematic analysis of problems as carried out by normally intelligent persons who apply intuition and judgment. Two distinctive features of this approach are uncertainty analysis and preference (or value or utility) analysis.

- *Decision Models* are in some sense a derived category as they combine the concept underlying the normative models and prescriptive models.

For further information, see [Mitra et al., 1988] and [Bell et al., 1988].

5.1.1 Organisation of the Chapter: Guided Tour

This chapter is organized as follows: In Section 5.2 we outline the historical market data and sentiment data for VIX that is employed in this research. This section also explains the news sentiment metadata and the micro-blogs sentiment data. In Section 5.3 we set out the most widely used neo-classical models in finance. In Section 5.4 we describe the ML models which are used by analysts; in particular, those which are utilized in this study. In Section 5.5 we construct a series of control variables and indicators and use them as input information to predict the direction movements of VIX. We then describe these investigations and their findings. A summary discussion of our work is presented in Section 5.6.

5.2 DATA: MARKET DATA, SENTIMENT DATA, AND ALTERNATIVE DATA

Artificial intelligence (AI) applications have significantly evolved over the past few years and have found its applications in almost every business sector. ML is a subset of artificial intelligence that helps you build AI-driven applications. Deep learning is a subset of ML that uses vast volumes of data and complex algorithms to train a model. AI models that are trained using vast volumes of data have the ability to make intelligent decisions. ML is a buzzword in the technology world right now and for good reason, it represents a major step forward in how computers can learn. The most common applications of ML across different industries are in finance, supply chain, consumer products, health care, retail, travel, media and mechanical engineering. The data that is used as input information in each application model is different, for example, in the consumer product models, the input variables are: (i) market data that will be the prices, (ii) the sentiment data which is the market sentiment data, (iii) and alternative data, typically, footfalls. In the financial models, the market data is the historical instrument prices, sentiment data is news metadata and the alternative data could be micro-blog sentiment data.

5.2.1 Market Data

Market data in financial applications, by definition, is the price and trade-related data for a financial instrument reported by a trading venue such as a stock exchange. The market data that is commonly utilised in the finance models (ML models or neo-classical models) is the daily or intraday open, high, low and closing prices, volumes, ... etc. of the stocks. There are other types of financial market data available for other asset classes, for example:

Commodities data: which is information on agricultural, mineral, metal, energy, and manufactured materials. Typical commodities data should contain accurate pricing for these items, their price rises or price falls, the current commodities market conditions and projected trends.

Fixed income data: which is information relating to municipal and corporate financial assets where an investor is paid a fixed interest or dividends rate. The fixed-income data feed will provide lots of different attributes. Typically, these include stock ticker, industry code and maturity. The most common fixed-income investments include Bonds, Treasury bills and Banker's acceptance.

Derivatives data: Derivatives are defined as Options and Futures and other derivative instruments relating to underlying assets, namely, equities, commodities and currencies [Hull, 2003]. They can be categorized as Options, Futures, Interest rate swaps and other derivative products. The related price data fall in the category of Market Data.

The derivatives (market) data used in this study is the daily open, high, low and closing prices of the Volatility Index (VIX) futures. The famous Chicago Board Options Exchange's Volatility Index (VIX), or "fear index", is widely used as a proxy of investors' sentiment and market volatility. VIX is a complex and well-defined derivative; for a formal definition of VIX given by CBOE [CBOE, 2016].

5.2.2 Sentiment Data and Alternative Data

News Sentiment Data: In recent years, commercial news analytics (meta) data providers have started computing quantitative measures called news sentiment scores for individual stocks; they also provide sentiment scores for macroeconomic news items [Mitra and Yu, 2016]. A sentiment score of a news item measures the emotional tone within the text and varies between positive, neutral and negative. In general, sentiment analysis is concerned with the analysis of direction-based text, that is, text containing opinions and emotions. Sentiment vendors have developed linguistic analytics which process the textual input of news stories to determine quantitative sentiment scores. Bloomberg, Thomson Reuters (Refinitiv), RavenPack and Alexandria Technology are leading players in this area. These companies apply NLP and classification engines to process textual information and elicit "sentiments" from news data sources. These are then distributed to market participants who in turn use these items of sentiment information in their respective decision engines.

Microblogs Sentiment (Alternative) Data

More recently, with the developments in NLP and computational linguistics, artificial intelligence and ML, researchers started looking at alternative sentiment measures like messages posted on social networks and blogs (for instance, Twitter and Facebook).

In this study, we have used micro-blog sentiment data to enhance the predictive ability of the models. The sentiment data is obtained from StockTwits which is a financial social media platform. Discussions in this platform are restricted to the domain of finance which makes it a rich and focused data source for financial models. This platform is different from textual news. These are individual views and opinions of investors and domain experts. Also, unlike newswire, StockTwits gets tweets posted 24 hours a day, 7 days a week. The historical data dates back to the year 2010. The data is streamed in JSON format. The various attributes of a sample sentiment stream of StockTwits include "message_id", "user_id", "sentiment_score", "symbol_id", "created_at", "exchange", "industry", "sector", "symbol" and "title".

5.3   NEO-CLASSICAL MODELS: AN OVERVIEW

In this section, we discuss a few widely used neo-classical finance models, namely, linear regression, LR, Auto-regressive Conditional Heteroskedastic (ARCH) and Generalized ARCH (GARCH). Depending upon the choice of the dependent variable, one can select one or a combination of the above models. This list is illustrative and not an exhaustive list of neo-classical models pertinent to finance.

For identification using regression models, we first need to define the variable for which we want to explain or predict. This variable is called the dependent or response variable. A response variable can be mainly classified into two categories: continuous variable and discrete variable, which can be further classified into binary and multi-class variables. If a response variable takes only two values, typically 0 and 1, it is referred to as a binary classifier. Whereas, if a response variable takes more than two discrete values, it is referred to as a multi-class classifier. Having defined the response variable, we need to identify a set of variables or predictors which significantly influences the response variable. This set of predictors is also called independent or control variables.

Linear regression is a useful tool for predicting a continuous response variable y (Please refer to [Stigler, 1986] for details on the history of regression analysis). Though it has been around for a long time, linear regression is still a useful and widely used statistical learning method. We can predict the value of y by using a set of explanatory variables X and finding a linear relationship between y and X. For instance, if we suppose that we have p distinct explanatory variables, then the linear regression model takes the following form:

$$Y = \beta_0 + \beta_1 x_1 + \beta_2 x_2 + \cdots + \beta_p x_p + \epsilon, \tag{5.1}$$

where x_j represents the jth explanatory variable and β_j quantifies the association between that variable and the response. We interpret β_j as the average effect on Y of a one unit increase in x_j, holding all other predictors fixed. The values of βs are determined by minimizing the square of the residual (ϵ).

Linear regression, however, does not perform very well if the dependent variable, y, is a discrete variable instead of a continuous variable. For these classification tasks, LR (also known as the logit model) is a more appropriate model. In LR, we learn a family of functions h from R to the interval $[0, 1]$ (Wooldridge, 2012). This regression estimates the probability of an event occurring, such as spam or non-spam email, based on a given dataset of independent variables. Since the outcome is a probability, the dependent variable is bounded between 0 and 1. In LR, a logit transformation is applied to the odds—that is, the probability of success divided by the probability of failure. This is also commonly known as the log odds or the natural logarithm of odds. In particular, the sigmoid function used in LR is the logit function, defined as:

$$logit(Z) = \frac{1}{[1 + exp(-Z)]} \tag{5.2}$$

$$ln(\frac{Z}{[1 + -Z]}) = \beta_0 + \beta_1 x_1 + \beta_2 x_2 + \cdots + \beta_p x_p \tag{5.3}$$

In this LR equation, $logit(Z)$ is the dependent or response variable and x is the independent variable. The beta parameters, or coefficient, in this model, are commonly estimated via maximum-likelihood estimation (MLE).

These two models work well while predicting daily returns or their direction. Nevertheless, when one is interested in predicting volatility, they are better off resorting to ARCH—introduced by [Engle, 1982]—and GARCH—introduced by [Bollerslev, 1986]. The discovery that common measures of risk and uncertainty exhibited strong time-varying variation led to the development of modelling time-variation in the second moment analysis. The stylized *volatility clustering* characteristics of a random variable, evolving through time, often refer to the momentum in conditional variance: the local variance is clustered in clumps of high/low values. Fluctuations with higher volatility often occur for longer periods, while small fluctuations tend to be followed by small ones. Statistically, volatility clustering implies time-varying conditional variance. These two models have been proven to be extremely useful for capturing the serial correlation and the momentum in conditional variance, which are commonly observed in empirical financial time series.

Let us denote Y_t as a time series of interest, e.g., the rate of return, the term $ARCH$ means that the conditional variance of Y_t evolves according to an autoregressive-type process. Formally, let ϵ_t be *i.i.d.* (independent and identically distributed) white noise with mean zero and variance of one. The time series Y_t is an $ARCH$ process of order p, $ARCH(p)$, if

$$Y_t = \sigma_t \epsilon_t \tag{5.4}$$

$$\sigma_t^2 = w + \sum_{i=1}^{p} \alpha_i Y_{(t-i)}^2 \tag{5.5}$$

where $w > 0$, $\alpha_i \geq 0, i = 1, \ldots, p$. The simplest $ARCH$ process is that of order $p = 1$. The $ARCH(1)$ process can be represented as

$$Y_t = \sigma_t \epsilon_t = \sqrt{w + \alpha_1 Y_{(t-1)}^2} \epsilon_t \tag{5.6}$$

with $w > 0$ and $\alpha_1 \geq 0$.

The estimation of ARCH models can be usually conducted using ordinary least squares (OLS) or MLE. However, one needs to determine the appropriate order p in an $ARCH(p)$ process. For some applications, a larger order p is required to capture all the serial correlation in Y_t^2, which makes the estimation procedure become unwieldy and numerically complicated. Furthermore, the parameters $w, \alpha_1, \ldots, \alpha_p$ in an $ARCH(p)$ adhere to a non-negative constraint. As the order p increases, such non-negative constraint will be more likely to be violated [Brooks, 2014a]. A parsimonious alternative to the ARCH process is the Generalised $ARCH(GARCH)$ process, which includes both the return as well as the past volatility of an underlying asset to forecast its volatility. The parameters of models can be estimated by the MLE method.

In sum, a linear model for regression or classification can be represented as a linear combination of a basis function $\Theta_j(x)$:

$$y(x, w) = f(\sum_{j=1}^{M} w_i \Theta_j(x)) \qquad (5.7)$$

where $f(.)$ tends to be a non-linear activation function for classification problems and is the identity for regression problems [Bishop and Nasrabadi, 2006]. In equation (5.7), x and w refer to the set of input parameters and their coefficients, respectively. Although neo-classical models have useful applications, they may not be appropriate for large-scale problems, which require adapting the basis function to data.

5.4 MACHINE LEARNING MODELS

The paradigm that deep learning provides for data analysis is very different from the traditional statistical modeling and testing framework. Traditional fit metrics, such as $R2$, $t - values$, $p - values$ and the notion of statistical significance are replaced in the ML literature understanding the bias-variance trade-off, i.e., the trade-off between a more complex model versus over-fitting. Usually, in the context of ML models, we divide the observations into training and testing samples. We train our model on training data, whereas we validate the accuracy of these models by checking their accuracies on testing samples.

In this section, we discuss the theoretical constructs of three ML models: **Artificial Neural Network (ANN), Decision Trees and Random Forest (RF)**. ML is an evolving area, and we observe that new models are introduced regularly. Typically, ML models have two versions—regressor and classifier. If the dependent variable is continuous, then one should use the regressor version, whereas for a discrete dependent variable, one should use the classifier version of these models.

ANN: is a model of computation inspired by the structure of neural networks in the brain—it consists of a large number of basic neurons that are connected to each other in a complex communication network, through which the brain is able to carry out highly complex computations. A neural network can be described as a directed graph whose nodes correspond to neurons and edges correspond to links between them. ANNs have been developed as generalizations of mathematical models of neural biology, based on the assumptions that:

1. Information processing occurs at many simple elements called neurons. Each neuron has an internal state, called its activation or activity level, which is a function of the inputs it has received.

2. A neuron sends its activation as a signal to several other neurons. Each connection link has an associated weight, which, in a typical neural net, multiplies the signal transmitted. It is important to note that a neuron can send only one signal at a time, although that signal is broadcast to several other neurons.

3. Each neuron applies an activation function to its net input to determine its output signal. The purpose of an activation function is to add non-linearity to the neural network.

ANN has multiple layers, namely, including input, hidden and output layers. Input layer units are set by some exterior function, which causes their output links to be activated at the specified level.

In the case of ANN each basis function, $\Theta_j(x)$, in equation (5.7) is a nonlinear function of a linear combination of the input variables (x) and the coefficients, w_i, are adaptive parameters. In other words, although ANN fixes the number of basis functions in advance, it allows them to be adaptive. This results in a basic neural network model that can be described as a series of functional transformations [Bishop and Nasrabadi, 2006]. Working forward through the network, the input function of each unit is applied to compute the input value. Usually, this is just the weighted sum of the activation on the links feeding into this node. We can use an L linear combination of input variables to construct activation functions, a_j, as follows:

$$a_j = \sum_{i=1}^{N} w_{ji}^{(1)} x_i + w_{j0}^{(1)} \tag{5.8}$$

where, $j = 1, \ldots, L$, and superscript (1) refer to the parameters in the *first* layer of the ANN. $w_{ji}^{(1)}$ and $w_{j0}^{(1)}$ correspond to weights and biases. Parameters, a_j are transformed using a differentiable, non-linear activation function, h(.). The activation function transforms this input function into a final value. Typically, this is a nonlinear function, often a sigmoid function corresponding to the "threshold" of that node. Using these activation functions yield the following:

$$Z_j = h(a_j) \tag{5.9}$$

This refers to the output of the basis function in (5.7) and in the context of ANN these are called hidden units. A linear combination of these values using equation (5.7) provides the output unit activation, as given by equation (5.10).

$$a_k = \sum_{j=1}^{L} w_{kj}^{(2)} z_j + w_{k0}^{(2)} \tag{5.10}$$

where, $k = 1, \ldots, K$, which is the total number of output variables. The superscript (2) refers to the *second* layer of the network. The output unit activations are transformed using an appropriate activation function, $\sigma(.)$, to predict the set of output variables, y_k, so that $y_k = \sigma(a_k)$.

For practical implementation of ANN, inputs are given random weights (usually between -0.5 and 0.5). If the weighted sum of the inputs to the neuron is above the threshold, then the neuron fires. Learning involves choosing values for the weights. An item of training data is presented. We use backpropagation [1] to minimize error or loss function based on the stochastic gradient-descent (SGD) method. ANNs are configured for pattern recognition or data classification through a learning process. The idea behind neural networks is that many neurons can be joined together by communication links to carry out complex computations.

Decision Tree: is a tree-structure model of regression as well as classification. The process includes the division of the predictor space generated from control variables, $x_1 + x_2 + \cdots + x_n$ into, say K, distinct and non-overlapping regions, $R_1, R_2, \ldots, R_K$. Such approach of stratification is called decision tree methods. To identify the strata of the response variable, the model typically uses the mean or mode of the given observations in various regions. In other words, for every observation that falls into the region R_j, we make the same prediction, which is the mean (or mode) of the response values for the training data in R_j.

A smaller tree with fewer splits (that is, fewer regions $R_1, R_2, \ldots, R_j$) might lead to lower variance and better interpretation at the cost of a little bias. Therefore, a better strategy is to grow a very large tree, T_0, and then prune it back in order to obtain a subtree. Rather than considering every possible subtree, we consider a sequence of trees indexed by a nonnegative tuning parameter α. We aim to select a sub-tree that results in the minimum standard error. Here, we follow the following algorithm:

---

[1]Backpropagation is the essence of neural network training. It is the method of fine-tuning the weights of a neural network based on the error rate obtained in the previous epoch (i.e., iteration). Proper tuning of the weights allows you to reduce error rates and make the model reliable by increasing its generalization.

1. Use recursive binary splitting to grow a large tree on the training data, stopping only when each terminal node has fewer than some minimum number of observations.

2. Apply pruning to the large tree to obtain a sequence of best subtrees, as a function α. For each value of α, we have a subtree $T \subset T_0$ such that:

$$\sum_{m=1}^{|T|} \sum_{x_i \in R_m} (y_i - \widehat{y_{R_m}})^2 + \alpha|T| \qquad (5.11)$$

3. Use k-fold cross-validation to choose α, i.e., divide the training sample into k-folds and for each k follow the following steps:

   [a] Repeat steps 1 and 2 on all but the $k - th$ fold of the training sample

   [b] Evaluate the mean squared error for predictions on $k - th$ fold as a function of α.

   [c] Estimate the mean of each α and choose α so that it minimizes the mean error.

4. Return the subtree from Step 2 that corresponds to the selected α.

As an EL technique, **Random Forests** are a collection of a large number of decision trees introduced during the training phase. [James et al., 2013] state "Random forests provide an improvement over bagged trees by way of a small tweak that decorrelates the trees". As in bagging, we build a number of decision trees on bootstrapped training samples. The final output allocated by the RF model is the classification given by the majority of the trees. The mean or average prediction of each individual tree is returned for regression tasks. A classification tree is very similar to a regression tree, except that it is a classification used to predict a qualitative response rather than a quantitative one. Unlike the regression RF model, where we predict response for an observation based on the average of the observations in the training model, for a classification tree. Although RF typically outperforms decision trees, gradient-boosted trees might be more accurate than random forests. Nonetheless, it is worth mentioning that the data quality has a significant bearing on the predictive accuracy of RF models.

5.4.1 Ensembling of Models

Ensemble essentially refers to combining the outputs of at least two analytical models to make the final prediction. These models can incorporate predictions from both neo-classical as well as ML models. In general, the predictions by ensembling exhibit higher accuracy than the individual accuracies of input

Conservative Approach

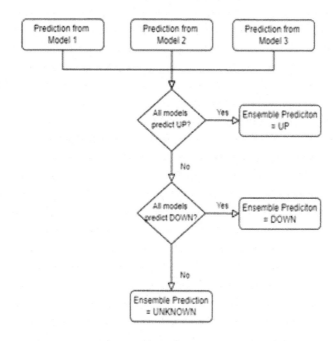

Figure 5.1   Conservative approach for voting techniques

analytical models. These techniques may include a maximum vote for classification and the average value for regression.

For instance, if the response variable is a binary variable, we can employ two voting techniques, which could be classed as "conservative" and "aggressive" approaches, to generate final predictions. The "conservative" approach implies that only if all models deployed predict UP (DOWN) on a given day would the final prediction be UP (DOWN). In all other cases, i.e., when there is at least one model that gives a different prediction, the ensemble prediction for the day was classified as "UNKNOWN". This effectively means that the ensemble prediction of UP or DOWN is made with utmost confidence.

On the other hand, the "aggressive" ensemble approach can be built on the maximum vote logic. In other words, the ensemble state (UP or DOWN) would be the direction predicted by at least two out of the three ML models. As an alternative to the "conservative" approach, this approach will not have any "UNKNOWN" predictions, but will produce more incorrect predictions and is therefore called the "aggressive" approach.

Another way of improving the accuracy of prediction is to combine multiple models; this is known as stacking. Stacking improves the accuracy of the predictions for response variables by combining the outputs of various models and running them through an additional model called a meta-learner; these

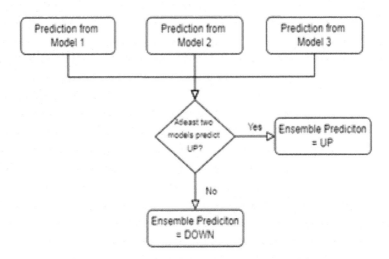

Figure 5.2 Aggressive approach for voting techniques

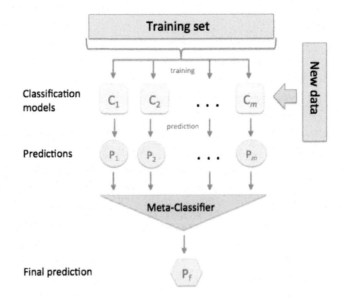

Figure 5.3 Stacking process

models can be either a neo-classical model or an ML model. The meta-learner attempts to minimize the error and maximize the accuracy of every individual model. The result is usually a very robust model that generalizes well on unseen data. Stacking is diagrammatically displayed and summarized (see Figure 5.3) [Stacking, 2020].

Although the ensemble method is very popular amongst data scientists, the stacking technique is gradually catching up. While we consider ensemble methods to improve our predictions, it might be a good idea to compare and contrast the findings of the ensemble with that of the stacking method.

5.5 APPLICATION IN FINANCE: INVESTIGATION OF AN EXEMPLAR PROBLEM

In this section, we discuss the motivation behind the selection of models based on our objective as well as the empirical results we document.

5.5.1 Problem statement

We aim to predict the directionality of VIX futures, that is, whether the VIX future closes above or below in comparison to the previous trading day. If the VIX future on the day t closes above that on the previous trading day $(t-1)$, then our dependent variable takes 1 and we call it the "UP" movement. On the other hand, if the VIX future on the day t closes below that on the previous day, then our dependent variable takes 0. We call it a "DOWN" movement. Since we are only interested in predicting the direction of the VIX futures, we are interested in using only those models—both neo-classical and ML models— that are appropriate for discrete dependent variables. In particular, we use LR, ANN classifier and RF classifier models to compare the outcome of the neo-classical model with that of the ML models.

5.5.2 Data: Market Data, Sentiment Data and Alternative Data

Our experimental dataset comprises two streams of time series data: the daily market (prices) data and the Micro-Blog metadata (quantified values of sentiment) of VIX. We collected the historical data set of VIX futures which consists of the daily open, high, low and closing prices with the period of 28 Jan 2015 – 07 Jan 2021 (n=1498). The market data is obtained from Refinitiv (Datascope). We considered the up and down movement changes of VIX futures closing prices from the previous day as the target variable of our prediction models and coded these as "UP" and "DOWN", respectively. The sentiment data employed in this study are the daily recorded sentiment scores retrieved from Stocktwits. This original data set consists of positive/negative sentiment scores/volume for S&P 500 and VIX indexes separately, with the period of 28 Jan 2015 – 07 Jan 2021.

For the purposes of model specification and forecasting, we divide the market dataset into two datasets; we reserve the first 2/3 of the full dataset for model training (in-sample data set) and used it as input information to estimate the parameters of the models. The test data set (out-of-sample) is

TABLE 5.1 Selected technical indicators

Name of indicators	Formulas
Simple n (3 here)—day Moving Average	$MA(n) = \dfrac{\sum_{i=1}^{n} C_{t-i+1}}{n}$
k (5 here)—day Exponential Moving Average	$EMA(k)_t = EMA(k)_{t-1} + \alpha \times (C_t - EMA(k)_{t-1})$
n (14 here)—day period Stochastic $K\%$	$K\% = \dfrac{C_t - LL_{t-(n-1)}}{HH_{t-(n-1)} - LL_{t-(n-1)}} \times 100$
Relative Strength Index (RSI)	$RSI = 100 - \dfrac{100}{1 + (\sum_{i=0}^{n-1} \frac{UP_{t-i}}{n})/(\sum_{i=0}^{n-1} \frac{DW_{t-i}}{n})}$

used to evaluate the forecasting performance of the models. The test data set goes from 15 Jan 2019 – 07 Jan 2021.

5.5.3 Model Construction

Active stock traders commonly integrate technical analysis into automated trading systems to forecast future price movements. The rapid growth of trading strategy systems over the past decade suggested that establishing adequate technical analysis methods could generate exceeding profit [Qin and Ren, 2019]. Technical analysis is based on the premise that the future behaviour of a financial time series can be predicted solely from its own past. Technical indicators are empirical tools for technical analyses and are rooted in mathematical formulas. Market participants ascribe these formulas in trading systems to generate timely buy or sell signals when there is an abundance of historical data available. As a result, technical indicators serve as a unified mathematical transformation over the raw price and volume data of each stock and play a crucial role in forming up profitable investment strategies. Technical indicators can be generally categorized into momentum measures, moving averages and stochastic oscillators.

5.5.3.1 Technical indicators

We use the closing, high and low prices for computing the selected technical indicators as reported in Table 5.1. For a detailed explanation of the selected technical indicators and their mathematical forms [Kara et al., 2011] in Table 5.1 below, the symbols are defined as C_t is the closing price at time t; α is a smoothing factor which is equal to $2/(k+1)$; LL_t and HH_t denote the lowest low and highest high in the last t days, separately; UP_t indicates upward price change while DW_t is the downward price change at time t.

Discrete transformation for technical indicators

As a general rule of thumb, a rising moving average indicates that the stock is in an uptrend, while a declining moving average indicates a downtrend. We convert the original 3-day MA and 5-day EMA values as follows. If the value of the moving average at time "t" is greater than the value at time "t-1", then the trend is "UP" and represented as "1"; otherwise, the trend is "DOWN" and represented as "0".

When Stochastic $K\%$ values are increasing, the stock prices are likely to go up and vice-versa. We conduct the following transformation. If the value of Stochastic $K\%$ at time "t" is greater than the value at time "t-1" then the trend is "UP" and represented as "1"; otherwise, the trend is "DOWN" and represented as "0".

RSI is generally used for identifying the overbought and oversold signals. It ranges between 0 and 100. The value of RSI exceeding 70 indicates that the stock is overbought and the price might go down in near future (recoding "0"); and the value of RSI under 30 means that the stock is oversold and the price might go up in near future (recoding "1"). For the values between (30, 70), if the RSI value at time "t" is greater than the RSI value at time "t-1", the trend is "UP" and represented as "1"; otherwise, the trend is "DOWN" and represented as "0".

5.5.3.2 Sentiment Features

We analyze the sentiment (meta) data and compute the following eight features as reported below in Table 5.2.

To refrain from any upfront importance imposed on the sentiment scores, we conduct feature scaling for all the aggregated positive and negative sentiment scores. Specifically, for positive scores "SP500_Pos_Aggregated" and "VIX_Pos_Aggregated", we consider the formula

$$Pos\_Sent_{normalized} = \frac{Pos\_Sent - min(Pos\_Sent)}{max(Pos\_Sent) - min(Pos\_Sent)} \quad (5.12)$$

and for negative scores "SP500_Neg_Aggregated" and "VIX_Neg_Aggregated", we employ

$$Neg\_Sent_{normalized} = \frac{Neg\_Sent - min(Neg\_Sen)}{max(Neg_{s}ent) - min(Neg\_Sent)} - 1 \quad (5.13)$$

The details of summary statistics for the scaled sentiment indexes are presented in Table 5.3.

TABLE 5.2 Description of eight sentiment features

Sentiment features	Description
SP500_Pos_Aggregated	Positive sentiment scores for all S&P 500 assets (active on the given day) aggregated over last 6 days
SP500_Neg_Aggregated	Negative sentiment scores for all S&P 500 assets (active on the given day) aggregated over last 6 days
SP500_Pos_Vol_Aggregated	Positive sentiment scores volume for all S&P 500 assets (active on the given day) aggregated over last 6 days
SP500_Neg_Vol_Aggregated	Negative sentiment scores volume for all S&P 500 assets (active on the given day) aggregated over last 6 days
VIX_Pos_Aggregated	Positive sentiment scores for VIX aggregated over last 6 days
VIX_Neg_Aggregated	Negative sentiment scores for VIX aggregated over last 6 days
VIX_Pos_Vol_Aggregated	Positive sentiment scores volume for VIX aggregated over last 6 days
VIX_Neg_Vol_Aggregated	Negative sentiment scores volume for VIX aggregated over last 6 days

TABLE 5.3 Summary statistics for scaled sentiment indexes

Sentiment features	1st Quantile	Median	Mean	3rd Quantile
SP500_Pos_Aggregated (Scaled)	0.1711	0.2325	0.2769	0.3425
SP500_Neg_Aggregated (Scaled)	−0.3121	−0.1899	−0.2407	−0.1316
VIX_Pos_Aggregated (Scaled)	0.1488	0.2051	0.2336	0.2876
VIX_Neg_Aggregated (Scaled)	−0.2133	−0.1479	−0.1729	−0.0987

5.5.4 Empirical Methodology

We employ the LR, random forest (RF) model and ANN technique for forecasting the VIX futures closing price movements. We divide the first 2/3 of the entire data as the training set ranging from 28 Jan 2015 to 14 Jan 2019 (n=998) and the remaining 1/3 as the testing set ranging from 15 Jan 2019

TABLE 5.4   Parameter settings for RF

Parameters	Level(s)
Number of trees to grow (ntree)	10, 20, …, 500
Maximum number of terminal nodes trees in the forest (maxnodes)	5, 6, …, 35
Number of variables randomly sampled as candidates at each split (mtry)	1, 2, …, 10

TABLE 5.5   Parameter settings for ANN

Parameters	Level(s)
size	1, 2, …, 10
decay	0.01, 0.02, …, 0.15

to 07 Jan 2021 (n=500). In the training set, we have 602 "DOWNs" and 396 "UPs"; for the hold-out set, we have 305 "DOWNs" and 195 "UPs", indicating that we have an imbalanced classification data set for the response variable. We employ the above-mentioned technical and sentiment indicators as the inputs for all the three models considered in the algorithm training process.

Logistic Regression

The LR is a popular technique to model the probability of discrete outcomes. This method is served as a benchmark model to compare with the performances of ML algorithms.

Random Forest Model

We conduct parameter tuning for RF to determine the best estimator. The parameter settings for RF are presented in Table 5.4.

Artificial Neural Network Model

We consider a single-hidden-layer neural network and specify two hyper-parameters: size and decay. Size is the number of units in the hidden layer, and decay is the regularization parameter to avoid over-fitting. The parameter settings for ANN are presented in Table 5.5.

A note on implementation

R and Python packages are used for LR models as well as the RF genre of models. Tensor flow and KERAS are regularly used by analysts to train and implement ANN classes of models. We have done all the implementation of the models described here using the packages available in R; for more information (see footnote [2]).

5.5.5 Experimental results

We employ accuracy and F-measure to evaluate the performance of considered models. The evaluation measures Precision and Recall are given using the True-Positive (TP), False-Positive (FP), True-Negative (TN) and False-Negative (FN):

$$Precision_{positive} = \frac{TP}{(TP + FP)}$$

$$Precision_{negative} = \frac{TN}{(TN + FN)}$$

$$Recall_{positive} = \frac{TP}{(TP + FN)}$$

$$Recall_{negative} = \frac{TN}{(TN + FP)}$$

Then the accuracy and F-measure are estimated by

$$\text{Accuracy} = \frac{TP + TN}{TP + FP + TN + FN}$$

$$\text{F-measure} = \frac{2 \times \text{Precision} \times \text{Recall}}{\text{Precision} + \text{Recall}}$$

Recall that in statistical analysis of binary classification, accuracy is not a desirable metric to employ when we have class imbalance, i.e., the data set contains many more examples of one class than the other. In such circumstances, the F-measure, which is calculated as the harmonic mean of precision and recall, serves as a helpful alternative when describing and comparing the model performances. The F-measure ranges between 0 and 1. The higher the precision and recall, the higher the F-measure, which implies a better model.

Table 6 and Table 7 present the best estimator in each parameter setting scenario for RF and ANN, respectively.

Table 8 indicates that the LR model provides the lowest performance at 0.1923 value of the F-measure, while RF and ANN algorithms present significantly better performances compared to the LR model. The RF technique has the highest accuracy of 0.630 and F-measure of 0.7587.

The experimental results reveal that for imbalanced classification problems, we might encounter the so-called "Accuracy Paradox" paradigm, where

---

[2]We implement RF using the randomForest package and ANN via the nnet package in R. Both were built under R version 4.2.2

TABLE 5.6 Best RF estimator in each parameter setting scenario

Scenario	ntree	maxnodes	mtry	Accuracy	F-measure
1	360	35	4	0.6280	0.7584
2	350	26	4	0.6300	0.7587
3	370	30	3	0.6300	0.7587

TABLE 5.7 Best ANN estimator in each parameter setting scenario

Scenario	size	decay	Accuracy	F-measure
1	4	0.01	0.6140	0.7402
2	4	0.12	0.6200	0.7545

TABLE 5.8 Performance comparison of the best models

	Accuracy	F-measure
RF	0.6300	0.7587
ANN	0.6200	0.7545
LR	0.5800	0.1923

traditional methodologies such as the logistic model may deliver undesirable predictive power and the emerging use of popular ML techniques serves as a remedy for such shortcomings of conventional models.

5.6 SUMMARY DISCUSSION

This chapter has been set out in the style of a tutorial and provides an overview of some common features of NCQ models and emerging paradigms of AI & ML models. We have included exemplar investigations to illustrate their use in a simple problem that requires a directional prediction. We use the directional movement of VIX, that is, whether VIX closes above or below the level of the previous day's close. We have used LR as a typical neo-classical model, since our response variable is a binary variable, it only takes the values of 0 and 1; 0 when VIX closes below and 1 when VIX closes above the previous day's level. We apply ANN and RF models as examples of ML models.

We observe that the predictive accuracy of both ML models significantly outperforms that of the LR approach for all the parameters we have used. In this chapter, we have not set out to compare the performance of these alternative models. So we point out that such a collection of models with variable accuracy can be used as an ensemble leading to a higher accuracy taken together than any one model taken individually. Thus an ensemble model combining the output of all three models, in general, will yield a better prediction than that of any single model.

REFERENCES

Abraham, R., Samad, M. E., Bakhach, A. M., El-Chaarani, H., Sardouk, A., Nemar, S. E., and Jaber, D. (2022). Forecasting a stock trend using genetic algorithm and random forest. *Journal of Risk and Financial Management*, 15(5):188.

Alberg, D., Shalit, H., and Yosef, R. (2008). Estimating stock market volatility using asymmetric garch models. *Applied Financial Economics*, 18(15):1201–1208.

Ali, S. S., Mubeen, M., Lal, I., and Hussain, A. (2018). Prediction of stock performance by using logistic regression model: evidence from pakistan stock exchange (psx). *Asian Journal of Empirical Research*, 8(7):247–258.

Amit, Y. and Geman, D. (1997). Shape quantization and recognition with randomized trees. *Neural computation*, 9(7):1545–1588.

Ampomah, E. K., Qin, Z., and Nyame, G. (2020). Evaluation of tree-based ensemble machine learning models in predicting stock price direction of movement. *Information*, 11(6):332.

Anand, C. (2021). Comparison of stock price prediction models using pre-trained neural networks. *Journal of Ubiquitous Computing and Communication Technologies (UCCT)*, 3(02):122–134.

Bell, D. E., Raiffa, H., and Tversky, A. (1988). *Decision Making: Descriptive, Normative, and Prescriptive Interactions*. Cambridge University Press.

Bishop, C. M. and Nasrabadi, N. M. (2006). *Pattern Recognition and Machine Learning*, volume 4. Springer.

Bollerslev, T. (1986). Generalized autoregressive conditional heteroskedasticity. *Journal of Econometrics*, 31(3):307–327.

Booth, A., Gerding, E., and McGroarty, F. (2014). Automated trading with performance weighted random forests and seasonality. *Expert Systems with Applications*, 41(8):3651–3661.

Bouras, C., Christou, C., Gupta, R., and Suleman, T. (2019). Geopolitical risks, returns, and volatility in emerging stock markets: evidence from a panel garch model. *Emerging Markets Finance and Trade*, 55(8):1841–1856.

Brahme, A. (2014). *Comprehensive biomedical physics*. Newnes.

Breiman, L. (1996). Bagging predictors. *Machine learning*, 24(2):123–140.

Breiman, L. (2001). Random forests. *Machine learning*, 45(1):5–32.

Brooks, C. (2014a). In C brooks. *Introductory Econometrics for Finance*, 331.

Brooks, C. (2014b). *Introductory Econometrics for Finance. 3rd Edition.* Cambridge University Press.

CBOE (2016). The VIX Index. https://www.cboe.com/tradable_products/vix/faqs/, Last accessed on 05 December 2022.

Chen, W., Jiang, M., Zhang, W.-G., and Chen, Z. (2021). A novel graph convolutional feature based convolutional neural network for stock trend prediction. *Information Sciences*, 556:67–94.

Chiang, T. C. and Doong, S.-C. (2001). Empirical analysis of stock returns and volatility: Evidence from seven asian stock markets based on tar-garch model. *Review of Quantitative Finance and Accounting*, 17(3):301–318.

Choudhry, T. (2000). Day of the week effect in emerging asian stock markets: evidence from the garch model. *Applied Financial Economics*, 10(3):235–242.

Engle, R. F. (1982). Autoregressive conditional heteroscedasticity with estimates of the variance of united kingdom inflation. *Econometrica: Journal of the Econometric Society*, pages 987–1007.

Fama, E. F. (1995). Random walks in stock market prices. *Financial Analysts Journal*, 51(1):75–80.

Freund, Y. (1995). Boosting a weak learning algorithm by majority. *Information and computation*, 121(2):256–285.

Gong, J. and Sun, S. (2009). A new approach of stock price prediction based on logistic regression model. In *2009 International Conference on New Trends in Information and Service Science*, pages 1366–1371. IEEE.

Gyamerah, S. A., Ngare, P., and Ikpe, D. (2019). On stock market movement prediction via stacking ensemble learning method. In *2019 IEEE Conference on Computational Intelligence for Financial Engineering & Economics (CIFEr)*, pages 1–8. IEEE.

Hastie, T., Tibshirani, R., Friedman, J. H., and Friedman, J. H. (2009). *The Elements of Statistical Learning: Data Mining, Inference, and Prediction*, volume 2.

Ho, T. K. (1995). Random decision forests. In *Proceedings of 3rd International Conference on Document Analysis and Recognition*, volume 1, pages 278–282. IEEE.

Ho, T. K. (1998). The random subspace method for constructing decision forests. *IEEE Transactions on Pattern Analysis and Machine Intelligence*, 20(8):832–844.

Hu, H., Tang, L., Zhang, S., and Wang, H. (2018). Predicting the direction of stock markets using optimized neural networks with google trends. *Neurocomputing*, 285:188–195.

Hull, J. C. (2003). *Options futures and other derivatives*. Pearson Education India.

Hung, J.-C. (2011). Applying a combined fuzzy systems and garch model to adaptively forecast stock market volatility. *Applied Soft Computing*, 11(5):3938–3945.

Jain, A. K., Mao, J., and Mohiuddin, K. M. (1996). Artificial neural networks: a tutorial. *Computer*, 29(3):31–44.

James, G., Witten, D., Hastie, T., and Tibshirani, R. (2013). *An Introduction to Statistical Learning*, volume 112. Springer.

Jiang, M., Liu, J., Zhang, L., and Liu, C. (2020). An improved stacking framework for stock index prediction by leveraging tree-based ensemble models and deep learning algorithms. *Physica A: Statistical Mechanics and its Applications*, 541:122272.

Kahneman, D. and Tversky, A. (2013). Prospect theory: an analysis of decision under risk. In *Handbook of the Fundamentals of Financial Decision Making: Part I*, pages 99–127. World Scientific.

Kahnemann, D. (1979). Prospect theory: an analysis of decision under risk. *Econometrica*, 47:263–292.

Kara, Y., Boyacioglu, M. A., and Baykan, Ö. K. (2011). Predicting direction of stock price index movement using artificial neural networks and support vector machines: the sample of the istanbul stock exchange. *Expert systems with Applications*, 38(5):5311–5319.

Lee, S. (2004). Application of likelihood ratio and logistic regression models to landslide susceptibility mapping using gis. *Environmental Management*, 34(2):223–232.

Lin, W., Wu, Z., Lin, L., Wen, A., and Li, J. (2017). An ensemble random forest algorithm for insurance big data analysis. *IEEE Access*, 5:16568–16575.

Lintner, J. (1969). The valuation of risk assets and the selection of risky investments in stock portfolios and capital budgets: a reply. *The Review of Economics and Statistics*, pages 222–224.

Lohrmann, C. and Luukka, P. (2019). Classification of intraday s&p500 returns with a random forest. *International Journal of Forecasting*, 35(1):390–407.

Luo, S.-S., Weng, Y., Wang, W.-W., and Hong, W.-X. (2017). L1-regularized logistic regression for event-driven stock market prediction. In *2017 12th International Conference on Computer Science and Education (ICCSE)*, pages 536–541. IEEE.

Markowitz, H. (1952). Portfolio selection. *The Journal of Finance*, 7(1):77–91.

Miller, M. H. and Modigliani, F. (1961). Dividend policy, growth, and the valuation of shares. *The Journal of Business*, 34(4):411–433.

Mitra, G., Greenberg, H. J., Lootsma, F. A., Rijkaert, M. J., and Zimmermann, H. J. (1988). *Mathematical Models for Decision Support*. Springer.

Mitra, G. and Mitra, L. (2011). *The Handbook of News Analytics in Finance*. John Wiley & Sons, Chichester, UK.

Mitra, G. and Yu, X. (2016a). *The Handbook of Sentiment Analysis in Finance*. Optirisk Systems in collaboration with Albury Books.

Mitra, G. and Yu, X. (2016b). *Handbook of Sentiment Analysis in Finance*. Optirisk Systems in collaboration with Albury Books.

Neumann, J. V. and Morgenstern, O. (1944). Theory of games and economic behavior. Princeton: Princeton Univ. Press.

Nti, I. K., Adekoya, A. F., and Weyori, B. A. (2020). A comprehensive evaluation of ensemble learning for stock-market prediction. *Journal of Big Data*, 7(1):1–40.

Nti, K. O., Adekoya, A., and Weyori, B. (2019). Random forest based feature selection of macroeconomic variables for stock market prediction. *American Journal of Applied Sciences*, 16(7):200–212.

Priddy, K. L. and Keller, P. E. (2005). *Artificial Neural Networks: An Introduction*, volume 68. SPIE press.

Qin, L. and Ren, R. (2019). Integrating market sentiment with trading rules-empirical study on china stock market. In *2019 International Conference on Pedagogy, Communication and Sociology (ICPCS 2019)*, pages 358–363. Atlantis Press.

Rakhra, M., Soniya, P., Tanwar, D., Singh, P., Bordoloi, D., Agarwal, P., Takkar, S., Jairath, K., and Verma, N. (2021). Crop price prediction using random forest and decision tree regression: a review. *Materials Today: Proceedings*.

Sadorsky, P. (2021). A random forests approach to predicting clean energy stock prices. *Journal of Risk and Financial Management*, 14(2):48.

Sagi, O. and Rokach, L. (2018). Ensemble learning: a survey wiley interdisciplinary reviews: data mining and knowledge discovery. 8(4):e1249.

Scholes, M. and Black, F. (1973). The pricing of options and corporate liabilities. *Journal of Political Economy*, 81(3):637–654.

SenRisk (2018). Sentiment analysis and risk assessment. http://senrisk.eu/. Last accessed on 05 December 2022.

Serrano, W. (2022). The random neural network in price predictions. *Neural Computing and Applications*, 34(2):855–873.

Sharpe, W. F. (1964). Capital asset prices: a theory of market equilibrium under conditions of risk. *The Journal of Finance*, 19(3):425–442.

Shastri, M., Roy, S., and Mittal, M. (2019). Stock price prediction using artificial neural model: an application of big data. *EAI Endorsed Transactions on Scalable Information Systems*, 6(20):e1–e1.

Smales, L. A. (2014). News sentiment in the gold futures market. *Journal of Banking & Finance*, 49:275–286.

Srinu Vasarao, P. and Chakkaravarthy, M. (2022). Time series analysis using random forest for predicting stock variances efficiency. In *Intelligent Systems and Sustainable Computing*, pages 59–67. Springer.

Stacking (2020). Ensembling technique to improve prediction. https://medium.com/ml-research-lab/stacking-ensemble-meta-algorithms-for-improve-predictions-f4b4cf3b9237. Last accessed on 07 December 2022.

Stigler, S. M. (1986). *The History of Statistics: The Measurement of Uncertainty Before 1900*. Harvard University Press.

Tan, Z., Yan, Z., and Zhu, G. (2019). Stock selection with random forest: An exploitation of excess return in the chinese stock market. *Heliyon*, 5(8):e02310.

Thakkar, A. and Chaudhari, K. (2020). Predicting stock trend using an integrated term frequency–inverse document frequency-based feature weight matrix with neural networks. *Applied Soft Computing*, 96:106684.

Thakkar, A. and Chaudhari, K. (2021). A comprehensive survey on deep neural networks for stock market: The need, challenges, and future directions. *Expert Systems with Applications*, 177:114800.

Weng, B., Lu, L., Wang, X., Megahed, F. M., and Martinez, W. (2018). Predicting short-term stock prices using ensemble methods and online data sources. *Expert Systems with Applications*, 112:258–273.

Wolpert, D. H. (1992). Stacked generalization. *Neural networks*, 5(2):241–259.

Xu, L., Kinkyo, T., and Hamori, S. (2018). Predicting currency crises: A novel approach combining random forests and wavelet transform. *Journal of Risk and Financial Management*, 11(4):86.

Xu, Y., Yang, C., Peng, S., and Nojima, Y. (2020). A hybrid two-stage financial stock forecasting algorithm based on clustering and ensemble learning. *Applied Intelligence*, 50(11):3852–3867.

Yegnanarayana, B. (2009). *Artificial Neural Networks*. PHI Learning Pvt. Ltd.

Zhang, D. and Lou, S. (2021). The application research of neural network and bp algorithm in stock price pattern classification and prediction. *Future Generation Computer Systems*, 115:872–879.

Zhou, F., Zhang, Q., Sornette, D., and Jiang, L. (2019). Cascading logistic regression onto gradient boosted decision trees for forecasting and trading stock indices. *Applied Soft Computing*, 84:105–747.

Zi, R., Jun, Y., Yicheng, Y., Fuxiang, M., and Rongbin, L. (2022). Stock price prediction based on optimized random forest model. In *2022 Asia Conference on Algorithms, Computing and Machine Learning (CACML)*, pages 777–783. IEEE.

(III)

HANDLING DIFFERENT ALTERNATIVE DATASETS

CHAPTER 6

Asset Allocation Strategies: Enhanced by Micro-Blog

Zryan Sadik

OptiRisk Systems Ltd, London, United Kingdom

Gautam Mitra

OptiRisk Systems Ltd and UCL, Department of Computer Science, London, United Kingdom

Shradha Berry

OptiRisk Systems Ltd, London, United Kingdom

Diana Roman

OptiRisk Systems Ltd and Brunel University London, Department of Mathematics, London, United Kingdom

CONTENTS

DOI: 10.1201/9781003293644-6

6.1 INTRODUCTION AND BACKGROUND

6.1.1 Literature Review

The rapid rise of social media communication has impacted the social milieu; thus many aspects of our Political, Economic, Social and Technological social and commercial life are affected. In particular, the rise of social media as the most preferred way of connecting people online has led to new models of information communication amongst the peers. Of these media, Twitter has emerged as a particularly strong platform and in the financial domain twits by market participants are of great interest and value. The provider of our Social media data, namely, StockTwits is a financial social media platform for traders, investors, media, public companies and investment professionals. Nowadays, with the rapid development of social media platforms, more and more financial market participants such as investors, analysts, traders, brokers and market makers prefer to communicate their respective perspectives about market as well as individual equities on social platforms. Our company and our quant analysts in an earlier study [Shi et al., 2017] used news and StockTwits data to enhance momentum strategy. The authors reported improved portfolios with news and StockTwits. The maximum portfolio returns were doubled with the introduction of StockTwits data reported in that study. Further that report made a strict distinction of sentiments expressed on social platforms, such as Twitter and StockTwits, from news sentiment obtained via newswires . This is because the messages shared publicly between traders are more germane and insightful than published news. Returning to Stocktwits the supplier of our social media data, they are a leading communication platform founded in 2008 with more than one million registered users, serves as a financial trending tool for regular investors and traders to share their opinions and learn from others about the market and stocks. Just like Twitter, messages shared on it are restricted to no more than 140 (now 250) characters, including ideas, charts, links and other forms of data [Oliveira et al., 2014]. The company, founded in 2008, was initially built to utilise Twitter's application programming interface (API). It has since grown in to a standalone micro-blogging platform for social media for finance. It is 2M registered members and gets about 4M monthly messages. Users can create free accounts in StockTwits and share messages on stocks with cash tags to identify them (example $AAPL for Apple Inc.) But what makes it extremely unique and interesting for financial market participants is that this platform focuses specifically on the field of financial markets and investment. Thus, a huge database with less noise than that collected from a more common social network is available. Lastly, the biggest feature of this platform is that people can directly see the level of bullishness and bearishness of a stock any time and this sentiment data is available as a chart. Based on the convenience and novelty of the StockTwits database, researchers are intrigued to find out the value within this dataset. Yet, to date, work in this area is not in abundance. [Oliveira et al., 2013] applied microblogging data to find a more robust evaluation method to forecast the following stock market variables: volume of trading, profits and volatility. Choosing five large US companies (AAPL, AMZN, GS, GOOG and IBM) and one market index (SPX), they obtained two kinds of daily data, namely, indicators for sentiment and the number of posts for each stock from June 2010 to November 2012. They explored several regression models but were unable to find a good predictive model that used sentiment indicators to predict the return and volatility. However, [Sprenger et al., 2014] found that public sentiment delivered via StockTwits is aligned with the movement of S&P 500 and is positively associated with trading volumes. [Hochreiter, 2016] uses StockTwits data for 30 listed companies in Dow Jones Industrial Average from 2010 to 2014. He applied evolutionary optimization methods to construct optimal rule-based trading strategies. The result was that the portfolio built with evolutionary optimization techniques outperforms the classical Markowitz optimal portfolio with reduced risks. Micro-blogging platforms such as Twitter and Weibo encourage short messages that are restricted in length and utilise tags to highlight the main topics. In turn this increases the speed at which users can create posts, and consequently the volume of posts, often resulting in the first release of particular information, for example, economic reports, earnings release and CEO departures. Obviously, this advantage in timing compared to classic newswires is contributed to the fact that information does not have to be verified by multiple sources. Over the past few years, the amount of academic literature associated with sentiment analysis has increased dramatically. In particular, the study by Pear Analytics [Kelly, 2009], Twitter conversations were analysed and classified to be 40% "Pointless babble", 38% conversational and the rest is either self-promotion, spam, pass along, or news. There are few studies that explored the StockTwits data, which is a micro-blogging platform exclusively dedicated to the stock market. [Al Nasseri et al., 2015] proposed a new intelligent trading support system based on sentiment prediction by combining text-mining techniques, feature selection and decision tree algorithms in an effort to analyze and extract semantic terms expressing a particular sentiment (sell, buy or hold) from StockTwits messages. They confirmed that

StockTwits postings contain valuable information and lead trading activities in capital markets

6.1.2 Asset Allocation by Second-Order Stochastic Dominance

SSD has a well recognised importance in portfolio selection, due to its connection to the theory of risk-averse investor behaviour and tail risk minimisation. Until recently, stochastic dominance models were considered intractable or at least very demanding from a computational point of view. Computationally tractable and scalable portfolio optimization models which apply the concept of SSD were proposed by [Dentcheva and Ruszczynski, 2006]; [Roman et al., 2006]; [Roman et al., 2013] and [Fabian et al., 2011]. These portfolio optimisation models assume that a benchmark, that is, a desirable "reference" distribution is available and a portfolio is constructed, whose return distribution dominates the reference distribution with respect to SSD. Index tracking models also assume that a reference distribution (that of a financial index) is available. A portfolio is then constructed, with the aim of replicating, or tracking, the financial index. Traditionally, this is done by minimising the tracking error, that is, the standard deviation of the differences between the portfolio and index returns. Other methods have been proposed (for a review of these methods, see for example [Beasley et al., 2003] and [Canakgoz and Beasley, 2009]. Recently further logical extensions of the SSD Long only models to Long/Short models have been proposed and formulated as Mixed Integer Programming (MIPs) [Kumar et al., 2008].

6.1.3 Micro-Blog Sentiment and the Impact of Micro-Blog Sentiment on Asset Return

Recently availability of high-performance computer systems has facilitated high frequency trading. Further the automated analysis of news feeds set the backdrop for computer automated trading which is enhanced by news (see e.g. [Mitra and Mitra, 2011] and [Mitra and Yu, 2016]). News sentiment is regarded to be unique to each individual and encompasses lots of emotions occurring during brief moments. For the financial domain, it is commonly known that investors in the bull market are positive and optimistic, while in the bear markets they seem relatively pessimistic and fear of loss. In other words, good sentiments are usually based on the rise in stock prices and can further stimulate a continued rise i.e. builds momentum. Therefore, relevant transaction data can be used to build sentiment indicators as one of the forecasting technologies of the future trend of stock price fluctuations (see, e.g. [Pring, 2002]). As pointed out in [Feuerriegel and Prendinger, 2016], sentiment analysis usually refers to the methods that judge the content of positive or negative through the relative details of text or other forms.

6.1.4 Micro-Blog Data for Trading and Fund Management

Micro-blogs in general and Tweets in particular stand apart as a platform for disseminating news. Because these are not subject to thorough editorial scrutiny and control they are far less reliable as a source of fact or evidence. On the other hand, these have the advantage of speed, that is, low latency of dissemination to a relatively larger readership. In this respect Twitter Data is ideal for enhancing Trading and Fund Management strategies. OptiRisk analytic team has been active in this domain and we have reported a number of studies. In 2015, for instance, Professor Hochreiter who is our associate [Hochreiter, 2016], published a paper discussing automated trading strategy using genetic algorithms and Micro-Blog sentiment data. Ms Shi an OptiRisk intern jointly with our analysts reported her findings of applying momentum strategy enhanced by StockTwits sentiment data in [Shi et al., 2017]. In the recently concluded study by Ms Berry [Berry et al., 2019] and co-workers have reported good results in volatility prediction which can be used for volatility trading and variance swaps.

6.1.5 Guided Tour

This chapter is structured as follows: In Section 6.2 we introduce the historical closing prices of the historical market data and the micro-blog meta data supplied by StockTwits. Section 6.3 sets out the asset allocation strategy used by OptiRisk; the details of the relevant SSD models are also included. In Section 6.4 we present a novel method of restricting the asset universe of choice using our proprietary method of constructing filters. In Section 6.5 we describe our investigations and the findings of our investigation. A discussion of our work and the conclusions are presented in Section 6.6. Finally, in the Appendix an expanded version of our back-testing results is presented.

6.2 MARKET DATA AND MICRO-BLOG DATA

The investigation reported in this chapter uses four-and-half years of historical data of the S&P 500 index. The data spans the period starting from 1 January 2014 to 1 July 2019. The daily closing prices of the market data and the micro-blog sentiment data supplied by Stocktwits, for each of the S&P 500 assets is included in this study.

6.2.1 Market Data

The time series data used in this study is the stock market daily closing price of the S&P 500 companies. We first filter the whole market database to contain only the daily prices of the assets from S&P 500 index covering from 1 January 2014 to 1 July 2019. This will produce the following eight columns:

TABLE 6.1 Market Data Sample

Date	Index	RIC	Open	High	Low	Close	Volume
20140102	S&P500	AMZN.O	398.8	399.36	394.02	397.97	2140246
20140103	S&P500	AMZN.O	398.29	402.71	396.22	396.44	2213512
20140106	S&P500	AMZN.O	395.85	397	388.42	393.63	3172207
20140107	S&P500	AMZN.O	395.04	398.47	394.29	398.03	1916684
20140108	S&P500	AMZN.O	398.47	403	396.04	401.92	2316903
20140109	S&P500	AMZN.O	403.71	406.89	398.44	401.01	2103793
20140110	S&P500	AMZN.O	402.53	403.764	393.8	397.66	2681701

TABLE 6.2 Sentiment Stream Attributes

message_id	user_id	sentiment_score	symbol_id	created_at	exchange	industry	sector	symbol	title
149482996	1695974	-0.1605	1	2019-01-02T 16:44:58.000Z	NASDAQ	Conglomerates	Conglomerates	ACC	AAC Holdings

Date, Index, RIC[1], *Open, High, Low* and *Close* prices in USD and *Volume*. Table 6.1 shows a sample of the market data.

The data was collected from Thomson Reuters Data Stream platform and adjusted to account for changes in index composition. This means that our models use no more data than was available at the time, removing susceptibility to the influence of survivor bias. For each asset, we compute the corresponding daily rates of return.

6.2.2 Micro-blog Data (StockTwits data)

Sentiment data is obtained from StockTwits which is a financial social media platform. The discussions in this platform are restricted to the domain of finance which makes it a rich and focused data source for financial models. This platform is different from textual news. These are individual views and opinions of investors and domain experts. Also, unlike news-wire, StockTwits gets tweets posted 24 hours a day, 7 days of the week. The data for this research is fetched from StockTwits Firestream API. The API allows licensed users with access to live and historical data on messages, activity and sentiment scores. The historical data dates back to the year 2010. The data is streamed in JSON format. Table 6.2 shows the various attributes of a sample sentiment stream. Table 6.3 describes each of these attributes.

---

[1]The Reuters instrument code (RIC) is a code assigned by Thomson Reuters to label each asset

TABLE 6.3   Description of each Attribute

Attribute	Description
message_id	Unique message ID
user_id	Unique user ID
sentiment_score	The sentiment score for this message. Range from −1 to 1 −1 implies very bearish +1 implies very bullish 0 implies neutral
symbol_id	Unique stock ID
created_at	Message timestamp
exchange	Exchange this stock is found at
industry	Stock's industry
sector	Stock's sector
symbol	Stock's symbol
title	Stock's title

Generating Micro-blog Impact Scores

In this study, we use the idea that was first proposed by [Yu, 2014] and used by [Yu et al., 2015]; [Sadik et al., 2018]; [Sadik et al., 2020] to construct micro-blog impact scores. These micro-blog impact scores can be used as proxies of firm-specific news impact in the new model. To calculate the micro-blog impact scores, the following steps have to be done:

1. The Timestamp for each micro-blog message has to be converted from UTC to EST time, which is the timing convention of the S&P 500 constituents from the New York Stock Exchange (NYSX).

2. Separating the positive and negative sentiment scores so that two different time series can be obtained.

3. After separating the scores, in a similar fashion to [Sadik et al., 2018], the positive and negative micro-blog impact scores for each sentiment score is calculated.

4. Finally, two daily time series are generated that represent the daily positive and negative micro-blog impact scores.

6.3 ASSET ALLOCATION STRATEGY

6.3.1 Portfolio Construction Models

The challenging problem of "active" portfolio selection is how to construct a portfolio such that its return at the end of the investment period is a maximum. Since portfolio returns are random variables, models that specify a preference relation among random returns are required. A portfolio is then chosen such that its return is non-dominated with respect to the preference relation that is under consideration; computationally, this is achieved using an optimisation model.

For portfolio selection, mean-risk models have been by far the most popular. They describe and compare random variables representing portfolio returns by using two statistics: the expected value (mean), where high values are desired and a risk value, where low values are desired. The first risk measure for portfolio selection was variance, proposed by Harry Markowitz [Markowitz, 1952]; he also introduced the concept of "efficient" portfolio, that is, one whose return has the lowest risk for a given mean value. A portfolio chosen for implementation should be "efficient" and is found via optimisation, where typically risk is minimised with a constraint on mean. Various risk measures, quantifying different "undesirable" aspects of return distributions, have been proposed in the literature, see for example [Fishburn, 1977]; [Ogryczak and Ruszczynski, 1999]; [Ogryczak and Ruszczynski, 2001]; [Rockafellar et al., 2000] and [Rockafellar and Uryasev, 2002].

Mean-risk models are intuitive and convenient from a computational point of view. However, they summarise a distribution with only two statistics; hence, a lot of information is overlooked and the resulting return distribution might still have undesirable properties.

Another paradigm in portfolio construction is Expected Utility Theory [Von Neumann et al., 2007]; here, random returns are compared by comparing their expected utilities (larger values are preferred). However, the expected utility values depend on the chosen utility function, which is a subjective choice. There are progressively stronger conditions on utility functions in order to correctly represent preference on wealth. The non-arguable requirement is that utility functions should be non-decreasing: higher wealth is preferred to lower wealth. Thus, non-decreasing utility functions represent rational behaviour. Furthermore, financial decision makers have been observed to be risk averse: the same increase in wealth is valued more at low wealth levels.

6.3.2 Second-Order Stochastic Dominance

Stochastic dominance provides a framework for comparing and ordering random variables (e.g. representing portfolio returns) that is closely connected to the expected utility theory, but it eliminates the need to explicitly specify a utility function (see, e.g. [Whitmore and Findlay, 1978] for a detailed description of stochastic dominance relations, [Kroll, 1980] for a review).

Progressively stronger assumptions about the form of utility functions used in investment lead to first, second and higher orders of SD. For example, first order stochastic dominance (FSD) is connected to "non-satiation" behaviour. A random return is preferred to another with respect to FSD if its expected utility is higher, for any non decreasing utility function. This is a strong condition and thus many random returns cannot be ordered with respect to FSD.

SSD has been widely recognised as the most sound framework for financial decision making, since it includes the preference of rational and risk averse investors, which is the observed attitude. A random return is preferred to another with respect to SSD if its expected utility is higher, for any non decreasing and concave utility function. There are equivalent definitions for SSD preference, underlying the close connection with tail risk measures such as Conditional Value-at-Risk (CVaR), proposed by [Rockafellar et al., 2000]. For $\alpha \in (0,1)$, The α-tail of a return distribution is approximately defined as the average or expected value of the worst $A\%$ of its outcomes, where $\alpha = A\%$ (e.g. $\alpha = 0.05$ corresponds to 5% tail.)

SSD involves comparison of increasingly higher portions of left tails: a random return is preferred to another with respect to SSD if its α-tail is higher, for all $\alpha \in (0,1)$. Equivalently, a random return is preferred to another with respect to SSD if its expected shortfall with respected to *any* target is lower. For rigorous definitions and treatment of SSD, please see, for example [Roman et al., 2006] and [Roman and Mitra, 2009].

SSD has always been regarded as a sound choice framework, as it does not lose information on the distributions involved. To obtain a portfolio whose return distribution is non-dominated with respect to SSD (thus attractive to all risk averse decision makers) has always been considered as highly desirable. Until recently, however, this was thought to be computationally intractable. Since the 2000s, there has been considerable research on computationally tractable optimisation models employing the SSD criterion. OptiRisk and its research team have been a leader in the field, producing several seminal papers and employing SSD commercially as its asset allocation engine. Two leading contributions are listed here [Dentcheva and Ruszczynski, 2006], [Fábián et al., 2011].

6.4 CONSTRUCTION OF FILTERS

We explain the rationale of why we need to use Filters and the advantage of using Filters. We then expand this section to describe the different type of filters which we have used. Our approach to incorporating filters for the choice of assets for inclusion in our portfolios has the aim of achieving only micro-blog-based choice or Market Price-based choice of asset allocation. In this approach we are able to choose (i) no influence of micro-blog or (ii) partial influence of micro-blog or (iii) even the extreme of Only Influenced by micro-blog and no Price Data influence.

6.4.1 Why Use Filters

The asset allocation strategy used by OptiRisk system uses the SSD model. The scenarios are the historical return data which captures accurately correlation structure of the constituent assets. As a consequence, the asset allocation is fairly robust in the long run and also achieves control of tail risk. The asset allocation is a static one period model; so it suffers from the draw back that the near term asset price movements are not taken into consideration. We have therefore introduced a method of restricting the asset universe for the choice of long assets and short assets which are to be held in the portfolio using a technique which we call asset filter. In the construction of the filter we take into account the near term behaviour of each of the assets in the asset universe. The near term behaviour is captured by the use of a well known technical indicator, namely, the relative strength index (RSI). We have extended this concept of applying the filter by taking into account the micro-blog sentiment and the impact of the micro-blog sentiment in respect of each asset. In rest of this Section 6.4 we describe the method by which we combine these two approaches.

6.4.2 Relative Strength Index

The Technical Indicator, Relative Strength Index (RSI), is an established momentum oscillator. The RSI compares the magnitude of a stock's recent gains to the magnitude of its recent losses and turns that information into a number that ranges from 0 to 100. The RSI indicator uses the daily closing prices over a given period is computed for each constituent asset of the market index under consideration. It is driven by the measure of the momentum of each asset. The RSI measure is expressed as:

$$RSI(t) = 100 - \frac{100}{1 + RS(t)};$$ (6.1)

The RSI and RS are re-expressed for the time bucket (t) as $RSI(t)$.

Computation of RSI

Formally, Relative Strength uses Exponential Moving Average (EMA); thus $RS(t)$ the relative strength is computed as the ratio of average gains and losses.

$$RS(t) = \frac{EMA(Gain_t)}{EMA(Loss_t)} \dots \text{calculated using market data of stock prices}$$

(6.2)

$$RSI(t) = 100 - \frac{100}{1 + RS(t)};$$ (6.3)

$$EMA(X_t) = \sum_{tn=1}^{N} e^{(-\lambda tn)} X_{tn}$$ (6.4)

where $Xt = Gain_t$ or $Loss_t$, λ = decay factor and N=RSI period.

The typical $RSI(t)$ value is calculated with average gains and losses over a period of N = 14 days (lookback period). The number of days N is a parameter in the RSI function and can be chosen in accordance with the characteristics of the data set. Secondly, the number of offset days can be varied. Gains and losses can therefore be daily gains and losses (days=1) or gains and losses over larger time intervals.

The $RSI(t)$ is considered to highlight overbought or oversold assets; when the $RSI(t)$ is above the thresholds of 70 and oversold when it is below the threshold of 30.

6.4.3 Micro-blog Relative Strength Index

We have introduced the concept of Micro-blog Relative Strength Index ($MRSI$). In this we extend the concept of RSI which is computed using Market Data by replacing it with the Impact of the streaming micro-blog sentiment Data.

Micro-blog RSI (MRSI) Computation

It is computed in a way comparable to that of **RSI(t)** whereby the up and down price movements are replaced by positive and negative micro-blog impact scores, respectively. Thus

$$MRS(t) = \frac{EMA(Positive\ Impact\ Scores)}{EMA(Negative\ Impact\ Scores))} \dots \text{is computed for each stock}$$

(6.5)

Hence,

$$MRSI(t) = 100 - \frac{100}{1 + MRS(t)};$$ (6.6)

Thus, the $MRSI(t)$ values range between $0 - 100$.

The micro-blog (sentiment) impact scores are computed in the same way as News (sentiment) impact score in $NRSI(t)$. An explanation of the computational model for $NRSI(t)$ and of News Impact scores the readers are referred to [Sadik et al., 2019] and [Yu, 2014] and [Yu et al., 2015]; [Sadik et al., 2018]; [Sadik et al., 2020], respectively.

6.4.4 Derived RSI (DRSI) Computation

We define the measure **Derived RSI** computation by taking a linear combination of $RSI(t)$ and $MRSI(t)$. So for the time bucket t, the measure Derived RSI $(DRSI(t))$ is defined as

$$DRSI(t) = \theta * RSI(t) + (1 - \theta) * MRSI(t), \qquad (6.7)$$

where $0 \leq \theta \leq 1$. The micro-blog impact scores are used to compute the $MRSI$. They reflect the same modelling paradigm of computing RSI. Thus, for the time bucket t we compute $RSI(t)$ and $MRSI(t)$ to calculate $DRSI(t)$:

$$DRSI(t) = \begin{cases} RSI(t) & \text{if } \theta = 1 \\ MRSI(t) & \text{if } \theta = 0 \\ DRSI(t) & \text{otherwise, that is, } 0 < \theta < 1 \end{cases} \qquad (6.8)$$

6.4.5 Applying the Filters: RSI (t), MRSI (t), DRSI (t)

As explained earlier the purpose of these filters are to restrict the choice of long and short positions of assets as they appear in the asset universe of the available assets. The choice is restricted in the following way: We apply a threshold of 70 to define the long and short bins; Long Bin is filled with the assets whose $RSI(t)$, or $MRSI(t)$, or $DRSI(t)$ values are below 70, and the Short Bin is filled with the assets whose $RSI(t)$, or $MRSI(t)$, or $DRSI(t)$ values are above 70. Finally the SSD method of asset allocation is applied to this restricted asset universe.

6.5 EMPIRICAL INVESTIGATION

6.5.1 The Framework

Our empirical investigation uses two time series datasets, namely, Market Data supplied by Thomson Reuters (Refinitiv) and Micro-blog Sentiment data supplied by StockTwits. We will use the following time-series in our experiments:

- 5 years of historical daily adjusted closing prices of the S&P 500 assets that covering the time period from 2014 to 2019.

- 5 years of daily micro-blog impact scores (positive and negative) are then derived from for each asset in the S&P 500 index. We set the looking back period of considering previous micro-blog items to 4320 minutes (3 days) and consider that the sentiment value decays to half of its initial value in 240 minutes.

TABLE 6.4 Parameter settings of the trading strategy

Parameter	Value	Parameter	Value
Index	SP500	Risk free rate	2.0% a year
Cash lend rate	2.0% a year	Cash borrow rate	2.0% a year
Proportion in long/short	100/0	Proportion in SSD cash	Up to 50%
Gearing	Not applied	Money mgmt. (prop. in cash)	No
Use lot sizing	Yes	Transaction costs	5 basis points
In-sample	500 days	SSD rebalancing frequency	3 days
UCITS compliant	No	Slippage	25 basis point
Cardinality constraints	Not enforced	Extra assets	Futures and VIX
Asset universe	Full OR Reduced	Stop Loss	Not enforced

6.5.2 Experimental Setup

Back-Testing study was carried out using the Framework described in Section 6.5.1, and the experimental setup is best explained by describing the Parameter Settings which are set out in Table 6.4. These Parameter Settings are used to control the trading that is done at each trading day of the Back-Testing period.

The following parameters set out in Table 6.4 were used in the investigation.

1. **Index:** S&P 500. <This model is generic apply to other indices: Hang Seng, Topix...>

2. **Risk free rate:** 2% <Risk free rate is set by default for the Market/Index geography>.

3. **Cash lend rate:** 2% <above comments apply>.

4. **Cash borrow rate:** 2% <above comments apply>.

5. **Proportion in long/short:** Long only...100/0 <specifies limits of long and short positions>.

6. **Proportion in SSD cash:** 0.5 [= 50%] <In the portfolio the Long position in cash is 0.5 or less of the Portfolio mark-to-market value>.

7. **Gearing:** <In some exchanges, namely, NIFTY or KOSPI traders instead of trading in underlying stock exploit gearing by using the futures contract>.

8. **Money management:** not applied here.

9. **Lot size:** traded only in available lot sizes.

10. **Transaction costs:** transaction costs is five basis point, that is, 0.0005.

11. **In-sample:** 500 <in sample trading days of historical data roughly two years worth of data>.

12. **SSD rebalancing frequency:** performs a rebalancing every 1, 2, ... days.

13. **UCITS complaint:** not applied here .

14. **Cardinality:** exact number of assets to be chosen in the portfolio.

15. **Extra assets:** Yes <whether to include index futures or/and VIX as an asset>.

16. **Asset universe:** whether or not a filter is applied.

17. **Stop Loss:** stop loss rule to be applied for individual assets.

The investigation is carried out using the concept of "rolling window"; in this case the window has a span of 500 trading days. The period investigated spans 5 years and 4 months (January 2014 to June 2019).

6.5.3 Back-testing Results

Back-testing was done using the experimental setup described in Section 6.5.2. In these tests three different strategies were applied and their outcomes examined and compared. The three strategies are

1. Full Asset Universe: This strategy takes into consideration the whole available assets from the S&P 500 index. It is included as a second benchmark, such that we are able to measure the improvements made by using different filters.

2. RSI: This is a momentum based strategy, which uses the RSI filter to restrict the choice of the long bins and the short bins of the asset universe.

3. DRSI: This is a derived strategy that combines RSI and MRSI which are defined in Section 6.4. In the results reported in this section, we have set the following value:

$$DRSI = \theta * RSI + (1 - \theta) * MRSI, \qquad (6.9)$$

where θ is set to $\theta = 0.4$.

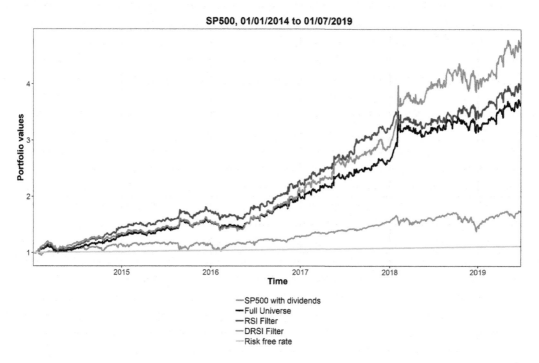

Figure 6.1   Comparison of the portfolio performance of two investment strategies

TABLE 6.5   Performance measurements for three portfolios

Portfolio	Final value	Excess RFR (%)	Sharpe ratio	Sortino ratio	Max draw- down (%)	Max. rec. days	Beta	Av. turnover	Wins	Losses
SP500 with dividends	1.75	8.76	0.62	0.86	19.49	218				
Full Universe	3.62	24.46	1.23	2.19	13.04	288	-0.05	6.84	662	720
RSI Filter	3.92	26.30	1.50	2.58	12.98	179	0.13	15.75	724	658
DRSI Filter	4.68	30.50	1.37	2.34	15.47	133	0.26	13.31	740	642

In Figure 6.1 the charts of the [− SP500 Index], [− Full Universe], [− RSI Filter],[− DRSI Filter] and [− Risk free rate] are presented. A complete set of charts which provide further detailed information about the portfolio composition such as SSD Cash, Cardinality and VIX position are supplied in the *Appendix*.

6.5.3.1   Analysis of Results

The performance of these three strategies is compared using the Industry Standard performance measures. These measures are tabulated and displayed in Table 6.5.

- Final value: Normalised final value of the portfolio at the end of the Back-testing period.

- Excess over RFR (%): Annualised excess return over the risk free rate. For S&P500 we used a yearly risk free rate of 2%.

- Sharpe ratio: Sharpe ratio computed using annualised returns.

- Sortino ratio: Sortino ratio computed using annualised returns.

- Max drawdown (%): Maximum peak-to-trough decline (as percentage of the peak value) during the entire Back-testing period.

- Max recovery days: Maximum number of days for the portfolio to recover to the value of a former peak.

- Beta: Portfolio beta when compared to the S&P500 index.

- Av. turnover: Av. turnover per day as a percentage of portfolio mark-to-market.

- Wins: Number of days that the portfolio makes profits throughout the Back-testing period.

- Losses: Number of days that the portfolio makes losses throughout the Back-testing period.

The relative performances can be compared and summarised under each headings of the Table 6.5.

Final value: The values as displayed in the Table show a steady improvement from the Index (1.75), ... until DRSI (4.68).

Excess RFR (%): As above and quite naturally, the values as displayed in the Table show a steady improvement from the Index (8.76), ... until DRSI (30.50).

Sharpe and Sortino ratios: We find these have improved: **Sharpe** Index (0.62), ... until DRSI (1.37) and **Sortino** Index (0.68), ... until DRSI (2.34). We observe the increase for Sharpe is not monotone which not surprising as the SSD asset allocation strategy minimises 'Tail Risk'.

Max draw- down (%) and Max. rec. days which are "Dynamic Risk" measures do not show consistent improvement. Our perspective on this is that on a long Back-Testing study spanning over multiple years **Max draw- down (%)** and **Max. rec. days** are less meaningful than if they are computed on a quarterly basis.

Beta and Av. Turnover are reasonable whereas **Wins** and **Losses** of DRSI is clearly best.

6.6 DISCUSSIONS AND CONCLUSION

In an associated report by [Berry et al., 2019], we have presented a descriptive analysis of Stocktwits sentiment data. In this study we have (i) processed the the StockTwits sentiment data and computed the impact on the returns for individual stocks in the S&P500 index. We have then (ii) created an asset filter $MRSI$ and then a derived filter $DRSI$. This filter is then applied to restrict the asset universe of choice. These filter restrictions which are based on tweets by Market participants prove to be beneficial and are seen to enhance our daily trading strategy. The back-testing results that we have presented vindicate our assertions.

We plan further work to explore how news sentiment time series data can be fused with micro-blog time series data. Since the information contents of these two sources are fairly different. Given the positive results that we have found in using this data, we wish to find other data sources from which we can obtain financial market tweets for other geographical trading venues and Indices such as TOPIX, Hang Seng, NIFTY and Euro STOXX.

APPENDIX

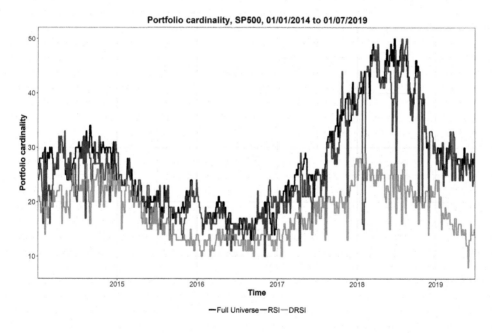

Figure 6.2   Portfolio Cardinality

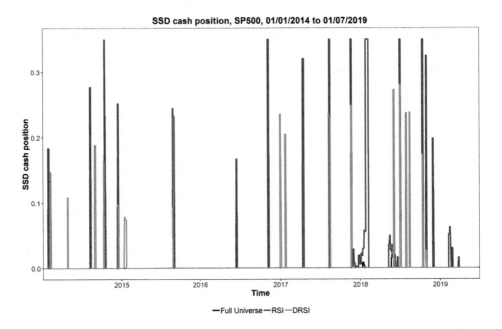

Figure 6.3   SSD Cash Position

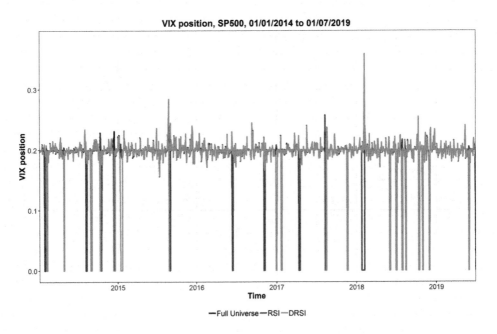

Figure 6.4   VIX Position

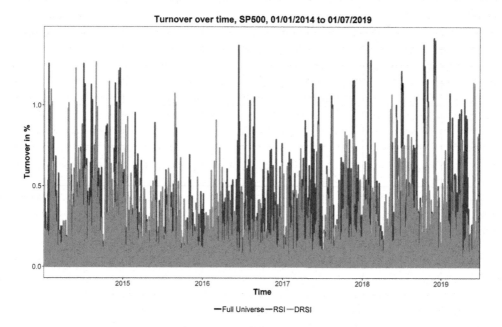

Figure 6.5   Turnover Over Time

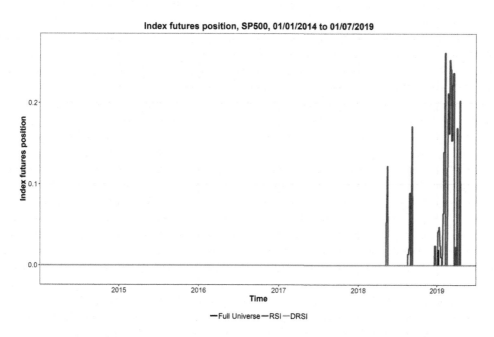

Figure 6.6   Index Futures Position

REFERENCES

Al Nasseri, A., Tucker, A., and de Cesare, S. (2015). Quantifying stock-twits semantic terms' trading behavior in financial markets: an effective application of decision tree algorithms. *Expert Systems with Applications*, 42(23):9192–9210.

Beasley, J. E., Meade, N., and Chang, T.-J. (2003). An evolutionary heuristic for the index tracking problem. *European Journal of Operational Research*, 148(3):621–643.

Berry, S., Mitra, G., and Sadik, Z. (2019). Improved volatility prediction and trading using stocktwits sentiment data. *White Paper, OptiRisk Systems.*

Canakgoz, N. A. and Beasley, J. E. (2009). Mixed-integer programming approaches for index tracking and enhanced indexation. *European Journal of Operational Research*, 196(1):384–399.

Dentcheva, D. and Ruszczynski, A. (2006). Portfolio optimization with stochastic dominance constraints. *Journal of Banking & Finance*, 30(2):433–451.

Fábián, C. I., Mitra, G., and Roman, D. (2011). Processing second-order stochastic dominance models using cutting-plane representations. *Mathematical Programming*, 130(1):33–57.

Fabian, C. I., Mitra, G., Roman, D., and Zverovich, V. (2011). An enhanced model for portfolio choice with SSD criteria: a constructive approach. *Quantitative Finance*, 11(10):1525–1534.

Feuerriegel, S. and Prendinger, H. (2016). News-based trading strategies. *Decision Support Systems*, 90:65–74.

Fishburn, P. C. (1977). Mean-risk analysis with risk associated with below-target returns. *The American Economic Review*, 67(2):116–126.

Hochreiter, R. (2016). Computing trading strategies based on financial sentiment data using evolutionary optimization. In *International Conference on Soft Computing-MENDEL*, pages 181–191. Springer.

Kelly, R. (2009). Twitter Study-August 2009. *San Antionio, TX: Pear Analytics.*

Kroll, Y. (1980). Stochastic Dominance: A Review and Some New Evidence. *Research in Finance*, 2:163–227.

Kumar, R., Mitra, G., and Roman, D. (2008). Long-short portfolio optimisation in the presence of discrete asset choice constraints and two risk measures. *Available at SSRN 1099926.*

Markowitz, H. (1952). Portfolio selection. *The Journal of Finance*, 7(1):77–91.

Mitra, G. and Mitra, L. (2011). *The Handbook of News Analytics in Finance*. John Wiley & Sons, Chichester, UK.

Mitra, G. and Yu, X. (2016). *The Handbook of Sentiment Analysis in Finance*. Optirisk Systems in Collaboration with Albury Books.

Ogryczak, W. and Ruszczynski, A. (1999). From stochastic dominance to mean-risk models: semideviations as risk measures. *European Journal of Operational Research*, 116(1):33–50.

Ogryczak, W. and Ruszczynski, A. (2001). On consistency of stochastic dominance and mean–semideviation models. *Mathematical Programming*, 89(2):217–232.

Oliveira, N., Cortez, P., and Areal, N. (2013). On the predictability of stock market behavior using stocktwits sentiment and posting volume. In *Portuguese Conference on Artificial Intelligence*, pages 355–365. Springer.

Oliveira, N., Cortez, P., and Areal, N. (2014). Automatic creation of stock market lexicons for sentiment analysis using stocktwits data. In *Proceedings of the 18th International Database Engineering & Applications Symposium*, pages 115–123. ACM.

Pring, M. (2002). *Technical Analysis Explained: The Successful Investor's Guide to Spotting Investment Trends and Turning Points*. McGraw-Hill Education.

Rockafellar, R. T. and Uryasev, S. (2002). Conditional value-at-risk for general loss distributions. *Journal of Banking & Finance*, 26(7):1443–1471.

Rockafellar, R. T., Uryasev, S., et al. (2000). Optimization of conditional value-at-risk. *Journal of Risk*, 2:21–42.

Roman, D., Darby-Dowman, K., and Mitra, G. (2006). Portfolio construction based on stochastic dominance and target return distributions. *Mathematical Programming*, 108(2-3):541–569.

Roman, D. and Mitra, G. (2009). Portfolio selection models: a review and new directions. *Wilmott Journal: The International Journal of Innovative Quantitative Finance Research*, 1(2):69–85.

Roman, D., Mitra, G., and Zverovich, V. (2013). Enhanced indexation based on second-order stochastic dominance. *European Journal of Operational Research*, 228(1):273–281.

Sadik, Z., Mitra, G., and Tan, Z. (2019). Asset allocation strategies: enhanced by news. *Whitepaper, OptiRisk Systems.*

Sadik, Z. A., Date, P. M., and Mitra, G. (2018). News augmented GARCH(1,1) model for volatility prediction. *IMA Journal of Management Mathematics*, 30(2):165–185.

Sadik, Z. A., Date, P. M., and Mitra, G. (2020). Forecasting crude oil futures prices using global macroeconomic news sentiment. *IMA Journal of Management Mathematics*, 31(2):191–215.

Shi, Y., Mitra, G., Arbex-Valle, C., and Yu, X. (2017). Using social media and news sentiment data to construct a momentum strategy. *Available at SSRN: https://ssrn.com/abstract=3406075 or http://dx.doi.org/10.2139/ssrn.3406075.*

Sprenger, T. O., Tumasjan, A., Sandner, P. G., and Welpe, I. M. (2014). Tweets and trades: the information content of stock microblogs. *European Financial Management*, 20(5):926–957.

Von Neumann, J., Morgenstern, O., and Kuhn, H. W. (2007). *Theory of Games and Economic Behavior (commemorative edition).* Princeton University Press.

Whitmore, G. A. and Findlay, M. C. (1978). *Stochastic Dominance: An Approach to Decision-Making Under Risk.* Lexington Books.

Yu, X. (2014). *Analysis of New Sentiment and Its Application to Finance.* PhD thesis, Brunel University London, Uxbridge, UK.

Yu, X., Mitra, G., Arbex-Valle, C., and Sayer, T. (2015). An Impact Measure for News: Its Use in Daily Trading Strategies. *Available at SSRN: https://ssrn.com/abstract=2702032 or http://dx.doi.org/10.2139/ssrn.2702032.*

CHAPTER 7

Asset Allocation Strategies: Enhanced by News

Zryan Sadik

OptiRisk Systems Ltd., London, United Kingdom

Gautam Mitra

OptiRisk Systems and UCL Department of Computer Science, London, United Kingdom

Ziwen Tan

OptiRisk Systems Ltd., London, United Kingdom

Christopher Kantos

Alexandria Technology, New York, United States

Dan Joldzic

Alexandria Technology, New York, United States

CONTENTS

DOI: 10.1201/9781003293644-7

7.1 INTRODUCTION AND BACKGROUND

7.1.1 Literature Review

News has always been a key source of investment information. The volumes and sources of news are growing rapidly. Published news in general influences the behaviour of institutional as well as individual investors. The financial market is a competitive venue; the investors and traders as well as other market participants such as dealers, brokers and market makers extract and analyse news relevant to their respective goals. Every party is motivated to take 'good' and timely decisions. The ease of accessibility to news and other information sources have become useful parameters for decision making by day traders. With news vendors (Alexandria Technology, Bloomberg, Thomson Reuters, Dow Jones, etc.) and brokers' platforms all disseminating news, the trading opportunities become endless. Based on such information making a human judgement and arriving at a decision naturally takes up time. Therefore, with an automated procedure of analysing sentiment meta data, the human decision-making process is improved in terms of time and efficiency.

Over the past few years, the amount of academic literature associated with news sentiment analysis has increased dramatically. In the beginning, the main research objective was to explore how sentiments explain the time series of returns. During this phase, the work of [De Long et al., 1990] needs to be highlighted as they were the great pioneers to notice that the more sentiments securities are exposed to, the higher unconditional expected returns could be achieved. There is also a growing body of research literature that argues news influences investor sentiment, hence asset prices, asset price volatility and risk (see, e.g. [Mitra and Mitra, 2011] and [Mitra and Yu, 2016]). Recently, some new research has been reported of studies which connect impacts of sentiments on liquidity which is an important financial attribute of an individual asset and that of investment portfolios (see, e.g. [Yu, 2014]).

In practice, people use sentiment analysis to extract subjective cognition from text resources to reasonably measure how market investors perceive and reflect news. For instance, machine translation has emerged as a scalable alternative for harmonizing information worldwide. Bloomberg, Thomson

Reuters (Refinitiv), RavenPack and Alexandria Technology are leading players in this area. These companies apply NLP and classification engines to process textual information and elicit 'sentiments' from the news data sources. These are then distributed to market participants who in turn use these items of sentiment information in their respective decision engines.

In this report we focus on the problem of asset allocation as faced by Day Traders, Quantitative Fund Managers and how their decision making can be enhanced by the application of News Sentiment. These Investment professionals are preoccupied with finding best investment strategies and trade signals.

Generally, fund managers follow one of the two approaches of active investment or passive investment. The active investment is supported by 'quant models' aided also by discretionary approach to trading. Active investment involves selecting securities within an asset class and selecting among asset classes. By contrast, the passive investment can be explained as: an investment portfolio which is based on the index itself or some replication of the index by a suitable analytic model. The recent research and contributions of Alexandria technology are noteworthy [Kantos et al., 2022]. In particular, they bring out how Loughran McDonald and FinBERT have taken a central position in this area.

Enhanced indexation models are related to index tracking, in the sense that they also consider the return distribution of an index as a reference. They, however, aim to outperform the index by generating 'excess' return (see, e.g. [DiBartolomeo, 2000]; [Scowcroft and Satchell, 2003]. Enhanced indexation is a new area of research and there is no generally accepted portfolio construction method in this field [Canakgoz and Beasley, 2009]. Although the idea of enhanced indexation was formulated as early as 2000, a paper explaining enhanced indexation and the application of Second Order Stochastic Dominance (SSD) as a scalable computational model is reported in [Roman et al., 2013].

7.1.2   Asset Allocation by Second-Order Stochastic Dominance

Second-order Stochastic Dominance (SSD) has a well-recognised importance in portfolio selection, due to its connection to the theory of risk-averse investor behaviour and tail risk minimisation. Until recently, stochastic dominance models were considered intractable or at least very demanding from a computational point of view. Computationally tractable and scalable portfolio optimisation models which apply the concept of SSD were proposed by [Dentcheva and Ruszczynski, 2006], [Roman et al., 2006, 2013] and [Fabian et al., 2011]. These portfolio optimisation models assume that a benchmark, that is, a desirable 'reference' distribution is available and a portfolio is constructed, whose return distribution dominates the reference distribution with respect to SSD. Index tracking models also assume that a reference distribution (that of a financial index) is available. A portfolio is then constructed, with the aim of replicating, or tracking, the financial index. Traditionally, this is done by minimising the tracking error, that is, the standard deviation of the differences between the portfolio and index returns. Other methods have been proposed (for a review of these methods, see for example [Beasley et al., 2003] and [Canakgoz and Beasley, 2009]. Recently further logical extensions of the SSD Long only models to Long/Short models have been proposed and formulated as Mixed Integer Programming (MIP) [Kumar et al., 2008].

7.1.3 News Sentiment and the Impact of News Sentiment on Asset Return

Recently availability of high-performance computer systems has facilitated high frequency trading. Further the automated analysis of news feeds set the backdrop for computer automated trading which is enhanced by news (see, e.g. [Mitra and Mitra, 2011]; [Mitra and Yu, 2016]). News sentiment is regarded to be unique to each individual and encompasses lots of emotions occurring during brief moments. For the financial domain, it is commonly known that investors in the bull market are positive and optimistic, while in the bear markets they seem relatively pessimistic and fear of loss. In other words, good sentiments are usually based on the rise in stock prices and can further stimulate a continued rise i.e. builds momentum. Therefore, relevant transaction data can be used to build sentiment indicators as one of the forecasting technologies of the future trend of stock price fluctuations (see, e.g. [Pring, 2002]). As pointed out in [Feuerriegel and Prendinger, 2016], sentiment analysis usually refers to the methods that judge the content of positive or negative through the relative details of text or other forms. In this report, the objective is to build news sentiment-based filters to choose a subset of the asset universe and utilize the SSD as a similar method of the index tracking model to construct different portfolios. We then investigate whether news sentiment data can enhance the performance of portfolios. The selection method of the asset universe of the portfolio is explained in Section 7.4.

7.1.4 Organisation of the paper: Guided Tour

This chapter is structured as follows: Section 7.2 introduces the historical closing prices of the historical market data and the news meta data supplied by Alexandria Technology. This section also outlines the four steps of classifying and ranking the news topics across the top 50 stocks by capitalisation belonging to the S&P 500 index. The experimental data analysis results are then presented as well, where we identify any recurring topics and those that have a significant impact on the return series and trading volume, as well as discovering the direction of change. Section 7.3 sets out the asset allocation strategy used by OptiRisk; the details of the relevant SSD models are also included. In Section 7.4 we present a novel method of restricting the asset universe of choice using our proprietary method of constructing filters. In Section 7.5 we describe our investigations and the findings of our investigation. A discussion of our work and the conclusions are presented in Section 7.6. The Chapter is concluded with an Appendix which contains a sample data set and a relatively complete explanation of the data fields of the Alexandria Technology News Meta Data.

7.2 MARKET DATA AND NEWS DATA

The investigation reported in this White Paper uses 5 years and 4 months of historical data relating to the S&P 500 index. The data spans the period starting 1 January 2017 to 30 June 2022. The daily adjusted closing prices and trading volume of the historical market data, as well as the news sentiment meta data supplied by Alexandria Technology, for each of the S&P 500 assets are included in this study.

7.2.1 Market Data

The time series data used in this study is the stock market daily closing price of the S&P 500 companies. We first filter the whole market database to contain only the daily prices of the assets from S&P 500 index covering from 1 January 2017 to 30 June 2022. This will produce the following eight columns: *Date*, *Index*, *RIC*, *Open*, *High*, *Low* and *Close* prices in USD. The Reuters instrument code (RIC) is a code assigned by Thomson Reuters to label each asset, and Table 7.1 shows a sample of the market data.

The data was collected from Thomson Reuters Data Stream platform and adjusted to account for changes in index composition. This means that our models use no more data than was available at the time, removing susceptibility to the influence of survivor bias. For each asset, we compute the corresponding daily rates of return.

7.2.2 Analysis of News Meta Data (Alexandria Technology)

News analytics data is presented in a metadata format where news is given characteristics such as relevance, novelty and sentiment scores according to an individual asset. The analytical process of producing such scores is fully automated from collecting, extracting, aggregating, to categorising and scoring. The result is an individual score assigned to each news article for each characteristic using different scales. A news sentiment score measures the emotional tone within a news item and varies between positive and negative. Sentiment

TABLE 7.1   Market Data Sample

Date	Index	RIC	Open	High	Low	Close
20170103	SP500	AMZN.O	37.896	37.938	37.385	37.6835
20170104	SP500	AMZN.O	37.9195	37.984	37.71	37.859
20170105	SP500	AMZN.O	38.0775	39.12	38.0128	39.0225
20170106	SP500	AMZN.O	39.118	39.972	38.924	39.7995
20170109	SP500	AMZN.O	39.9	40.0887	39.5885	39.846
20170110	SP500	AMZN.O	39.83	39.9	39.47715	39.795
20170111	SP500	AMZN.O	39.683	39.975	39.4755	39.951
20170112	SP500	AMZN.O	40.0155	40.7065	39.975	40.682
20170113	SP500	AMZN.O	40.716	41.0825	40.57	40.857
20170117	SP500	AMZN.O	40.785	40.8	40.172	40.486
20170118	SP500	AMZN.O	40.475	40.5865	40.2135	40.374

score is a value falling within a range consisting of a minimum and maximum depicting the overall tone of a news article.

The Alexandria database provides sample historical data regarding news sentiment. The whole database contains over 10 millions data points across 752 companies from the S&P 500 index. For further information on Alexandria Technology news meta data, readers may refer to their website (www.alexandriatechnology.com).

We analyse the Alexandria news meta data since we want to analyse this data in the form of an event study, that is, a statistical investigation into the effect of different events on a company's value. We follow the steps below in search for the most influential news types on stock returns:

1. Choose a small sample of six companies from the S&P 500 index whose historical price series shows strong volatility between the year 2017 and 2022.

2. Identify any recurring news topics around extreme returns.

3. Expand the investigation to the top 50 S&P 500 assets by capitalization.

4. If possible, rank the topics by their scale of impact.

Six companies are chosen for a preliminary study. Their historical prices all exhibit highly volatile behaviour. The purpose of this step is to narrow the scope of our investigation and get a sense of which news topics occur around extreme price movement–since naturally, news topics that happen around extreme events should have a stronger impact. The selected companies are: AT&T, Tractor Supply Co., Pfizer Inc., Ford Motor Co., Verizon Communications Inc. and Wells Fargo & Co..

Identifying Potentially impactful Topics

We first calculate the daily return on day t for each of the six assets. In this study, daily returns (r_t) are calculated as the continuously compounded returns which are the first difference in logarithm of closing prices of the asset of successive days:

$$r_t = \log(P_t) - log(P_{t-1}) \tag{7.1}$$

where P_t and P_{t-1} are the daily closing price of the asset at current and previous days, respectively. Next, looking at the top five extreme returns of each asset, we examine the news sentiment of every topic around these dates in a three-day window, that is, two days leading up to the events plus the event date. Recall from Figure 7.3 (in the *Appendix 1*) that the sentiment score of the Alexandria data comes in three possible values: 1, 0 or –1. We notice that there are a large number of neutral news which does not contribute much value to the current research and introduce unnecessary noise to the analysis. Therefore, we transform the trinary scores into numbers ranging between 0 and 100 using the following equation:

$$\hat{S}Sent = 100 * (Prob(positive) + \frac{1}{2}Prob(neutral)), \tag{7.2}$$

where $Prob(positive)$ and $Prob(neutral)$ are the probability that the sentiment is positive and neutral, receptively. We then subtract 50 from all the obtained values from equation 7.2 to obtain scores between -50 and +50, as explained in [Yu, 2014]. In this way, almost all of the previously neutral news is eliminated and re-categorized into positive/negative sentiment and now only news with exactly 0 score would represent neutral items. The number of positive, neutral and negative news is plotted for each topic. Figure 7.1 shows an example of such plot, where green represents positive news, yellow represents neutral news and red represents negative news. The topics which do not appear on the graph means no news from that category were found in the three-day window. From these graphs, we are able to observe certain recurring topics based on their frequency of appearance. The results are explained in more details in the rest of this section.

Directional Results and Trends for 50 Assets

After we identify any recurring or high-frequency topics, we test the results across the top 50 S&P 500 companies (by capitalisation) and determine the

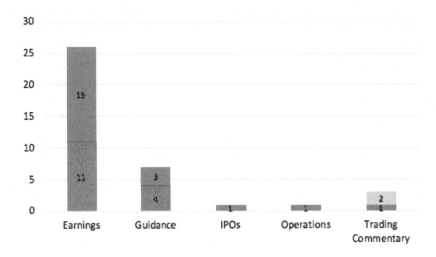

Figure 7.1 Number of news with different polarity–an example

following points: (1) if the topics stay consistent; (2) if the sentiment scores of these topics are truly impactful in moving the return. Further manipulation on the Alexandria database is required before we can make such statistical inference. There are two complications arising from the date the news are published: (1) multiple news could be published on the same day for the same asset; (2) news could be released on the weekend. To resolve these two issues, recall that we have transformed the original sentiment to lie between −50 and +50, now, we aggregate the positive and negative sentiment scores, respectively, for all news found on the same day to obtain a single daily score. This is a reasonable way to combine the scores as it implies the compounding effect of news in a day, that is, the higher the score, the more extreme the polarity. Similarly, we reassign the weekend scores to the following Monday.

Next, for each of the 31 topics, we perform linear regression on the positive and negative sentiment against the return values asset-by-asset and evaluate the level of significance. We only do so if the number of data points is greater than 30 after the scores have been aggregated so to ensure the accuracy of the test. The linear relationship between the scores and return is given by the following equation:

$$Return = \beta_0 + \beta_1 Positive + \beta_2 Negative \tag{7.3}$$

A natural question to ask at this point is at what direction (upward or downward) is the topics moving the return? It is possible for a positive sentiment on a piece of news to have a negative effect on the return; and vice versa. Thus, by looking at the signs of the coefficients, we can determine the trajectory of the return given the sentiment.

Impactful Topics on Return and Trading Volume

From the top five extreme returns of the six initial companies, we observe certain recurring topics based on their frequency of appearance as well as the number of news items belonging to that category in the three-day window. The top 20 results are listed in Table 7.4 in the *Appendix 1*. We can see that, for example, within the 30 selected days of extreme return, 28 of them contain at least one piece of news on *Earnings* and there are altogether 752 news items on *Earnings* found across this period. The distribution of news density around these extreme returns shown in Table 7.4 is a potential implication of the scale of impact from each topic.

In this section, we present statistically tested results to verify if the recurring topics identified above is indeed impacting the return, that is, if the *p*-values of the coefficients computed from equation (7.3) are less than 0.05. Table 7.5 (in the *Appendix 1*) shows the number of assets whose positive or negative sentiment has a significant effect on the return values. For example, we can see that there are 32 out of 50 assets whose positive sentiment on *Earnings* significantly move the prices, and 30 out of 50 assets experience impact from negative news on *Earnings*.

Observe that the top eight news types from Table 7.5 can be found in Table 7.4, suggesting that the news groups observed around the extreme returns are a good indication of the overall price movement and reinforcing the significance of these impactful topics.

We further test this result on the daily trading volume as an additional evaluation metric using the following equation (7.4) and arrive at similar results demonstrated in Table 7.6 in the *Appendix 1*.

$$Trading\ Volume = \beta_0 + \beta_1 Positive + \beta_2 Negative \qquad (7.4)$$

Likewise in Table 7.6, the *Positive* and *Negative* columns represent the number of companies whose positive/negative sentiment significantly drives trading volume. For instance, the trading volume would be significantly increased for 21 out of 50 assets if there is a positive *Earnings* news, and trading volume would be significantly decreased if a negative *Earnings* news is found for 37 out of 50 assets.

Notice that the top four news topics are consistent between Tables 7.5 and 7.6, that is, *Earnings*, *Estimates*, *Research Ratings* and *Trading Commentary*; and the number of influenced assets are evidently higher in these topics than others.

We record the number of companies whose positive and negative sentiments exhibit a significant effect on the return (i.e. *p*-value < 0.05) and rank them in descending order. Consequently, the topic that produces the highest number of significant sentiments would be regarded as the most impactful news group.

We rank the four topics according to the effect of their positive sentiment on return, that is, by the number of impacted assets, as follows: (i) *Earnings*, (ii) *Research Ratings*, (iii) *Trading Commentary* and (iv) *Estimates*.

And the order ranked according to the impact of negative sentiment is: (i) *Trading Commentary*, (ii) *Earnings*, (iii) *Research Ratings* and (iv) *Estimates*.

Similarly, positive news on *Research Ratings* seem to have the most significant influence on the trading volume and so are negative news on *Earnings*.

Finally, focusing on the four topics above, we only find one news item on *Earnings*, *Research Rating* and *Estimates*, respectively, whose positive sentiments induce a negative return, and one additional news items on *Earnings* that results in a positive return given a negative sentiment. In other words, almost all positive news lead to a positive return and negative news to a negative return.

Choosing Alexandria News Event

In an earlier version of this white-paper we reported our findings re: the scope of choosing a smaller but more useful set of News Events for our study. In the trial data set with which the initial research was carried out most of the Alexandria news items were released on different days and at different frequencies. We now use the enhanced version of news release: news data is now published within milliseconds of release from Dow Jones and pushed to users.

The reader is referred to the Alexandria Technology manual for further information on how the data is structured. An explanation of how the News Data used in this study is extracted is outlined below.

1. Alexandria data is categorising news articles as Auto Message events, i.e. scheduled or unscheduled news event. If the announcement of an event was anticipated, arranged, or planned according to some schedule or timetable, then the value will be TRUE otherwise it will be FALSE. This study considered only the scheduled news from Alexandria data as a part of the input.

2. The fact that each Alexandria news item is given characteristics such as relevance, novelty, and sentiment scores according to an individual asset. The result for relevance is an individual score assigned to each news article using scales from 0.1 to 1.00, which indicates how strongly related the company is to the underlying news story. The chosen data is further filtered in such a way that only the news items with relevance of 0.5 or greater are taken into consideration in this research.

Generating News Impact Scores

After extracting the news data that is related to the asset price movements in the first stage, one can select the news sentiment score for each news items based on their relevance and novelty scores. In this study, rather than using the Alexandria news sentiment score itself we use the idea that was first proposed by [Yu, 2014] and used by [Yu et al., 2015] and [Sadik et al., 2018, 2020] to construct news impact scores. We construct News Impact Scores that can be used as proxies of firm-specific news impact in the new model. It is worth recalling that the sentiment score of Alexandria data comes in three possible values: 1, 0, or –1. We notice that there are a large number of neutral news which does not contribute much value to the current research and introduce unnecessary noise to the analysis. Therefore, we transform the trinary scores into numbers ranging between 0 and 100 using equation 7.2 and subtract 50 from all values to obtain scores between –50 and +50, as explained in [Yu, 2014]. In this way, almost all of the previously neutral news is eliminated and re-categorized into positive/negative sentiment and now only news with exactly 0 score would represent neutral items. To calculate the news impact scores, the following steps have to be done:

1. The 'Timestamp' for each news item has to be converted from UTC to EST time, which is the timing convention of the S&P 500 constituents from the New York Stock Exchange (NYSE).

2. Separating the positive and negative sentiment scores so that two different time series can be obtained.

3. After separating the scores, in a similar fashion to [Sadik et al., 2018], the positive and negative news impact scores for each sentiment score is calculated.

4. Finally, two daily time series are generated that represent the daily positive and negative news impact scores.

7.3 ASSET ALLOCATION STRATEGY

7.3.1 Portfolio Construction Models

The challenging problem of 'active' portfolio selection is how to find a portfolio x such that its return at the end of the investment period R_x is 'maximised'. Since portfolio returns are random variables, models that specify a preference relation among random returns are required. A portfolio x is then chosen such that its return R_x is non-dominated with respect to the preference relation that is under consideration; computationally this is achieved using an optimisation model. For portfolio selection, mean-risk models have been by far the most popular. They describe and compare random variables using two statistics: the expected value (mean) and a risk value. Various risk measures have been proposed in the literature, see for example [Markowitz, 1952], [Fishburn, 1977], [Ogryczak and Ruszczynski, 1999, 2001], [Rockafellar et al., 2000] and [Rockafellar and Uryasev, 2002]. Mean-risk models are convenient from a computational point of view and have an intuitive appeal, but their approach is somewhat oversimplified. Expected utility theory [Von Neumann et al., 2007] compare random returns by comparing their expected utilities (larger value preferred). However, the expected utility values depend on the utility function that is used; the choice of a specific utility function is somewhat subjective.

7.3.2 The Second-Order Stochastic Dominance

In contrast to the classical approaches explained above, stochastic dominance (SD) has been recognised as a sounder model of choice, as it exploits 'the three p's: price, probability and preference' [Lo, 1999]. It is closely connected to the expected utility theory, but it eliminates the need to explicitly specify a utility function (see, e.g. [Whitmore and Findlay, 1978] for a detailed description of stochastic dominance relations, [Kroll, 1980] for a review). With stochastic dominance, random variables are compared by pointwise comparison of functions constructed from their distribution functions. There are progressively stronger assumptions about the form of utility functions used in investment, which lead to first, second and higher orders of SD. For example, first-order stochastic dominance (FSD) is connected to 'non-satiation' behaviour. A random return is preferred to another with respect to FSD relation if its expected utility is higher, for any increasing utility function. This is a strong condition and thus many random returns cannot be ordered with respect to FSD.

In the domain of asset allocation in the context of finance the SSD has found increasing use as a model for portfolio construction. Despite of these attractive advantages of the model, it had posed computational difficulty in scaling up the portfolio construction. Typically this happens for moderate to large market indices such as S&P 500 and Russel 2000 market indices. For a full discussion of computational issues (see, e.g. [Vickson, 1975]). OptiRisk and its research team has addressed this computational issue adequately and it is one of OptiRisk's USP in the domain of computational solution of SSD. [Fabian et al., 2011] solved the computational problem to reduce the constraints in the process by using a cutting-plain approach. In general the SSD Long only models are formulated as Linear Programs (LP); but Long/Short models are formulated as Mixed Integer Programming (MIP), (see, e.g. [Kumar et al., 2008]). An explanation regarding the details of SSD and its application in finance can be found in [Roman et al., 2013].

7.4 CONSTRUCTION OF FILTERS

We explain the rationale of why we need to use Filters and the advantage of using Filters. We then expand this section to describe the different type of filters which we have used. Our approach to incorporating filters for the choice of assets for inclusion in our portfolios has the aim of achieving only News based choice or Market Price based choice of asset allocation. In this approach we are able to choose (i) no influence of News or (ii) Partial influence of News or (iii) even the extreme of Only Influenced by News and no Price Data influence.

7.4.1 Why Use Filters

The asset allocation strategy used by OptiRisk system uses the SSD model. The scenarios are the historical return data which captures accurately correlation structure of the constituent assets. As a consequence, the asset allocation is fairly robust in the long run and also achieves control of tail risk. The asset allocation is a static one period model; so it suffers from the draw back that the near term asset price movements are not taken into consideration. We have therefore introduced a method of restricting the asset universe for the choice of long assets and short assets which are to be held in the portfolio using a technique which we call asset filter. In the construction of the filter, we take into account the near term behaviour of each of the assets in the asset universe. The near term behaviour is captured by the use of a well known technical indicator, namely, the relative strength index (RSI). We have extended this concept of applying the filter by taking into account the news sentiment and the impact of the news sentiment in respect of each asset. In rest of this section, we describe the method by which we combine these two approaches.

7.4.2 Relative Strength Index

The Technical Indicator, Relative Strength Index (RSI), is an established momentum oscillator. The RSI compares the magnitude of a stock's recent gains to the magnitude of its recent losses and turns that information into a number that ranges from 0 to 100. The RSI indicator uses the daily closing prices over a given period computed for each constituent asset of the market index under consideration. It is driven by the measure of the momentum of each asset. The RSI measure is expressed as:

$$RSI(t) = 100 - \frac{100}{1 + RS(t)};$$ (7.5)

The RSI and RS are re-expressed for the time bucket (t) as $RSI(t)$.

Computation of RSI

Formally, Relative Strength uses Exponential Moving Average (EMA); thus $RS(t)$ the relative strength is computed as the ratio of average gains and losses.

$$RS(t) = \frac{EMA(Gain_t)}{EMA(Loss_t)} \ldots \text{calculated using market data of stock prices}$$
(7.6)

$$RSI(t) = 100 - \frac{100}{1 + RS(t)};$$
(7.7)

$$EMA(X_t) = \sum_{tn=1}^{N} e^{(-\lambda tn)} X_{tn}$$
(7.8)

Where $X_t = Gain_t$ or $Loss_t$, $\lambda = $ decay factor and N=RSI period.

The typical $RSI(t)$ value is calculated with average gains and losses over a period of $N = 14$ days (lookback period). The number of days N is a parameter in the RSI function and can be chosen in accordance with the characteristics of the data set. Secondly, the number of offset days can be varied. Gains and losses can therefore be daily gains and losses (days=1) or gains and losses over larger time intervals.

The $RSI(t)$ is considered to highlight overbought or oversold assets; when the $RSI(t)$ is above the thresholds of 70 and oversold when it is below the threshold of 30.

7.4.3 News Relative Strength Index (NRSI)

We have introduced the concept of News Relative Strength Index ($NRSI$). In this we extend the concept of RSI which is computed using Market Data by replacing it with the Impact of the streaming News Sentiment Data.

News RSI (NRSI) Computation

It is computed in a way comparable to that of **RSI(t)** whereby the up and down price movements are replaced by positive and negative *news impact scores*, respectively. Thus

$$NRS(t) = \frac{EMA(Positive\ Impact\ Scores)}{EMA(Negative\ Impact\ Scores))} \ldots \text{is computed for each stock}$$
(7.9)

Hence,

$$NRSI(t) = 100 - \frac{100}{1 + NRS(t)};$$
(7.10)

Thus, the $NRSI(t)$ values ranges range from 0 to 100.

For an explanation of the news sentiment impact score the readers are referred to the works of OptiRisk analysts please see: [Yu, 2014], [Yu et al., 2015] and [Sadik et al., 2018, 2020].

7.4.4 Derived RSI (DRSI) Computation

We define the measure **Derived RSI** computation by taking a linear combination of $RSI(t)$ and $NRSI(t)$. So for the time bucket t, the measure Derived RSI ($DRSI(t)$) is defined as

$$DRSI(t) = \theta * RSI(t) + (1 - \theta) * NRSI(t), \tag{7.11}$$

where $0 \leq \theta \leq 1$. The news impact scores are used to compute the $NRSI$. They reflect the same modelling paradigm of computing RSI. Thus, for the time bucket t we compute $RSI(t)$ and $NRSI(t)$ to calculate $DRSI(t)$:

$$DRSI(t) = \begin{cases} \text{RSI(t)} & \text{if } \theta = 1 \\ \text{NRSI(t)} & \text{if } \theta = 0 \\ \text{DRSI(t)} & \text{otherwise, that is, } 0 < \theta < 1 \end{cases} \tag{7.12}$$

7.4.5 Applying the Filters: RSI(t), NRSI(t), DRSI(t)

As explained earlier the purpose of these filters are to restrict the choice of long and short positions of assets as they appear in the asset universe of the available assets. The choice is restricted in the following way: We apply a threshold of 70 to define the long and short bins; Long Bin is filled with the assets whose $RSI(t)$, or $NRSI(t)$, or $DRSI(t)$ values are below 70, and the Short Bin is filled with the assets whose $RSI(t)$, or $NRSI(t)$, or $DRSI(t)$ values are above 70. Finally the SSD method of asset allocation is applied to this restricted asset universe.

7.5 EMPIRICAL INVESTIGATION

7.5.1 The Framework

Our empirical investigation uses two time series datasets, namely, Market Data supplied by Thomson Reuters (Refinitiv) and News Sentiment data supplied by Alexandria Technology. We will use the following timeseries in our experiments:

- 5 and a half years of historical daily adjusted closing prices of the S&P 500 assets that covering the time period from 2017 to mid-2022.

- 5 and a half years of daily news sentiments classified as probabilities and explained in Section 7.2.2 using equation (7.2). These are then converted to positive and negative sentiment scores for each asset in the S&P 500 index.

TABLE 7.2 Parameter settings of the trading strategy

Parameter	Value	Parameter	Value
Index	SP500	Risk free rate	2.0% a year
Cash lend rate	2.0% a year	Cash borrow rate	2.0% a year
Proportion in long/short	100/0	Proportion in SSD cash	Up to 50%
Gearing	Not applied	Money mgmt. (prop. in cash)	No
Use lot sizing	Yes	Transaction costs	5 basis points
In-sample	500 days	SSD rebalancing frequency	3 days
UCITS compliant	No	Slippage	25 basis point
Cardinality constraints	Not enforced	Extra assets	Futures and VIX
Asset universe	Full OR Reduced	Stop Loss	Not enforced

- 5 and a half years of daily news impact scores (positive and negative) are then derived from for each asset in the S&P 500 index. We set the looking back period of considering previous news items to 4320 minutes (3 days) and consider that the sentiment value decays to half of its initial value in 240 minutes.

7.5.2 Experimental Setup

Back-Testing study was carried out using the Framework described in Section 7.5.1, and the experimental setup is best explained by describing the Parameter Settings which are set out in Table 7.2. These Parameter Settings are used to control the trading that is done at each trading day of the Back-Testing period.

The following parameters set out in Table 7.2 were used in the investigation.

1. **Index:** S&P 500 <This model is generic apply to other indices: Hang Seng, Topix...>.

2. **Risk free rate:** 2% <Risk free rate is set by default for the Market/Index geography>.

3. **Cash lend rate:** 2% <above comments apply>.

4. **Cash borrow rate:** 2% <above comments apply>.

5. **Proportion in long/short:** Long only...100/0 <specifies limits of long and short positions>.

6. **Proportion in SSD cash:** 0.5 [= 50%] <In the portfolio the Long position in cash is 0.5 or less of the Portfolio mark-to-market value>.

7. **Gearing:** <In some exchanges, namely, NIFTY or KOSPI traders instead of trading in underlying stock exploit gearing by using the futures contract>.

8. **Money management:** not applied here.

9. **Lot size:** traded only in available lot sizes.

10. **Transaction costs:** transaction costs is five basis point, that is, 0.0005.

11. **In-sample:** 500 <in sample trading days of historical data roughly two years worth of data>.

12. **SSD rebalancing frequency:** performs a rebalancing every 1, 2, ... days.

13. **UCITS complaint:** not applied here.

14. **Cardinality:** exact number of assets to be chosen in the portfolio.

15. **Extra assets:** Yes <whether to include index futures or/and VIX as an asset>.

16. **Asset universe:** whether or not a filter is applied.

17. **Stop Loss:** stop loss rule to be applied for individual assets.

The investigation is carried out using the concept of 'rolling window'; in this case the window has a span of 500 trading days. The period investigated spans 5 and a half years (January 2017 to 30 June 2022).

7.5.3 Back-testing Results

Back-testing was done using the experimental setup described in Section 7.5.2. In these tests three different strategies were applied and their outcomes examined and compared. The three strategies are

1. Full Asset Universe: This strategy takes into consideration the whole available assets from the S&P 500 index. It is included as a second benchmark, such that we are able to measure the improvements made by using different filters.

2. RSI: This is a momentum-based strategy, which uses the *RSI* filter to restrict the choice of the long bins and the short bins of the asset universe.

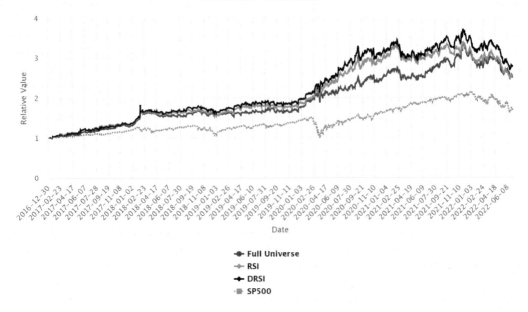

Figure 7.2 Comparison of the portfolio performance of three investment strategies

3. DRSI: This is a derived strategy that combines RSI and $NRSI$ which are defined in Section 7.4. In the results reported in this Section, we have set the following value:

$$DRSI = \theta * RSI + (1 - \theta) * NRSI, \qquad (7.13)$$

where θ is set to $\theta = 0.4$.

In Figure 7.2 the charts of the [− SP500 Index], [− Full Universe], [− RSI] and [− DRSI] are presented.

7.5.4 Analysis of Results

The performance of these three strategies is compared using the Industry Standard performance measures. These measures are tabulated and displayed in Table 7.3.

- Final value: Normalised final value of the portfolio at the end of the Back-testing period.

- Excess over RFR (%): Annualised excess return over the risk free rate. For S&P500 we used a yearly risk free rate of 2%.

- Sharpe ratio: Annualised Sharpe ratio computed using annualised returns.

TABLE 7.3 Performance measurements for three portfolios

Portfolio	Final value	Excess over RFR (%)	Sharpe Ratio	Sortino Ratio	Average turnover (%)	Max Draw-Down (%)	Max Recover Days
SP500	1.69	8.08	0.48	0.66		33.92	145
Full Universe	2.51	16.24	0.81	1.37	26.67	25.55	486
RSI	2.56	16.66	0.83	1.42	42.44	27.66	313
DRSI	2.80	18.65	0.90	1.54	22.84	27.50	303

- Sortino ratio: Annualised Sortino ratio computed using annualised returns.

- Max draw-Down (%): Maximum peak-to-trough decline (as percentage of the peak value) during the entire Back-testing period.

- Max Recover Days: Maximum number of days for the portfolio to recover to the value of a former peak.

- Average turnover: Av. turnover per day as a percentage of portfolio mark-to-market.

The relative performances can be compared and summarised under each headings of the Table 7.3.

Final value: The values as displayed in the Table show a steady improvement from the Index (1.69), ... until DRSI (2.80).

Excess RFR (%): As above and quite naturally, the values as displayed in the Table show a steady improvement from the Index (8.08), ... until DRSI (18.65).

Sharpe and Sortino ratios: We find these have improved: **Sharpe** Index (0.58), ... until DRSI (0.90) and **Sortino** Index (0.66), ... until DRSI (1.54). We observe the increase for Sharpe is not monotone which not surprising as the SSD asset allocation strategy minimises 'Tail Risk'.

Max draw- Down (%) and Max Recovery Days which are 'Dynamic Risk' measures do not show consistent improvement. Our perspective on this is that on a long Back-Testing study spanning over multiple years **Max Draw-Down (%)** and **Max Recovery Days** are less meaningful than if they are computed on a quarterly basis.

Average Turnover is reasonable and improved in DRSI strategy.

7.6 DISCUSSIONS AND CONCLUSION

In this report we have first presented our experimental statistical analysis of News Data supplied by Alexandria Technology. We have found that of the two categories of news namely, Calendar (Scheduled) News and Non-Calendar

(Unscheduled) News, the former influences asset returns with relatively more impact.

We have then introduced the concept of Asset Filters which take in to consideration short term asset price movement or the impact of news. The Filters are used to restrict Asset Universe that is used for portfolio choice. Trade portfolios constructed by using Filters in this way improves the dynamic risks measured as Draw Down. We have described momentum based Filter (RSI), News Impact Filters ($NRSI$) and their combination Derived Filters ($DRSI$). The superior $DRSI$ results vindicate our approach of combining Market Information (Market Data) and News Sentiment Information .

Future research is planned to examine if applied to other Global Market Indices such as TOPIX, Hang Seng, NIFTY, Euro STOXX and FTSE 100; this approach leads to commensurate improvements.

APPENDIX

	RpCode	year	date	hour	sentiment	confidence	novelity	relevance	marketImpactScore	prob_POS	prob_NTR	prob_NEG	headlineOnly	autoMessage	events	source	country	ISIN
1	0157B1	2013	20130102	132543	0	0.9623101	1	0.0222200	-0.1733560	0.0069395	0.9748734	0.0181871	F	F	AA@ALEX,AA@RAT	DJN	USA	US0231351067
2	0157B1	2013	20130102	235340	-1	0.8918847	1	0.5000000	-1.8612827	0.0060372	0.0660396	0.9279231	F	F	AA@ALEX,AA@ERN	WSJ	USA	US0231351067
3	0157B1	2013	20130103	153423	-1	0.9791408	1	1.0000000	-1.2060564	0.0047960	0.0091101	0.9860938	T	F	AA@ALEX,AA@LGL,AA@REG	BNZ	USA	US0231351067
4	0157B1	2013	20130103	140752	-1	0.9700000	1	0.1666700	-0.2258692	0.0100000	0.0100000	0.9800000	F	F	AA@ALEX,AA@ERN	DJN	USA	US0231351067
5	0157B1	2013	20130103	142252	-1	0.9700000	1	0.1666700	-0.2258692	0.0100000	0.0100000	0.9800000	F	F	AA@ALEX,AA@ERN	DJN	USA	US0231351067
6	0157B1	2013	20130103	144502	-1	0.9700000	1	0.1666700	-0.0180408	0.0100000	0.0100000	0.9800000	F	F	AA@ALEX,AA@ERN	DJN	USA	US0231351067
7	0157B1	2013	20130103	161718	0	0.8344641	1	0.5000000	-0.2987756	0.0328141	0.8896427	0.0775431	F	F	AA@ALEX,AA@EST	DJN	USA	US0231351067
8	0157B1	2013	20130103	163021	-1	0.9700000	15	0.1666700	0.0016585	0.0100000	0.0100000	0.9800000	F	F	AA@ALEX,AA@ERN	WSJ	USA	US0231351067
9	0157B1	2013	20130103	192704	-1	0.9700000	2	0.3333300	3.4518851	0.0100000	0.0100000	0.9800000	T	F	AA@ALEX,AA@ERN	DJN	USA	US0231351067
10	0157B1	2013	20130103	223427	-1	0.8436873	1	1.0000000	-2.1172666	0.0261300	0.0780785	0.8957915	T	F	AA@ALEX,AA@REG	DJN	USA	US0231351067

Figure 7.3 Alexandria Data Sample

TABLE 7.4 Recurring News Topics

Topics	No. of Occurrence	No. of News Items
Earnings	28	752
Operations	27	176
Trading Commentary	23	94
Research Ratings	17	106
Guidance	12	54
Government Interaction	12	47
Regulatory Affairs	12	49
Dividends	10	25
Bond Ratings	8	57
Estimates	8	36

TABLE 7.5 Significant News Topics on Return Series

Code	Description	Positive	Negative
AA@ERN	Earnings	32	30
AA@TRD	Trading Commentary	23	36
AA@RAT	Research Ratings	28	26
AA@EST	Estimates	19	19
AA@GDC	Guidance	6	7
AA@DIV	Dividends	8	5
AA@OPS	Operations	4	6
AA@GOV	Government Interaction	6	4
AA@BUY	Buy Backs	2	4
AA@REG	Regulatory Affairs	2	4
AA@HFD	Hedge Fund Trades	3	3
AA@LGL	Legal	5	1
AA@MGT	Management	5	1
AA@OWN	Equity Ownership	4	1
AA@INS	Insider Transactions	1	3
AA@FDA	Clinical Trials	1	2
AA@MNA	Mergers & Acquisitions	2	1
AA@IPO	IPOs	0	2
AA@CBT	Bond Ratings	1	1
AA@CPG	Corporate Governance	1	1
AA@DRV	Equity Derivative	1	1
AA@BNQ	Bankruptcy	0	1
AA@DEB	Debt Financing	1	0
AA@IMB	Trade Imbalances	1	0
AA@ACC	Accounting	0	0
AA@AST	Asset Purchases/Sales	0	0

(*Continued on next page*)

TABLE 7.5 (Continued)

Code	Description	Positive	Negative
AA@CPA	Corporate Actions	0	0
AA@DMNA	Dissolved Mergers	0	0
AA@ETF	Index Reconstitution	0	0
AA@MFD	Mutual Fund Trades	0	0
AA@PRE	Private Equity	0	0

TABLE 7.6 Significant News Topics on Trading Volume

Code	Description	Positive	Negative
AA@ERN	Earnings	21	37
AA@EST	Estimates	25	27
AA@RAT	Research Ratings	33	26
AA@TRD	Trading Commentary	18	24
AA@OPS	Operations	6	11
AA@INS	Insider Transactions	8	8
AA@GDC	Guidance	5	7
AA@GOV	Government Interaction	5	7
AA@REG	Regulatory Affairs	1	5
AA@DRV	Equity Derivative	1	4
AA@MFD	Mutual Fund Trades	2	4
AA@MGT	Management	3	4
AA@OWN	Equity Ownership	3	4
AA@CBT	Bond Ratings	1	3
AA@DIV	Dividends	5	3
AA@FDA	Clinical Trials	2	3
AA@HFD	Hedge Fund Trades	0	3
AA@MNA	Mergers & Acquisitions	1	3

(*Continued on next page*)

TABLE 7.6 (Continued)

Code	Description	Positive	Negative
AA@LGL	Legal	4	2
AA@BUY	Buy Backs	2	1
AA@CPG	Corporate Governance	1	1
AA@IMB	Trade Imbalances	1	1
AA@IPO	IPOs	2	1
AA@ACC	Accounting	0	0
AA@AST	Asset Purchases/Sales	0	0
AA@BNQ	Bankruptcy	0	0
AA@CPA	Corporate Actions	0	0
AA@DEB	Debt Financing	0	0
AA@DMNA	Dissolved Mergers	0	0
AA@ETF	Index Reconstitution	0	0
AA@PRE	Private Equity	0	0

REFERENCES

Beasley, J. E., Meade, N., and Chang, T.-J. (2003). An evolutionary heuristic for the index tracking problem. *European Journal of Operational Research*, 148(3):621–643.

Canakgoz, N. A. and Beasley, J. E. (2009). Mixed-integer programming approaches for index tracking and enhanced indexation. *European Journal of Operational Research*, 196(1):384–399.

De Long, J. B., Shleifer, A., Summers, L. H., and Waldmann, R. J. (1990). Noise trader risk in financial markets. *Journal of Political Economy*, 98(4):703–738.

Dentcheva, D. and Ruszczynski, A. (2006). Portfolio optimization with stochastic dominance constraints. *Journal of Banking & Finance*, 30(2):433–451.

DiBartolomeo, D. (2000). The enhanced index fund as an alternative to indexed equity management. *Northfield Information Services, Boston*.

Fabian, C. I., Mitra, G., Roman, D., and Zverovich, V. (2011). An enhanced model for portfolio choice with SSD criteria: a constructive approach. *Quantitative Finance*, 11(10):1525–1534.

Feuerriegel, S. and Prendinger, H. (2016). News-based trading strategies. *Decision Support Systems*, 90:65–74.

Fishburn, P. C. (1977). Mean-risk analysis with risk associated with below-target returns. *The American Economic Review*, 67(2):116–126.

Kantos, C., Joldzic, D., Mitra, G., and Hoang Thi, K. (2022). Comparative Analysis of NLP Approaches for Earnings Calls. *White Paper Available on https://optirisk-systems.com/publications/whitepapers/*.

Kroll, Y. (1980). Stochastic dominance: a review and some new evidence. *Research in Finance*, 2:163–227.

Kumar, R., Mitra, G., and Roman, D. (2008). Long-short portfolio optimisation in the presence of discrete asset choice constraints and two risk measures. *Available at SSRN 1099926*.

Lo, A. W. (1999). The three p's of total risk management. *Financial Analysts Journal*, 55(1):13–26.

Markowitz, H. (1952). Portfolio selection. *The Journal of Finance*, 7(1):77–91.

Mitra, G. and Mitra, L. (2011). *The Handbook of News Analytics in Finance*. John Wiley & Sons, Chichester, UK.

Mitra, G. and Yu, X. (2016). *The Handbook of Sentiment Analysis in Finance*. Optirisk Systems in Collaboration with Albury Books.

Ogryczak, W. and Ruszczynski, A. (1999). From stochastic dominance to mean-risk models: semideviations as risk measures. *European Journal of Operational Research*, 116(1):33–50.

Ogryczak, W. and Ruszczynski, A. (2001). On consistency of stochastic dominance and mean–semideviation models. *Mathematical Programming*, 89(2):217–232.

Pring, M. (2002). *Technical Analysis Explained: The Successful Investor's Guide to Spotting Investment Trends and Turning Points*. McGraw-Hill Education.

Rockafellar, R. T. and Uryasev, S. (2002). Conditional Value-at-Risk for General Loss Distributions. *Journal of Banking & Finance*, 26(7):1443–1471.

Rockafellar, R. T., Uryasev, S., et al. (2000). Optimization of conditional value-at-risk. *Journal of Risk*, 2:21–42.

Roman, D., Darby-Dowman, K., and Mitra, G. (2006). Portfolio construction based on stochastic dominance and target return distributions. *Mathematical Programming*, 108(2-3):541–569.

Roman, D., Mitra, G., and Zverovich, V. (2013). Enhanced indexation based on second-order stochastic dominance. *European Journal of Operational Research*, 228(1):273–281.

Sadik, Z. A., Date, P. M., and Mitra, G. (2018). News augmented GARCH(1,1) model for volatility prediction. *IMA Journal of Management Mathematics*, 30(2):165–185.

Sadik, Z. A., Date, P. M., and Mitra, G. (2020). Forecasting crude oil futures prices using global macroeconomic news sentiment. *IMA Journal of Management Mathematics*, 31(2):191–215.

Scowcroft, A. and Satchell, S. (2003). *Advances in Portfolio Construction and Implementation*. Elsevier.

Vickson, R. (1975). Stochastic dominance tests for decreasing absolute risk aversion. i. discrete random variables. *Management Science*, 21(12):1438–1446.

Von Neumann, J., Morgenstern, O., and Kuhn, H. W. (2007). *Theory of Games and Economic Behavior (commemorative edition)*. Princeton University Press.

Whitmore, G. A. and Findlay, M. C. (1978). *Stochastic Dominance: An Approach to Decision-Making Under Risk*. Lexington Books.

Yu, X. (2014). *Analysis of New Sentiment and Its Application to Finance*. PhD thesis, Brunel University London, Uxbridge, UK.

Yu, X., Mitra, G., Arbex-Valle, C., and Sayer, T. (2015). An Impact Measure for News: Its Use in Daily Trading Strategies. *Available at SSRN: https://ssrn.com/abstract=2702032 or http://dx.doi.org/10.2139/ssrn.2702032*.

CHAPTER 8

Extracting Structured Datasets from Textual Sources – Some Examples

Matteo Campellone

Executive Chairman and Head of Research, Brain

Francesco Cricchio

CEO and CTO, Brain

CONTENTS

DOI: 10.1201/9781003293644-8

8.1 INTRODUCTION

Recent years have been witnessing a continuous search from active investors for additional sources of information on financial markets. The number of alternative data providers and of platforms that aggregate and integrate alternative data for asset managers has consequently been steadily increasing, creating a large and fairly variegate ecosystem.

Alternative data itself is an increasingly wider term, embracing satellite images originated data, credit card transactions, weather data, socials, news and corporate documents derived data. In an intuitive way alternative data could perhaps be defined as *all data that does not belong to the financial and economic data traditionally used by investors and that could be of interest to exploit in investment strategies.* A general and updated review of many currently available alternative datasets can be found here [Alternative Data Org, 2022].

A wide portion of the current available alternative data is extracted from textual documents through Natural Language Processing (NLP) tools, an area which is also developing fast [Alternative Data Org, 2022]; [Alpha, 2021].

In this chapter we will provide a general overview of Brain approaches to the creation of alternative datasets based on a number of language metrics on textual documents such as news, SEC Filings and earning call transcripts.

Outline

The remainder of this paper is organized as follows:

- Section 8.2 provides an overview of some, amongst many, useful sources of unstructured financial text: public newsflow, company filings and earnings calls transcripts.

- Section 8.3 shows how Natural Language Processing techniques can be used to define some measurable quantities on the above unstructured sources thus creating structured datasets that can then be used as building blocks for various analyses in the financial domain and investment strategies. Some results and case studies are shown.

- Section 8.4 provides some conclusions.

8.2 SOURCES OF TEXTUAL INFORMATION

8.2.1 Newsflow collection

A relevant source of information is public news. Generally speaking, if one adheres to the efficient market hypothesis (EMH), public news-flow should not be considered a useful source for trading, since its content will be quickly captured by market prices, unless one aims at beating the market on the speed to react to the information contained in the single news item.

EMH has been widely challenged over several years starting from the 80s, with a large number of studies highlighting inconsistencies with observed data [Malkiel, 2003; Bouchaud, 2022]. More recent works suggest very interesting interpretations of these inconsistencies through order flows dynamics [Bouchaud, 2022; Gabaix and Koijen, 2021].

So, with a less EMH-believer point of view, one can suppose that changes in news-flow may trigger trades which could have a relevant impact on the market, or that could bear a longer effect on the evolution of prices. In fact the analysis of correlation between sentiment movements and market performance is a currently widely studied topic, as well as the analysis of sentiment correlations among stocks [Tetlock, 2007; Wan et al., 2021; Ahelegbey et al., 2022; Mitra and Yu, 2016; Mitra and Mitra, 2011].

It may therefore make sense to assess, at least for descriptive purposes for the strict EMH believer, or with more predictive ambitions for others, how is the "mood" of the news-flow on each single asset, by systematically measuring quantities such as aggregated sentiment on various time scales, volume of news and related quantities.

8.2.2 SEC Filings: Form 10-Ks and Form 10-Qs

An interesting source of potentially relevant information are Form 10-K and Form 10-Q documents, that are compulsorily required by the U.S. Securities and Exchange Commission (SEC) to most listed U.S. companies, respectively, annually and quarterly. These documents, especially the 10-Ks, contain detailed information on the company's history, organization, performance and related potential risks.

Notwithstanding the fact that these documents are public and fairly easily accessible to most, some interesting studies in literature claim the existence of relevant inefficiencies in the way the market captures the information available on these reports, maybe due to their increased complexity and length during the years [Cohen et al., 2020; Loughran and McDonald, 2011; Padysak, 2020; Hanicova et al., 2021].

To better quantify this last point, we performed the following analyses for the 10-K company filings of the past 10 years:

- Figure 8.1(a) shows an increasing trend of the company filings length measured in terms of the total number of sentences.

- Figure 8.1(b) shows an increasing trend of the readability score calculated as average of various readability tests such as Gunning Fog index [Gunning et al., 1952], Flesch Kincaid grade level

[Kincaid et al., 1975] and others. These metrics are used to quantify the ease with which a reader can understand a text. The readability score is typically measured as the US grade needed to be able to read the text; therefore the higher the score, the more difficult the text is to read.

- Figure 8.1(c) shows the increasing trend of the mean length of sentences in terms of number of words.

8.2.3 Earnings Calls Transcripts

Earnings calls are conference calls held by listed companies where the management presents to analysts, media and investors the results of the previous quarter or fiscal year, also discussing the future outlook. The calls are held by the top-line managers of the company such as the CEO, the CFO or other executives. The last part of the call is a Q&A section where questions can be asked to the management on specific topics that are relevant for understanding of the company's results and future outlook. While 10-K and 10-Q company filings are only related to US companies, earnings calls transcripts are structured using a similar format for global stocks. There are several papers that highlight the earnings calls content as a source of valuable information for investors [Kimbrough, 2005; Matsumoto et al., 2011; Mayew and Venkatachalam, 2012; Frankel et al., 2017].

There are different providers that enable a relatively quick access to earnings calls transcripts. Some of them use a complete manual procedure for transcription while others combine software with manual review to facilitate the process. However, given the complexity and the specificity of the language used in such texts, a complete automatic transcription process seems currently not feasible. The transcripts can be available from few minutes to few hours after the event depending on the provider.

8.2.4 Comparison of Metrics between Company Filings and Earning Call transcripts

We compared some specific metrics calculated on 10-K company filings and earnings call transcripts of past 10 years (from January 2012 to October 2021) for a universe of approximate 5000 US stocks.

We selected the following metrics for comparison:

1. The sentiment measured using Brain algorithm (details are provided in Section 8.3.1) with some modification for application to company filings and earnings calls.

2. The amount of positive language in the text ("score positive"); this metric corresponds to the percentage of "financial positive" words based on

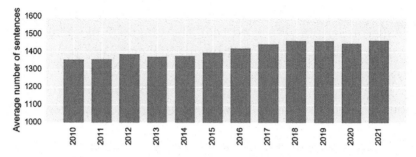

(a) Average number of sentences in 10-K reports from 2010 to 2021.

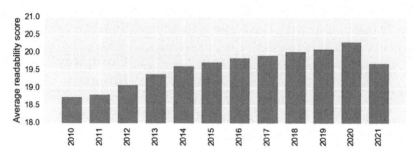

(b) Average readability score of 10-K reports from 2010 to 2021. The score is calculated as average of various readability tests such as Gunning Fog index [Gunning et al., 1952], Flesch Kincaid grade level [Kincaid et al., 1975] and others. The readability score is typically measured as the US grade needed to be able to read the text; therefore the higher is the score, the more difficult is the text to read.

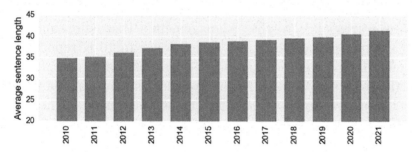

(c) Average sentence length in 10-K reports from 2010 to 2021.

Figure 8.1 Various metrics of complexity applied to 10-K company filings. For each year, an average on all available stocks is performed.

a dictionary specific to Finance; the percentage is calculated with the respect to all words present in the dictionary.

3. The amount of negative language in the text ("score negative"); this metric corresponds to the percentage of "financial negative" words based on a dictionary specific to Finance; the percentage is calculated with the respect to all words present in the dictionary.

TABLE 8.1 Comparison of various sentiment metrics for company filings and earnings call transcripts. We note that the language of earnings calls transcripts is more positive (higher sentiment, higher "score positive" and lower "score negative") compared to the language used in regulatory filings. The sentiment metric is calculated using the Brain Sentiment Indicator (BSI) algorithm described in Section 8.3.1 with some modifications for application to company filings and earnings call transcripts; the BSI score ranges from -1 (most negative) to $+1$ (most positive). The metrics "score positive" and "score negative" correspond to the percentage of "financial positive" and "financial negative" words, respectively, based on a dictionary specific to Finance; such percentage is calculated with the respect to all words present in the dictionary.

Metric (scale)	Section	Company filings	Earnings calls
Sentiment $(-1, +1)$	Whole document	0.20	0.35
Score positive $(0, 1)$	Whole document	0.13	0.37
Score negative $(0, 1)$	Whole document	0.30	0.24
Sentiment $(-1, +1)$	Management Discussion	0.25	0.36
Score positive $(0, 1)$	Management Discussion	0.13	0.40
Score negative $(0, 1)$	Management Discussion	0.26	0.20

Such comparison was performed for the whole document and for the Management Discussion (MD) section. More precisely the MD section under scrutiny corresponds to the "Management's Discussion and Analysis of Financial Condition and Results of Operations" in 10-K reports and to the "Management Discussion" section in the case of the earnings calls transcripts.

From the results in Table 8.1, we note that the language of earnings calls transcripts is on average more positive (higher sentiment, higher "score positive" and lower "score negative") compared to the language used in regulatory filings. This is not surprising given the different context of the two sources of information, since the managers during the earnings calls are likely to use a more positive language regarding the business compared to regulatory documents where a more objective language must be used. As expected, this effect is even more evident if we focus the analysis on the Management Discussion section.

In Figure 8.2 we show the average correlations of some language metrics of the MD&A section of quarterly company filings (10-Ks and 10-Qs) and earnings calls transcripts. Typically the company filings are published the same day of the earnings call; however, when this was not the case we were careful in the alignment of dates such that the reference period of the company filing corresponded to the one of the earning call. As it is reasonable to expect the correlations are negligible except from some positive correlations between

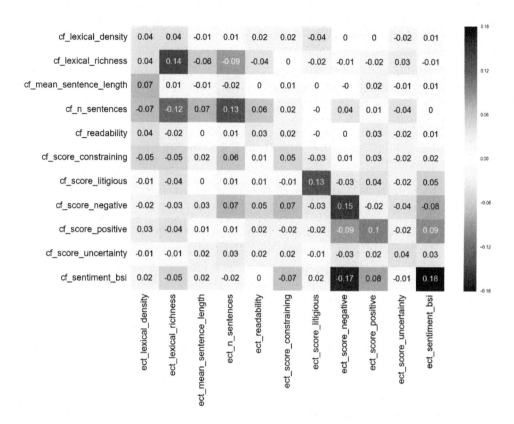

	ect_lexical_density	ect_lexical_richness	ect_mean_sentence_length	ect_n_sentences	ect_readability	ect_score_constraining	ect_score_litigious	ect_score_negative	ect_score_positive	ect_score_uncertainty	ect_sentiment_bsi
cf_lexical_density	0.04	0.04	-0.01	0.01	0.02	0.02	-0.04	0	0	-0.02	0.01
cf_lexical_richness	0.04	0.14	-0.06	-0.09	-0.04	0	-0.02	-0.01	-0.02	0.03	-0.01
cf_mean_sentence_length	0.07	0.01	-0.01	-0.02	0	0.01	0	-0	0.02	-0.01	0.01
cf_n_sentences	-0.07	-0.12	0.07	0.13	0.06	0.02	-0	0.04	0.01	-0.04	0
cf_readability	0.04	-0.02	0	0.01	0.03	0.02	-0	0	0.03	-0.02	0.01
cf_score_constraining	-0.05	-0.05	0.02	0.06	0.01	0.05	-0.03	0.01	0.03	-0.02	0.02
cf_score_litigious	-0.01	-0.04	0	0.01	0.01	-0.01	0.13	-0.03	0.04	-0.02	0.05
cf_score_negative	-0.02	-0.03	0.03	0.07	0.05	0.07	-0.03	0.15	-0.02	-0.04	-0.08
cf_score_positive	0.03	-0.04	0.01	0.01	0.02	-0.02	-0.02	-0.09	0.1	-0.02	0.09
cf_score_uncertainty	-0.01	-0.01	0.02	0.03	0.02	0.02	-0.01	-0.03	0.02	0.04	0.03
cf_sentiment_bsi	0.02	-0.05	0.02	-0.02	0	-0.07	0.02	-0.17	0.08	-0.01	0.18

Figure 8.2 Correlation matrix between the some language metrics calculated on company filings and earnings calls transcripts. This matrix is calculated as an average over all the stocks of the correlation between the metrics time series of each stock over the whole time histor. The metrics on y-axis marked with prefix `cf` refer to language metrics calculated on company filings and described in Section 8.3.2, while the metrics on x-axis marked with prefix `ect` refer to language metrics calculated on earnings calls transcripts and described in Section 8.3.4.

the sentiment of company filings and the sentiment of the earning calls and by a negative correlation between the amount of negative language used in the earnings call and the sentiment of the corresponding company filing. A positive correlation seems to be present between the amount of litigious language of filings and earnings calls.

8.3 STRUCTURED DATASETS BASED ON UNSTRUCTURED SOURCES

As opposed to "structured data" which are to be intended as organized, typically quantitative data, and organized to be processed my a computer, an enormous amount of information available is in unstructured form, generally definable as data in a native form, not organized or in a form easily processed by a computer. Textual documentation that we shall be discussing in this work such as news, corporate documents, transcripts, belong to this second category. Intuitively speaking Natural Language Processing is a very general area which deals with enabling computers to process and measure quantities on texts. See for example [Wikipedia, 2019]

8.3.1 Measuring Sentiment and Attention on the Corporate Newsflow – The Brain Sentiment Indicator

Measuring sentiment on texts is a quite common activity nowadays, with a wide range of applications, from assessing of marketing and political campaigns, or product launches to monitoring news and the content of corporate documents [Alpha, 2021; Mitra and Yu, 2016; Mitra and Mitra, 2011].

There is a large number of providers that calculate sentiment on listed companies, each based on different choices of sources, such as news, socials and documents [Alternative Data Org, 2022; Alpha, 2021].

Also in academic literature, the relation between stock market and media has been widely explored [Tetlock, 2007; Wan et al., 2021; Mitra and Yu, 2016; Mitra and Mitra, 2011].

There is no unique algorithm of "sentiment", and each sentiment measure depends on the assumptions and the specific methodological choices used, not only in the scoring algorithm but also on the overall process, starting from the definition of the sources, the parsing approach or the tagging and aggregation phase just to mention some.

We hereby present an overview of the general process defined to calculate the "Brain Sentiment Indicator" (BSI) [Brain, 2022] and discuss some specific methodological choices that have been made in its determination.

The BSI algorithm is based on a fairly simple approach that aims at simplicity and explicability. It uses a combination of semantic rules for financial news classification and dictionary-based approach (Bag of Words – BOW). The sentiment is calculated from the news headlines that typically in financial news well summarize the overall meaning of the text while limiting the noise that would be borne by a long text. The overall logic of the calculation runs as described in the following paragraphs. First, some general integration and preprocessing steps are performed:

1. news integration from several sources (currently approximately 2000 global sources);

2. text cleaning and normalization;

3. tagging each news item to one or more specific financial asset, e.g. a stock, an FX, a commodity or a cryptocurrency.

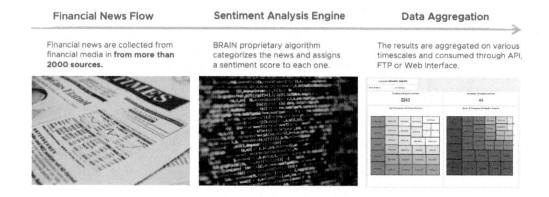

Financial News Flow	Sentiment Analysis Engine	Data Aggregation
Financial news are collected from financial media in **from more than 2000 sources.**	BRAIN proprietary algorithm categorizes the news and assigns a sentiment score to each one.	The results are aggregated on various timescales and consumed through API, FTP or Web Interface.

Figure 8.3   Workflow overview for the "Brain Sentiment Indicator"

Then the sentiment of each news item is measured. To do so Brain combined approach follows these steps:

1. News item headlines are assigned to specific categories using a series of semantic rules for each asset class. Each category has a predefined value of sentiment.

2. If no category is identified in step (1) then a Bag of Words approach is used, based on a proprietary dictionary customized for Finance. Negations and adverbs to make the sentence stronger or weaker are handled.

3. Aggregation of the sentiment of the single news item by similar news and/or by related asset.

Finally, the sentiment scores for each news item are aggregated over a specific time window (e.g. last 7 or 30 days) to calculate the BSI score at asset level. These steps are summarized in Figure 8.3.

As an example we show the "Brain Sentiment Indicator" on some selected stocks during the outburst of the COVID-19 pandemic in 2020:

- Figure 8.4(a) shows an example for "Apple Inc.": major tech stocks showed a quick recovery in sentiment after the pandemic outbreak.

- Figure 8.4(b) shows an example for "Salesforce.com": the sentiment of companies offering cloud based services stayed positive even during the pandemic since many companies accelerated the digitalization processes.

- Figure 8.4(c) shows an example for "United Airlines": the sentiment of stocks related to travel sector switched to negative shortly after the the pandemic outbreak and after 6 months still did not recover.

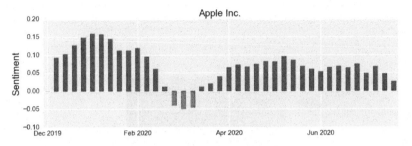

(a) Apple Inc.: major tech stocks showed a quick recovery in sentiment after the pandemic outbreak.

(b) Salesforce.com: the sentiment of companies offering cloud-based services stayed positive even during the pandemic since many companies accelerated the digitalization processes.

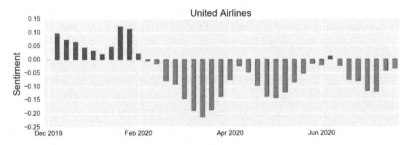

(c) United Airlines: the sentiment of stocks related to travel sector switched to negative during the pandemic and after 6 months from the outbreak still did not show any sign of recovery.

Figure 8.4 "Brain Sentiment Indicator" for selected stocks during the pandemic.

Finally, when calculating metrics related to news volume, one has to consider that the average volume of news depends on the specific company (e.g. typically increasing with company size for example). Therefore, it is useful to define a normalized measure of volume. A possible choice, often called "buzz", is to consider the volume distribution of past previous months. The "buzz" is then obtained as the difference of current news volume and past volume

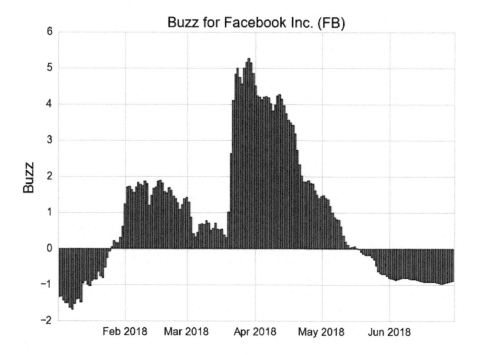

Figure 8.5 Normalized volume (buzz) of Facebook Inc. (FB) stock during the first half of 2018. The "buzz" is calculated as the difference between the current news volume (in this example it is calculated using a time window of 30 days) and the average of past volume in a time window of several months. We note the large peak during March-April in correspondence to the outbreak of the "Cambridge Analytica scandal" [Confessore, 2018].

average in units of standard deviations. A value close to 0 means that the stock not is receiving particular attention by media compared to its average, while a value significantly larger than 0 highlights a larger than normal attention. An example for Facebook Inc. (FB) stock is shown in Figure 8.5 where we note the large peak in correspondence to the outbreak of the "Cambridge Analytica scandal" [Confessore, 2018].

8.3.2 Measures on Corporate Documents – The Brain Language Metrics on Company Filings Dataset

One may be interested in creating a framework to systematically exploit at least some of the information available in the 10-K and 10-Q reports. Following the ideas of reference [Cohen et al., 2020] one can sort reports by how similar (or different) they are with respect to the previous one. Specifically by "previous report", we mean the report referring to the same period of the preceding year.

If the similarity among the total reports can already be a source of insights and potentially used for systematic models, it could be even more insightful and interesting to measure different language characteristics on the above reports such as, for example, the sentiment, the number of positive/negative words, or the weight of different language styles such as "litigiousness" or "constraining" language [Loughran and McDonald, 2011; Padysak, 2020]. Also other characteristics of the language used could potentially be a source of information, such as readability scores or lexical richness or lexical density [Hanicova et al., 2021]

These measures can be used either in absolute terms or by looking at variations (using differences in the scores, or vector similarity measures depending on the case) with respect to the previous reports.

For instance, one can focus on differences with respect to various language styles and systematically go long or short (or avoid) stocks based on any of these metrics, e.g. avoid stocks for which litigious language has increased the most or similar analyses.

Aiming at reducing noise and focusing on more dense in relevant information text, metrics can be calculated also on specific sections, such as the "Management's Discussion and Analysis of Financial Condition and Results of Operations" (MD&A) section or the "Risk Factor" sections.

With the purpose of providing a base for possible strategies based on the above thoughts, the "Brain Language Metrics on Company Filings" dataset [Brain, 2022] measures several language metrics on 10-Ks and 10-Qs company reports, on a daily basis, for approximately 6000+ US stocks.

In particular, the dataset is made of two parts; the first one includes the language metrics of the most recent 10-K or 10-Q report for each firm, namely:

1. financial sentiment, calculated using a similar same approach used in the BSI, with some differences due to the different nature of the sources (news vs longer corporate documents)

2. percentage of words belonging to financial domain classified by language types: "constraining", "interesting", "litigious" and "uncertain" language;

3. lexical metrics such as readability, lexical density and lexical richness;

4. text statistics such as the report length and the average sentence length.

The second part includes the differences between the two most recent 10-Ks or 10-Qs reports of the same period for each company, namely:

1. difference of the various language metrics (e.g. delta of sentiment, delta of readability etc.);

2. similarity metrics between documents, also with respect to a specific language type (for example similarity with respect to "litigious" language or "uncertainty" language), based on typical similarity measure between texts (e.g. cosine similarity).

The dataset includes the metrics and related differences both for the whole report and for specific sections; for example "Risk Factors" and "MD&A" sections.

One of the language metrics included in the dataset is the similarity between subsequent 10-K reports as shown in the following figures:

- Figure 8.6(a) shows the similarity between subsequent 10-K filings of Procter & Gamble (PG) stock by considering all sections in the report and all types of financial language.

- Figure 8.6(b) shows the similarity between subsequent 10-K filings of Apple Inc. (AAPL) stock by focusing on the negative financial language of the "MD&A" section. We note that the similarity between the reports of 2021 and 2020 is much lower compared to the values of the previous years. In Table 8.2 we provide a sample of the words and sentences that contributed mostly to such difference.

Finally, we performed a correlation analysis of the language metrics of last 10-K and 10-Q report for each stock. The correlation matrix shown in Figure 8.7 is calculated by performing an average over all the stocks of the correlation matrix between the time series metrics of each stock over the whole time history. The high correlations found between certain metrics are not surprising, and some of them may be trivial as implied by construction, but it makes sense to be aware of them when using them as input to any model.

1. The sentiment score (`sentiment_bsi`) is positively correlated with the amount of positive words (`score_positive`) and negatively correlated with the amount of negative words (`score_negative`).

2. The mean sentence length metric (`mean_sentence_length`) is, consistently with its definition, highly correlated with the readability score (`readability_score`), meaning that the longer are the sentences the lesser is the readability (the readability score is measured as the US grade needed to be able to read the text; the higher is the score, then the more difficult is the text to read [Gunning et al., 1952; Kincaid et al., 1975]).

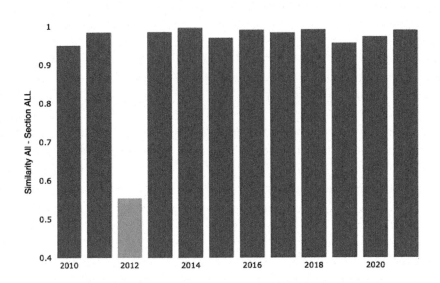

(a) Similarity between 10-K filings of Procter & Gamble (PG) stock by considering all sections in the report and all types of financial language.

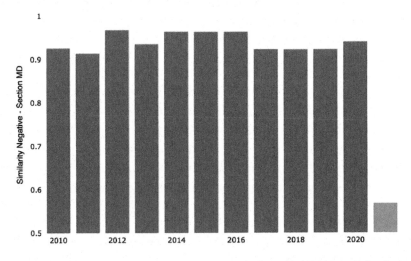

(b) Similarity between 10-K filings of of Apple Inc. (AAPL) stock by focusing on the negative financial language of the MD&A section

Figure 8.6 Examples of similarity metrics between 10-K reports for selected stocks.

3. A relevant negative correlation is found between the number of sentences n_sentences and the lexical richness (field lexical_richness measured as type-tokens ratio [Johansson, 2008]) meaning that the longer is the

TABLE 8.2 Examples of words and sentences that provided the largest contribution to the dissimilarity between Apple Inc. 10-K reports published in 2020 and 2021 (see Figure 8.6(b)).

Year	Word(s)	Sentence Extract
2021	Disruptions	"similar **disruptions** could occur in the future, the extent of the continuing impact of the covid-19 pandemic on the company's operational and financial performance is uncertain and will depend on many factors outside the company's control, including the timing, extent, trajectory and duration of the pandemic ..."
2021	Disruptions, shortages	"during the fourth quarter of 2021, certain of the company's component suppliers and logistical service providers experienced **disruptions**, resulting in supply **shortages** that affected sales worldwide."
2020	Abrupt, decline	"if there is an **abrupt** and substantial **decline** in estimated demand for one or more of the company's products, a change in the company's product development plans, or an unanticipated change in technological requirements for any of the company's products, the company may be required to record write-downs or impairments of manufacturing-related assets or accrue purchase commitment cancellation fees."
2020	Weakness, unfavorable	"the **weakness** in foreign currencies relative to the u.s. dollar had an **unfavorable** impact on greater china net sales during 2020."

text then the less varied is the vocabulary used by the author. The inverse correlation between type-tokens ratio and text length is a well-known effect [Johansson, 2008].

4. A negative correlation is measured between the number of sentences (n_sentences) and the mean sentence length (mean_sentence_length), meaning that more sentences are in the text, then on average the shorter are the sentences.

8.3.3 Some Case Studies based on Measures on Corporate Documents

In this section we show some case studies obtained using the metrics of the dataset "Brain Language Metrics on Company Filings". More specifically we

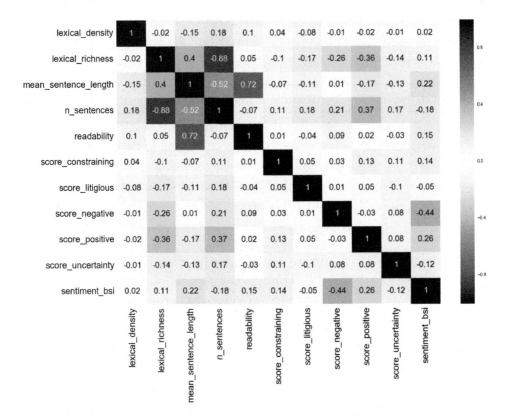

Figure 8.7 Correlation matrix between some language metrics of company filings from the dataset "Brain Language Metrics on Company Filings". This matrix is calculated as an average over all the stocks of the correlation between the metrics time series of each stock over the whole time history.

ranked the stocks according to selected metrics and we performed a backtest of each quintile of such ranking.

The general setup is the following:

1. the validation interval ranges from January 2010 to July 2021;

2. the reference universe is constituted by the largest 3000 US stocks by market cap; this universe is updated at the beginning of each year using the data of previous year to avoid any survival bias;

3. the quintiles are updated each quarter with uniform weights;

4. no transaction costs are included; this is due to the fact that at this stage we are only trying to assess the relevance of the features and not focusing or suggesting on any implementation strategy.

The following metrics showed the most promising results in terms of separation of quintile performance:

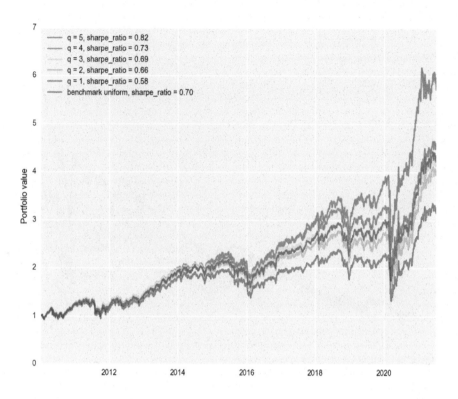

Figure 8.8 Quintile backtesting based on the metric "similarity uncertainty" measuring the similarity with focus on "uncertainty" language between one report and the previous one of the same period. The reference universe is constituted by the most liquid 3000 US stocks (dynamical universe updated each year). The interval is January 2010 – July 2021, no transaction costs are included and the quintiles are updated each quarter with uniform weights. The blue line corresponds to the benchmark constituted by the whole universe with uniform weights.

1. **Similarity among reports based on uncertainty language**: here the hypothesis is that the easiest path for company filings authors is to retain much of the structure and the content in their newest filings when the underlying business is strong and growing. When the business exhibits weakness, the author needs to provide more explanations. One way this can manifest is via textual amendments and additions. In this use case we focus on the similarity of language specifically related to the "uncertainty language" between one report and the previous one of the same period to avoid seasonal effects (e.g. 10-K of 2021 compared with 10-K of 2020 or 10-Q of Q1/2021 compared with 10-Q of Q1/2020). As shown in Figure 8.8 we find that, in the time interval of analysis,

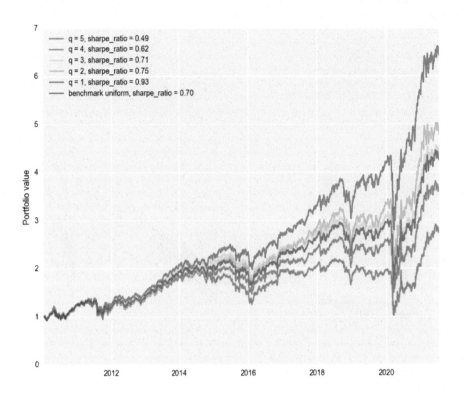

Figure 8.9 Quintile backtesting based on the metric "constraining language score" measuring the percentage of constraining financial language in each report. The reference universe is constituted by the most liquid 3000 US stocks (dynamical universe updated each year). The interval is January 2010–July 2021, no transaction costs are included and the quintiles are updated each quarter with uniform weights. The blue line corresponds to the benchmark constituted by the whole universe with uniform weights.

the top quintile (dark green line, Sharpe ratio 0.82) corresponding to reports that are changing less the uncertainty language in the whole report (more similar) outperforms the bottom quintile (red line, Sharpe ratio 0.58) corresponding to reports that are changing more the uncertainty language in the whole report (less similar). The five quintiles show a performance that reflects their inverse order.

2. **Amount of constraining financial language in the report**: we find that, in the analyses time interval, the stocks that contain less "constraining" financial language in their annual 10-K reports outperform in the following period the stocks that contain more "constraining" financial language. As shown in Figure 8.9, the bottom quintile corresponding

to reports that contain less constraining language (red line, Sharpe ratio 0.93) outperforms the top quintile (dark green line, Sharpe ratio 0.49) corresponding to reports that contain more constraining language. The five quintiles show a performance that reflects their order.

Some recent papers have studied other metrics of the "Brain Language Metrics on Company Filings" dataset:

- The work by M. Padysak [Padysak, 2020] focused on the metrics **"similarity of positive language"** and shows that the less positive similarity stocks outperform the most positive similarity stocks.

- The study by D. Hanicova et al. [Hanicova et al., 2021] tested possible investment strategies based on the metrics **"lexical richness"** related to how many unique words are used by the author and **"lexical density"** (a high lexical density indicates a large amount of information-carrying words).

8.3.4 Measures on Earning Call Transcripts – The Brain Language Metrics on Earnings Calls Transcripts dataset

One can apply the same analyses and metrics used for company filings (Section 8.3.2) to Earnings Calls Transcripts (ECT), that also are full of relevant information regarding the performance of the company, the outlook and the potential risks.

While the SEC forms are redacted mainly for regulatory purposes and therefore are necessarily skewed towards highlighting risks and possible threats, in the context of ECT, the discussion is possibly more balanced discussing results, business opportunities and risks also perhaps given the more discursive setting of the disclosures. See also the analysis of Section 8.2.4.

For ECT therefore the investment idea behind simply distinguishing changers and non-changers would probably be less relevant than other investment ideas or systematic patterns that could involve the relative "sign" of the differences or the absolute values of some specific features.

As example of calculation of possible measure on the earning calls, we provide some details of the "Brain Language Metrics on Earnings Calls Transcripts" dataset [Brain, 2022] that has the objective of monitoring several language metrics for the quarterly earnings call transcripts of 4500+ US stocks. Similarly to the dataset for Company Filings, the dataset is composed of two parts. Part one includes several language metrics for the most recent earnings call transcript for each stock, namely:

1. financial sentiment;

2. percentage of words belonging to financial domain classified by language types: "constraining", "litigious" and "uncertainty" language;

3. lexical metrics such as lexical density and richness of text;

4. text statistics such as the transcript length.

Part two measures the changes between the most recent earnings call transcript and the previous one:

1. Difference of the various language metrics (e.g. delta sentiment, delta readability score, delta percentage of a specific language type etc.)

2. Similarity metrics between documents, also with respect to a specific language type (for example similarity with respect to "litigious" language or "uncertainty" language)

The metrics calculation is reported separately for the following sections of the transcript:

a. Management's Discussion (MD)

b. Analysts' Questions (AQ)

c. Management Answers to Analysts' Questions (MAAQ)

The dataset is updated with a daily frequency since new earnings calls transcripts can be published every day for some of the universe stocks. Clearly the data for each stock will change on a quarterly basis when new earnings calls are published. An example of sentiment calculation on some sentences of Apple Inc. earnings call of October 2021 is provided in Table 8.3.

One of the dataset metrics is the similarity between the most recent earnings call transcript and the previous one for various sections. We show some examples in the following figures:

1. Figure 8.10(a) shows the similarity between the "Management's Discussion" section of subsequent earnings calls of Microsoft with focus on generic financial domain language.

2. Figure 8.10(b) shows the similarity between the "Management's Discussion" section of subsequent earnings calls of Microsoft with focus on "negative" financial domain language. In this case, for example, we note a larger difference (less similarity) between the second and third quarter of 2020. The larger difference seems caused by changes in the business related to the pandemic outbreak at the beginning of 2020. This is shown in Table 8.4 where we provide a sample of the words and sentences that contributed mostly to such difference.

TABLE 8.3 Examples of sentences with positive and negative sentiment for various sections of Apple Inc. (AAPL) earnings call of October 2021.

Section	Sentence Extract	Sentiment
Management Discussion	"This level of sales performance, combined with the unmatched loyalty of our customers and the strength of our ecosystem, drove our installed base of active devices to a new all-time record."	Positive
Management Discussion	"Customers are loving the ninth generation iPad, which features a beautifully sharp display and twice the storage of the previous generation, as well as the new iPad mini, with its ultra portable design and impressive speed and performance."	Positive
Management Discussion	"A year ago, i spoke to you about the atmosphere of uncertainty in which we were living and the way it had come to define our daily experience, both as people and as a company."	Negative
Analysts' Questions	"How do you think about balancing the regulators push for more choice with a customer base that is happy with the existing experience?"	Positive
Analysts' Questions	"How do you think about the infrastructure and visibility to sort of rebound and sort of handle sort of these disruptions that seem to crop up from time to time? "	Negative
Management Answers to Questions	"And so, what we're doing is working with our partners on making sure that they have supply that we need and making sure that our demand statements are accurate as we see them and so forth."	Positive
Management Answers to Questions	"If you look at Q4 for a moment, we had about $6 billion in supply constraints and it affected the iPhone the iPad and the Mac."	Negative

Another metric included in the dataset is the difference of sentiment score measuring the tone of the "Analysts' Questions" section; one example for Apple Inc. stock is shown in Figure 8.11.

Finally, we performed a correlation analysis of the language metrics of last earnings calls transcript for each stock. The correlation matrix shown

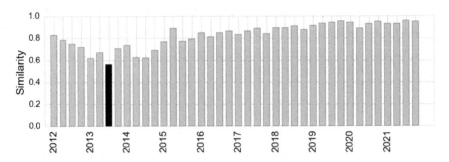

(a) Similarity between the "Management's Discussion" section of subsequent earnings calls of Microsoft with focus on generic financial domain language.

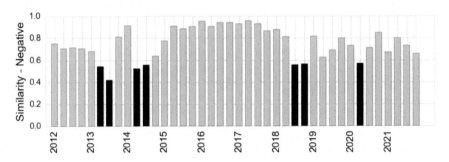

(b) Similarity between the "Management's Discussion" section of subsequent earnings calls of Microsoft with focus on "negative" financial domain language.

Figure 8.10 Examples of similarity metrics between subsequent earnings calls transcripts of Microsoft (MSFT).

in Figure 8.12 is calculated by performing an average over all the stocks of the correlation matrix between the time series metrics of each stock over the whole time history. Similarly to the case of the Company Filings, we note the following points:

1. As expected the sentiment score (`sentiment_bsi`) is positively correlated with the amount of positive words (`score_negative`) and negatively correlated with the amount of negative words (`score_positive`).

2. Similarly the amount of "uncertainty" words (`score_uncertainty`) is correlated with the amount of negative words (`score_positive`); clearly many "uncertainty" words shows also have a negative connotation.

3. The mean sentence length (`mean_sentence_length`) is highly correlated by construction with the readability score (`readability_score`). This is because the formula of readability scores explicitly depends on the mean sentence length [Gunning et al., 1952; Kincaid et al., 1975]); the longer

TABLE 8.4 Examples of words and sentences that provided the largest contribution to the dissimilarity between Microsoft (MSFT) earnings call transcript of third and second quarter of 2020 (see Figure 8.10(b)). Some of the differences seem related to consequences of the pandemic outbreak.

Quarter	Year	Word(s)	Sentence Extract
3	2020	Weak	"however, we expect the sales dynamics from march to continue, including a significant impact in Linkedin from the **weak** job market and increased volatility in new longer lead time deal closures"
3	2020	Volatility	"the remaining 20%, which is primarily made up of new annuity agreements, transactional licensing and enterprise services consulting revenue, is subject to more **volatility**."
3	2020	Slowdown	"we saw a **slowdown** in our transactional business across segments but particularly in small and medium businesses."
3	2020	Delay	"and in enterprise services, we expect a low single-digit revenue decline, driven by continued **delays** in our consulting business."; "in enterprise services, growth rates slowed as consulting projects were delayed".
2	2020	Challenge	"Surface revenue increased 6% and 8% in constant currency, lower-than-expected as continued strong momentum in the commercial segment was partially offset by execution **challenges** in the consumer segment."
2	2020	Cybercrime	"now to security, **cybercrime** will cost businesses, governments and individuals $1 trillion this year."

are the sentences the lesser is the readability (the readability score is measured as the US grade needed to be able to read the text; the higher the score, the more difficult the text is to read [Gunning et al., 1952; Kincaid et al., 1975]).

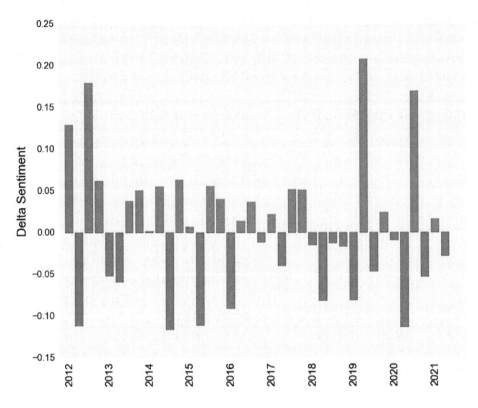

Figure 8.11 Difference of sentiment score in "Analysts' Questions" section between subsequent earnings calls transcripts of Apple Inc. (AAPL) stock. We consider the metrics related to the Analyst questions to be "handled" with greater care since we imagine that the tone of a question is generally less defined than in the case of a direct statement.

4. A relevant negative correlation is found between the number of sentences n_sentences and the lexical richness (field lexical_richness measured as type-tokens ratio [Johansson, 2008]) meaning that the longer is the text then the less varied is the vocabulary used by the author. The inverse correlation between type-tokens ratio and text length is a well-known effect [Johansson, 2008].

8.3.5 Some Case Studies based on Measures on Earning Call Transcripts

In this section we show some case studies obtained using the metrics of the dataset "Brain Language Metrics on Earnings Calls Transcripts". More specifically we ranked the stocks according to selected metrics and we performed a backtesting of each quintile of such ranking.
The general setup is the following:

- the validation interval ranges from January 2012 to July 2021;

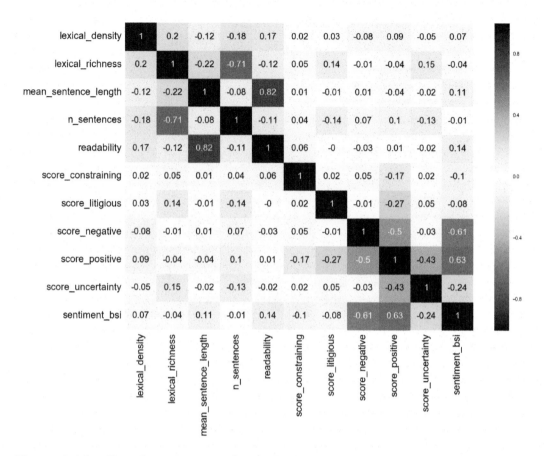

Figure 8.12 Correlation matrix between some metrics of earnings calls transcripts from the dataset "Brain Language Metrics on Earnings Calls Transcripts". This matrix is calculated as an average over all the stocks of the correlation between the metrics time series of each stock over the whole time history.

- the reference universe is constituted by the largest 3000 US stocks by market cap; this universe is updated at the beginning of each year using the data of the previous year to avoid any survival bias;

- the quintiles are updated each quarter with uniform weights;

- no transaction costs are included; this is due to the fact that at this stage we are only trying to assess the relevance of the features and not focusing on or suggesting any implementation strategy.

We performed two tests:

1. For each date we ordered the stocks by **financial sentiment of the Management Discussion section of the last earnings call**. We

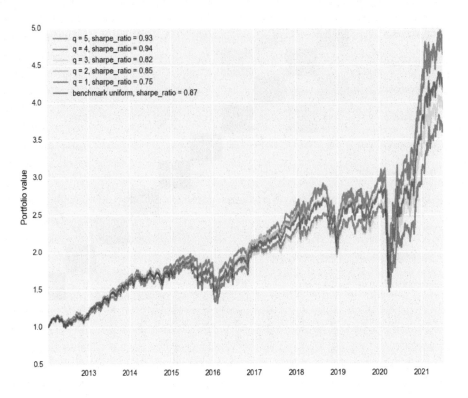

Figure 8.13 Quintile backtesting based on the sentiment of the "Management Discussion" section of the last earnings call transcript. The reference universe is constituted by the largest 3000 US stocks by market cap (dynamical universe update every year). The blue line corresponds to the benchmark constituted by the whole universe with uniform weights.

wanted to test if the higher sentiment stocks outperform the lower sentiment ones on a quarterly time horizon. We then performed a backtesting of each quintile of the ranking with the setup described above. As shown in Figure 8.13, the top quintile (dark green line, Sharpe ratio 0.96) corresponding to the stocks with highest sentiment of the "Management Discussion" section of the last earnings call outperforms the bottom quintile (red line, Sharpe ratio 0.76) corresponding to stocks with lowest sentiment. The five quintiles show a performance that reflects their order.

2. We analysed a use case similar to the one tested for the dataset "Brain Language Metrics on Company Filings" (BLMCF) (see Section 8.3.3) by looking at the similarity between subsequent earnings calls with focus

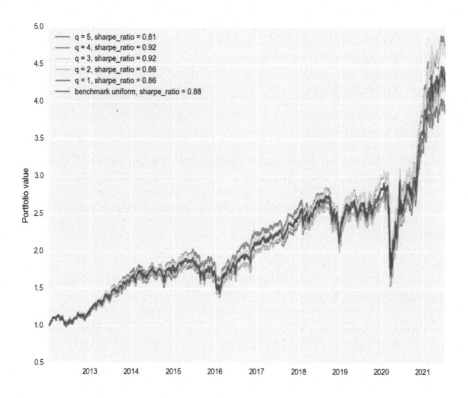

Figure 8.14 Quintile backtesting based on the similarity of the "Management Discussion" section of the last earnings call transcript. The reference universe is constituted by the largest 3000 US stocks by market cap (dynamical universe update every year). The blue line corresponds to the benchmark constituted by the whole universe with uniform weights.

on the "uncertainty" language. We then performed a backtesting of each quintile of the ranking with the setup described above. As shown in Figure 8.14, the quintiles do not separate and behave in a similar manner. Differently than in BLMCF case the similarity-uncertainty metric does not seem to produce interesting results.

8.3.6 Combination of Features based on Machine Learning

Here we focus on the dataset "Brain Language Metrics on Company Filings" and we combine the dataset features using a machine learning approach to generate a daily stock ranking based on the predicted future returns for the next 3 months for a universe of 3000 US stocks. This is interesting for two reasons; the first is that we build a model that combines more than one feature at the time, also able to capture nonlinear effects. The second reason is that, while in the preceding case studies the selection of the feature was done *a-posteriori*, meaning that we were looking at a possible explanation of the results, but with a feature-selection bias that was implicit in the analysis, in the machine learning model it is the training phase that selects the most relevant features for the validation phase. The used machine learning model implements a voting scheme of machine learning classifiers. More precisely the model setup is the following

1. the time horizon for the model prediction corresponds to the 3 months following the time of prediction;

2. the out-of-sample validation interval ranges from January 2012 to September 2021;

3. the reference universe is constituted by a dynamic universe updated each year to include the top 3000 stocks by dollar volume of the previous year to avoid survival bias;

4. the model is built as a classifier trying to predict if in the next 3 months a stock will underperform (class 0) or outperform (class 1) the median of the universe.

5. specific "embargo" techniques are used to be sure not to contaminate the training data with the validation data

6. the model is re-trained every 6 months with an expanding window starting from January 2007. Each trained model is then used to generate out-of-sample predictions for the following 6 months in a "walking-forward" manner.

7. a subset of features for which the history was available is from January 2007 is used. This corresponds to metrics and differences on the whole content of 10-K reports described in Section 8.3.2.

The top features selected by the model in the last training interval (January 2007 - June 2021) are shown in Figure 8.15; the top feature is represented by the amount of "constraining" language present in the report. This is consistent with the fact that the use case based on this single feature (see Section 8.3.3) produced promising results. However, the machine learning approach is logically more robust since it selects the most relevant features on subsequent training intervals in a "walking-forward" manner and producing prediction on the following out-of-sample period.

The model predictions are used to build an out of sample quintile backtest, where each quintile is selected according to the model prediction. The following setup is used:

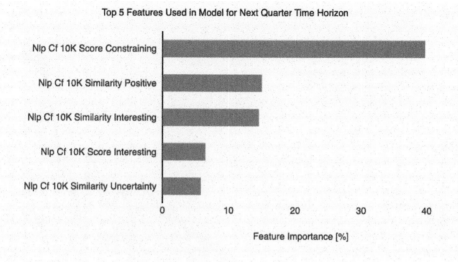

Figure 8.15 Top features selected by the machine classifier using the language metrics on company filings as input.

1. the out-of-sample validation interval ranges from January 2012 to September 2021;

2. the reference universe is constituted of a dynamic universe updated each year to include the top 3000 stocks by dollar volume of the previous year to avoid survival bias;

3. the quintiles are updated each quarter with uniform weights;

4. no transaction costs are included; this is due to the fact that at this stage we are only trying to assess the relevance of the features and not focusing or suggesting on any implementation strategy.

In Figure 8.16 we can see that the top quintile of stocks according to the model prediction (dark green line, Sharpe ratio 0.94) outperforms the bottom quintile of stocks according to the model prediction (red line, Sharpe ratio 0.52). More specifically, the five quintiles show a performance that reflects their order. In Table 8.5 we report the corresponding backtesting metrics for the top and bottom quintile, and for a benchmark constituted of all stocks belonging to the universe with uniform weights.

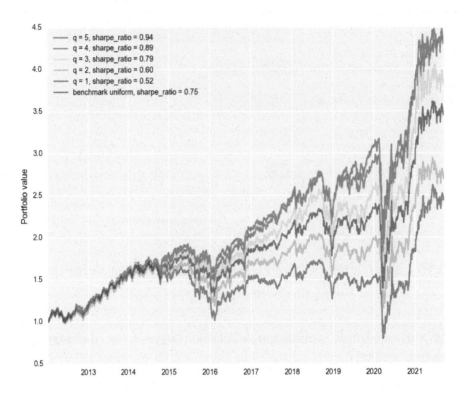

Figure 8.16 Out-of-sample quintile backtesting based on the prediction of a machine learning classifier using the language metrics on company filings as input.

TABLE 8.5 Metrics for out-of-sample backtesting based on the prediction of a machine learning classifier using the language metrics on company filings as input. Such metrics are reported the top and bottom quintile, and for a benchmark constituted of all stocks belonging to the universe with uniform weights.

Metric	Top Quintile	Bottom Quintile	Benchmark
Annualized Return [%]	16.0	9.8	13.5
Volatility [%]	17.9	22.8	19.8
Sharpe Ratio	0.94	0.52	0.75
Max Drawdown [%]	-38.2	−54.2	−42.0

8.4 CONCLUSIONS

We have presented some approaches to extract information in the form of usable structured datasets based on textual sources such as news, SEC regulatory filings, and earnings calls. In particular we discussed some Brain proprietary datasets based on the above sources, with some case studies. For the "Brain Language Metrics on Company Filings" dataset, we also showed the setup of a machine learning model that provides a periodically updated stock ranking based on its confidence in over/under performance of the investment universe. The model is dynamic in training and uses an expanding window approach that combines the most relevant features selected in the training phase. The model results are promising; the top quintile of stocks according to the model prediction outperforms the bottom quintile of stocks according to the model prediction and more specifically the five quintiles show a performance that reflects their order.

8.5 DISCLAIMER

REFERENCES

Ahelegbey, D. F., Cerchiello, P., and Scaramozzino, R. (2022). Network based evidence of the financial impact of Covid-19 pandemic. *International Review of Financial Analysis*, 81:102101.

Alpha, E. (2021). Eagle alpha's 1st annual alternative data report – 2021. https://www.eaglealpha.com/2021/11/18/eagle-alphas-1st-annual-alternative-data-report-2021/.

Alternative Data Org (2022). Alternative data database. https://alternativedata.org/. Accessed: 2022-06-23.

Bouchaud, J.-P. (2022). The inelastic market hypothesis: a microstructural interpretation. *Quantitative Finance*, pages 1–11.

Brain (2022). Brain website. https://braincompany.co. Accessed: 2022-06-23.

Cohen, L., Malloy, C., and Nguyen, Q. (2020). Lazy prices. *The Journal of Finance*, 75(3):1371–1415.

Confessore, N. (2018). Cambridge analytica and facebook: The scandal and the fallout so far. *The New York Times*, 4:2018.

Frankel, R. M., Jennings, J. N., and Lee, J. A. (2017). Using natural language processing to assess text usefulness to readers: the case of conference calls and earnings prediction. *Available at SSRN 3095754*.

Gabaix, X. and Koijen, R. S. (2021). In search of the origins of financial fluctuations: the inelastic markets hypothesis. Technical report, National Bureau of Economic Research.

Gunning, R. et al. (1952). *Technique of Clear Writing. McGraw-Hill*, pages 36–37.

Hanicova, D., Kalus, F., and Vojtko, R. (2021). How to use lexical density of company filings. *Available at SSRN 3921091*.

Johansson, V. (2008). Lexical diversity and lexical density in speech and writing: a developmental perspective. *Working papers/Lund University, Department of Linguistics and Phonetics*, 53:61–79.

Kimbrough, M. D. (2005). The effect of conference calls on analyst and market underreaction to earnings announcements. *The Accounting Review*, 80(1):189–219.

Kincaid, J. P., Fishburne Jr, R. P., Rogers, R. L., and Chissom, B. S. (1975). Derivation of new readability formulas (automated readability index, fog count and flesch reading ease formula) for navy enlisted personnel. Technical report, Naval Technical Training Command Millington TN Research Branch.

Loughran, T. and McDonald, B. (2011). When is a liability not a liability? Textual analysis, dictionaries, and 10-Ks. *The Journal of Finance*, 66(1):35–65.

Malkiel, B. G. (2003). The efficient market hypothesis and its critics. *Journal of Economic Perspectives*, 17(1):59–82.

Matsumoto, D., Pronk, M., and Roelofsen, E. (2011). What makes conference calls useful? The information content of managers' presentations and analysts' discussion sessions. *The Accounting Review*, 86(4):1383–1414.

Mayew, W. J. and Venkatachalam, M. (2012). The power of voice: managerial affective states and future firm performance. *The Journal of Finance*, 67(1):1–43.

Mitra, G. and Mitra, L. (2011). *The Handbook of News Analytics in Finance.* John Wiley & Sons.

Mitra, G. and Yu, X. (2016). *Handbook of Sentiment Analysis in Finance.* Optirisk Systems in Collaboration with Albury Books.

Padysak, M. (2020). The positive similarity of company filings and the cross-section of stock returns. *Available at SSRN 3690461.*

Tetlock, P. C. (2007). Giving content to investor sentiment: The role of media in the stock market. *The Journal of Finance,* 62(3):1139–1168.

Wan, X., Yang, J., Marinov, S., Calliess, J.-P., Zohren, S., and Dong, X. (2021). Sentiment correlation in financial news networks and associated market movements. *Scientific Reports,* 11(1):1–12.

Wikipedia (2019). Natural language processing. https://en.wikipedia.org/wiki/Natural_language_processing. Accessed 2022-06-23.

CHAPTER 9

Comparative Analysis of NLP Approaches for Earnings Calls

Christopher Kantos

Alexandria Technology, London, United Kingdom

Dan Joldzic

Alexandria Technology, London, United Kingdom

Gautam Mitra

OptiRisk Systems and UCL Department of Computer Science, London, United Kingdom

Kieu Thi Hoang

OptiRisk Systems, London, United Kingdom

CONTENTS

DOI: 10.1201/9781003293644-9

9.1 INTRODUCTION

9.1.1 An Overview

In this chapter, we review a few natural language processing methods mainly in the context of 'Earnings Call' of the companies/stocks under consideration. The methods selected show an evolution of methods which have matured in the recent years.

There are many approaches of natural language processing (NLP) to extract quantified sentiment scores from texts. These sentiment scores are used by financial professionals in creating portfolio construction or risk control models and thereby add value to the investment process. We start by considering and critically analysing the recent history of three well-known approaches, namely, LM, FinBERT and Alexandria Technology.

9.1.2 Loughran McDonald

Loughran McDonald (LM), developed by Tim Loughran and Bill McDonald of Notre Dame in 2011, is arguably the most popular financial lexicon method available. Lexicon, often called the "bag of words" approach to NLP use a dictionary of words or phrases that are labelled with sentiment. The motivation for the LM dictionary was that in testing the popular and widely used Harvard Dictionary, negative sentiment was regularly mislabeled for words when applied in the financial context, see [Loughran and McDonald, 2011]. The LM dictionary was built from a large sample of 10 Q's and 10 K's from the years 1994 to 2008. The sentiment was trained on approximately 5,000 words from these documents and over 80,000 words from the Harvard Dictionary. The dictionary is updated on a periodic basis to capture new words, phrases and context that enter the business lexicon, specifically the financial domain.

9.1.3 FinBERT

FinBERT is a specialized version of the machine learning model BERT (Bidirectional Encoder Representations from Transformers) developed by Google in 2018. The BERT model is pretrained on both language modelling and next sentence prediction, which provides a framework for providing contextual relationships among words [Araci and Genc, 2019]. FinBERT was developed in 2019 to extend the BERT model for better understanding within financial contexts. FinBERT uses the Reuters TRC2-financial corpus as the pre-training dataset. It is a subset of the broader Reuters TRC2 dataset, which consist of 1.8 million of articles that were published by Reuters between 2008 and 2010. This dataset is filtered for financial keywords and phrases to improve relevance and suitability with the computer power available. The resulting dataset is 46,143 documents consisting of 29 million words and

400,000 sentences. For sentiment training, FinBERT uses Financial Phrase-bank, developed in 2014 by Malo et al, consisting of 4,845 sentences in the LexisNexis database labelled by 16 financial professionals. Lastly, the FiQA Sentiment dataset which consists of 1,174 financial news headlines completes the FinBERT sentiment training [Araci and Genc, 2019]. FinBERT is an open-source project that is freely available for download on Github.

9.1.4   Alexandria Technology

Alexandria Technology uses a unique approach to NLP that is an ensemble of pure machine learning techniques without the need for dictionaries, rules, or price trend training. Developed in 2007, the model's initial application was to classify DNA sequences for genomic functionality. The underlying NLP technology was then tailored for applications in the institutional investment industry.

Alexandria's language model uses dynamic sequencing to identify phrase length. A phrase can be one, two, three or more words depending on the conditional probability of the string. The strings become stand- alone features which are then re-analysed to determine which features occur together within a body of text. If two or more features have a high probability of occurring together, the features are joined to form higher order concepts. For earnings calls, the Alexandria language model was created from millions of sentences using the FactSet XML Transcripts corpus.

Once the language model is formed, training can begin for features such as sentiment and themes. For sentiment, Alexandria's training function is senti-ment labels on sentences from GICS sector analysts. A sentence is reviewed and given a label for topic and sentiment for that particular topic. The earn-ings call sentiment training sample is over 200,000 unique sentence labels from earnings calls.

9.1.5   Chapter Structure: A Guided Tour

The rest of this chapter is organized in the following sections. In Section 2 we consider and describe the Data Sources. We discuss Earnings Call Transcripts, Sentiment Data and our approach to NLP models. In Section 3 we describe the models and methods used by us. We consider NLP for Earnings Calls and describe our method of Sentiment Classification. The investigations and the results are set out in Sections 4 and 5, respectively. In Section 6 we present discussions and conclusions in a summary form.

9.2 DATA SOURCES AND MODELS

9.2.1 Earnings Call Transcripts

In our analysis, we use the FactSet Earnings Call Transcript database, which covers earnings calls of companies around the world. Earnings call is a conference call between the management of a public company, analysts, investors and the media to discuss the company's financial results during a given reporting period, such as a quarter or a fiscal year [Chen, 2021]. An earnings call is usually preceded by an earnings report. This contains summary information on financial performance for the period. FactSet is a data vendor who provides transcripts of these earning calls. Their coverage begins in 2003 for earnings calls, of which we look at a period from 2010-2021. Our subset of this global dataset focuses on the S&P 500 equities.

Earnings calls have always been a source of integral information for investors. During these calls, companies discuss financial results, operations, advancements and hardships that will steer the course and future of the company. This information is a key resource for analysts to forecast the ongoing and future financial performance of the company. The calls are split up between Management Discussion (MD) where executives present financial results and forecasts, followed by Questions and Answers (QA) where participants such as investors and analysts can ask questions regarding the MD results.

Using NLP and ML, we can analyse these calls in near real time. By parsing and scoring them for sentiment, we have an overall view of thousands of calls as well as a topic-by-topic (or sentence-by-sentence) detailed view of what was said, who said it and what the sentiment was without ever having to dial in or read through the transcripts. Armed with the quantitative information derived from these calls, we can then explore various ways to incorporate and enhance our investment processes.

9.2.2 Sentiment Data

The sentiment data used in our testing is an output of the three NLP approaches. We use these three NLP methods to build classification models which create sentiment score from the earnings call transcripts dataset. All three approaches use a trinary system of -1 (negative sentiment), 0 (neutral sentiment) and 1 (positive sentiment) to score each section of the earnings call.

LM method uses an alternative negative word list of the widely used Harvard Dictionary words list, along with five other word lists, that better reflect tone in financial text. This is proved when they research a large sample of 10-Ks during 1994 to 2008, almost three-fourths of the words [Loughran and

McDonald, 2011] identified as negative by the widely used Harvard Dictionary are words typically not considered negative in financial contexts.

FinBERT is a pre-trained NLP model to analyse sentiment of financial text based on BERT. BERT, short for Bidirectional Encoder Representations from Transformers, is a Machine Learning (ML) model for NLP. BERT consists of a set of transformer encoders stacked on top of each other. BERT can be used on several language tasks, such as sentiment analysis, text prediction, text generation, summarization and so on. FinBERT is built by further training the BERT language model in the finance domain, using a large financial corpus and thereby fine-tuning it for financial sentiment classification. FinBERT improved the state-of-the-art performance by 15 percentage points for a financial sentiment classification task in FinancialPhrasebank dataset [Araci and Genc, 2019].

Different from Loughran McDonald method and FinBERT method, Alexandria Technology's method does not use any dictionaries, rules, or price training trends, but pure machine learning techniques to produce sentiment score from texts. Alexandria's language model was created from millions of sentences using the FactSet XML Transcripts corpus. The training function in this approach is sentiment labels on topics and sentences from GICS sector analysts. A section of text is analysed and labelled with a theme and sentiment. The earnings call sentiment training sample is over 200,000 unique labels from earnings calls.

9.2.3   NLP Models

For our comparative analysis of NLP approaches for classifying and scoring sentiment for earnings calls, we use open-source libraries for the LM and FinBERT approaches and the proprietary machine learning approach of Alexandria Technology. More information about the three sources is found below and in the bibliography section of this paper.

9.3   METHODOLOGY OF THE THREE APPROACHES

9.3.1   Natural Language Processing for Earnings Calls

NLP for financial text has a wide array of applications for institutional investors. Previous research shows that using NLP for financial news has applications in risk management, asset allocation, alpha generation, among many others. It is only recently that we have begun to look at using the same techniques that we use for financial news for other sources such as earnings calls.

As early as 2012, studies looked at large stock price movements, and if they correlated to analyst earnings revisions [Govindaraj et al., 2013]. More recent studies have shown that using sentiment on earnings transcripts can lead to outperformance that is not explained by traditional risk and return factors such as Momentum and EPS Revisions [Jha et al., 2015]. Further study by Societe Generale in 2019 shows that sentiment on earnings calls actually outperformed traditional factors such as Value, Profitability, Quality, Momentum and Growth as well, see [Short, 2016]. In a study by S&P in 2018, they show that market sentiment surrounding earnings calls not only has low correlation with other signals but amplifies the effectiveness of other earnings transcript-based signals [Tortoriello, 2018]. Finally, Alexandria's internal research shows that alpha signals from earnings calls have a longer decay rate and drift compared to shorter-term signals such as news and social media.

It is clear much research has been done to show the value add in analysing the sentiment of earnings calls and in the next section, we will extend on the research to explore what effect different NLP approaches have using a homogeneous set of earnings call transcript data as inputs.

9.3.2 Sentiment Classification

For our sentiment classification, we use transcript data provided by FactSet. FactSet's coverage includes over 1.7 million global company corporate events, including earnings calls, conference presentations, guidance calls, sales & revenue calls and investor meeting notes [fac, 2019].

We first use the three distinct models to apply sentiment labels to the S&P500 over a time period of 2010-2021. Among individual section classifications (ie. Topics, Sentences), we found the highest correlation to be between LM and FinBERT at 0.38. Alexandria had the lowest correlations with both LM and FinBERT at 0.14 and 0.17 respectively. We then aggregated the individual sections into net sentiment values for each security in the sample. Unsurprisingly, when aggregated the correlations rose, with the highest correlation being between LM and Alexandria at 0.45, the lowest being FinBERT and Alexandria at 0.30 and FinBERT and LM having correlation of 0.41.

Correlations

TABLE 9.1   Section Classification Correlation

	Alexandria Technology	Loughran McDonald
Loughran McDonald	0.14	
FinBERT	0.17	0.38

TABLE 9.2   Net Sentiment Correlation

	Alexandria Technology	Loughran McDonald
Loughran McDonald	0.45	
FinBERT	0.30	0.41

Sentiment Classification: Illustrated with Examples

We provide several examples of how each approach classifies sections of an earnings call for sentiment. The differences are indicative of how each approach does or does not take context and word order into question.

Example 1: Turning to the Google Cloud segment, revenues were $4.6 billion for the second quarter, up 54%. GCP's revenue growth was again above Cloud overall, reflecting significant growth in both infrastructure and FinBERT platform services. Once again, strong growth in Google Workspace revenues was driven by robust growth in both seats and average revenue per seat. Google Cloud had an operating loss of $591 million. As to our Other Bets, in the first (sic) [second] quarter, revenues were $192 million. The operating loss was $1.4 billion.

Example 2: Turning to the balance sheet, total spot assets were $1.1 trillion and standardized RWAs increased to $454 billion, reflecting high levels of client activity and the closing of E TRADE. Our standardized CET1 FinBERT ratio was flat to the prior quarter at 17.4%.

Example 3: Non-GAAP operating expenses increased 9% year-over-year and 2% sequentially, slightly above the high end of our guidance range, primarily due to higher variable compensation related to better-than-expected order momentum.

TABLE 9.3   Sentiment Classification: Illustrated with Examples

Example	Method	Sentiment Score
Example 1	Loughran McDonald	−1
	Alexandria	1
	FinBERT	−1
Example 2	Loughran McDonald	−1
	Alexandria	1
	FinBERT	−1
Example 3	Loughran McDonald	1
	Alexandria	−1
	FinBERT	1

9.4 INVESTIGATION

For comparison, we run a monthly long/short simulation where we are long the top quintile and short the bottom quintile based on our sentiment classifications.

The earnings calls are separated into two sections, the MD and the QA. Each component of MD and QA is then totaled for each transcript to get a net sentiment of positive, neutral or negative $(1, 0, -1)$. We then use the average of (MD + QA) as our sentiment score. Our final net sentiment score is the log of positive over negative sentiment counts.

$$NetSentiment = log_{10} \frac{CountPositive + 1}{CountNegative + 1} \tag{9.1}$$

To avoid concerns of liquidity, we use the S&P 500 as our universe and split the data into quintiles on a monthly basis using the net sentiment scores, with Q1 being the highest sentiment in the sample, and Q5 being the lowest. We use an aggregation window of six months to ensure each security in our universe has sample data over two unique earnings calls (Earnings calls are required quarterly for United States public companies.). Once we have our quintiles, we rebalance on the last trading day of the month going long quintile 1 and short quintile 5. Quintiles 2-4 (generally the most neutral sentiment) will be ignored.

We run our simulation monthly, from January 2010 to September of 2021, on Loughran McDonald Monthly S&P500 Universe 2010-2021.

In the sample period, Alexandria outperforms LM and FinBERT in all years except 2010 and 2011. Alexandria performs positively in every year apart from 2016, and in that year showed the least loss compared to the other approaches. Over the course of the sample period, Alexandria had the best performance, with cumulative P&L of 221.87%, followed by FinBERT with 16.77%, and Loughran McDonald at 19.81%.

TABLE 9.4 Loughran McDonald Q1–Q5 Annual Returns

Quintile	Avg Stocks	Annual Returns	Standard Deviation	Sharpe Ratio	vs. S&P500
1	108	14.36%	15.05%	0.95	2.18%
2	107	13.46%	16.05%	0.84	1.27%
3	108	12.95%	15.34%	0.84	0.76%
4	107	12.14%	16.85%	0.72	−0.05%
5	107	12.09%	16.92%	0.71	−0.10%
Long/Short	215	1.56%	5.82%	0.27	

TABLE 9.5   Loughran Monthly S&P500 Universe 2010–2021(09)

	Annual	Std Dev	Sharpe	SP 500
2010	6.75%	7.52%	0.9	12.78%
2011	5.74%	6.65%	0.86	0.00%
2012	−5.29%	4.89%	−1.08	13.41%
2013	1.93%	3.12%	0.62	29.60%
2014	1.02%	4.49%	0.23	11.39%
2015	9.30%	6.62%	1.41	−0.73%
2016	−6.53%	4.99%	−1.31	9.54%
2017	3.48%	4.56%	0.76	19.42%
2018	−0.32%	4.98%	−0.06	−6.24%
2019	2.35%	5.24%	0.45	30.30%
2020	4.57%	8.97%	0.51	8.34%
2021	−4.64%	6.71%	−0.69	14.68%

TABLE 9.6   Alexandria Q1–Q5 Annual Returns

Quintile	Avg Stocks	Annual Returns	Standard Deviation	Sharpe Ratio	vs. S&P500
1	108	19.07%	15.06%	1.27	6.88%
2	107	13.41%	15.42%	0.87	1.22%
3	108	13.33%	16.07%	0.83	1.14%
4	107	12.32%	16.73%	0.74	0.13%
5	107	7.03%	17.45%	0.4	−5.16%
Long/Short	213	10.54%	7.83%	1.35	

TABLE 9.7   Alexandria Monthly S&P500 Universe 2010–2021(09)

	Annual	Std Dev	Sharpe	SP 500
2010	3.60%	5.68%	0.63	12.78%
2011	5.47%	7.95%	0.69	0.00%
2012	11.26%	6.69%	1.68	13.41%
2013	8.69%	3.08%	2.82	29.60%
2014	16.15%	8.38%	1.93	11.39%
2015	28.81%	10.61%	2.72	−0.73%
2016	−4.04%	9.22%	−0.44	9.54%
2017	22.88%	8.22%	2.78	19.42%
2018	14.24%	8.60%	1.66	−6.24%
2019	4.03%	5.75%	0.7	30.30%
2020	10.69%	8.87%	1.2	8.34%
2021	6.66%	7.64%	0.87	14.68%

TABLE 9.8   FinBERT Q1–Q5 Annual Returns

Quintile	Avg Stocks	Annual Returns	Standard Deviation	Sharpe Ratio	vs. S&P500
1	108	14.59%	14.09%	1.04	1.66%
2	107	13.66%	15.66%	0.87	0.73%
3	108	12.62%	16.14%	0.78	−0.31%
4	107	12.05%	16.69%	0.72	−0.88%
5	107	11.97%	18.11%	0.66	−0.96%
Long/Short	215	1.34%	8.41%	0.16	

TABLE 9.9   FinBERT Monthly S&P500 Universe 2010–2021(09)

	Annual	Std Dev	Sharpe	SP 500
2010	−1.94%	9.05%	−0.21	12.78%
2011	4.43%	4.80%	0.92	0.00%
2012	0.01%	5.99%	0	13.41%
2013	3.52%	3.80%	0.93	29.60%
2014	1.51%	5.32%	0.28	11.39%
2015	20.37%	7.42%	2.75	−0.73%
2016	−13.82%	7.84%	−1.76	9.54%
2017	9.88%	4.14%	2.39	19.42%
2018	4.17%	6.60%	0.63	−6.24%
2019	−4.38%	9.86%	−0.44	30.30%
2020	3.78%	16.41%	0.23	8.34%
2021	−10.44%	11.83%	−0.88	14.68%

9.5   SUMMARY RESULTS

Annual Returns

TABLE 9.10   Annual Returns

	LM	**Alexandria**	**FinBERT**
2010	**6.75%**	3.60%	−1.94%
2011	**5.74%**	5.47%	4.43%
2012	−5.29%	**11.26%**	0.01%
2013	1.93%	**8.69%**	3.52%
2014	1.02%	**16.15%**	1.51%
2015	9.30%	**28.81%**	20.37%
2016	−6.53%	**−4.04%**	−13.82%
2017	3.48%	**22.88%**	9.88%
2018	−0.32%	**14.24%**	4.17%
2019	2.35%	**4.03%**	−4.38%
2020	4.57%	**10.69%**	3.78%
2021	−4.64%	**6.66%**	−10.44%

Cumulative Returns

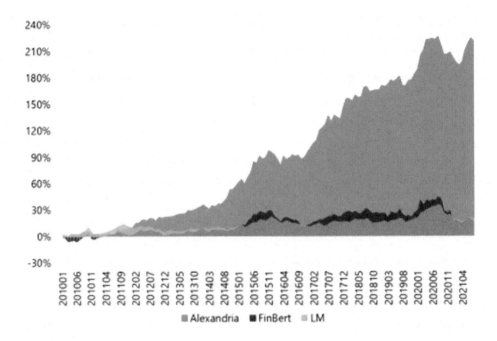

Figure 9.1   Cumulative Returns

9.6   DISCUSSION AND CONCLUSION

As our analysis shows, not all NLP methods are the same when applied to earnings calls. We see a stark difference between the sentiment labelling between the three methods as shown from the low correlations among them, and subsequently the performance when applied to a trading strategy over the past decade.

The result shows that using sentiment for earnings calls can generate alpha not explained by traditional risk and return factors. Furthermore, when using the three different NLP methods to generate sentiment from earning calls, there is a low correlation of sentiment labeling between Alexandria and both LM and FinBERT at the individual classification level. Higher correlations are found at the aggregate security sentiment level but remain low. As a result, low correlations and large performance differences arise from the distinct language models, sentiment training and NLP technology of each approach. The research also suggests that Alexandria Technology method significantly outperforms both other NLP approaches over the period of 2010–2021. Earnings call MD and QA sentiment have been the best-performing factors against traditional style factors over the past 10 years. See [Short, 2016].

We find Alexandria's NLP method to perform the best during the sample by a significant margin. Alexandria's unique approach of using domain experts to train their model appears to be more robust compared to the rigid technique of a dictionary-based approach, and even to the non-earnings specific deep learning methods as well.

REFERENCES

(2019). Factset research systems 2019. https://investor.factset.com/static-files/83ca372c-8283-419f-a8c7-40c8e0daab51.

Araci, D. F. and Genc, Z. (2019). Financial sentiment analysis with pre-trained language models. *arXiv preprint arXiv:1908.10063.*

Chen, J. (2021). Earnings call. https://www.investopedia.com/terms/e/earnings-call.asp.

Govindaraj, S., Livnat, J., Savor, P. G., and Zhao, C. (2013). Large price changes and subsequent returns. *Journal Of Investment Management.*

Jha, V., Blaine, J., and Montague, W. (2015). Finding value in earnings transcripts data with alphasense. *Extractalpha.com.*

Loughran, T. and McDonald, B. (2011). When is a liability not a liability? Textual analysis, dictionaries, and 10-Ks. *The Journal of Finance,* 66(1):35–65.

Short, J. (2016). Societe generale cross asset research/equity quant. https://silo.tips/download/societe-generale-cross-asset-research.

Tortoriello, R. (2018). Their Sentiments Exactly: Sentiment Signal Diversity Creates Alpha Opportunity. *S&P Global.*

CHAPTER 10

Sensors Data

Alexander Gladilin
OptiRisk Systems

Kieu Thi Hoang
OptiRisk Systems

Gareth Williams
Transolved Ltd

Zryan Sadik
OptiRisk Systems

CONTENTS

10.1 INTRODUCTION

What is sensor data?

Networked sensors embedded in a broad range of devices are among the most rapidly growing data sources, driven by the proliferation of smartphones and the reduction in the cost of satellite technologies. This category of alternative data is typically very unstructured and often significantly larger in volume than data generated by individuals or business processes, and it poses much tougher processing challenges. Sensors, among many other things, capture economic activity through images from satellites or security cameras, or through movement patterns such as cell phone towers. Key alternative data sources in this category include

- Satellite imaging to monitor economic activity, such as construction, shipping, or commodity supply.

- Geolocation data to track traffic in retail stores, such as using volunteered smartphone data, or on transport routes, such as on ships or trucks.

- Cameras positioned at a location of interest.

- Weather and pollution sensors.

The Internet of Things (IoT) will further accelerate the large-scale collection of this type of alternative data by embedding networked microprocessors into personal and commercial electronic devices, such as home appliances, public spaces and industrial production processes. Sensor-based alternative data that contains satellite images, mobile app usage or cellular location tracking is typically available with a 3- to 4-year history.

10.2   SATELLITE DATA

Earth observation is an everyday use for satellites providing data and information about weather changes, environmental monitoring and other planet parameters. Companies can use satellite data to determine the population, location and behavioural patterns of consumers from different parts of the world. Satellite data is also useful in controlling mining, production, construction and shipments. Asset managers, who are interested in longer-term strategic allocation, evaluate economic conditions by studying satellite images, which are available at scale and high frequency. This data proxies economic activity arguably better than GDP and in real time rather than quarterly [Henderson et al., 2012].

The resources and timelines required to launch a geospatial imaging satellite have dropped dramatically; instead of tens of millions of dollars and years of preparation, the cost has fallen to around $100,000 to place a small satellite as a secondary payload into a low Earth orbit. With the vast amount of geospatial imaging data available and even with the cost of launching geospatial satellites continuing to fall it is prudent to examine the use of machine learning techniques with such alternative data sources.

Satellite data can provide information on many scales, from the use of hydrogen spectral analysis to determine the composition of astronomical phenomena to the monitoring of human activity to inform financial decisions. Satellite images can be used to count cars in parking lots, a potential source of insight into sales activity for retailers or output at factories. For instance, data from the satellite imagery which counts the number of vehicles in shopping mall car parks uses it as a measurement for retail sales activity and geospatial

Figure 10.1 The locations from which testing and training patches were extracted from an overhead image of Toronto. Blue areas are training patch areas while red areas are testing patch areas [Mundhenk et al., 2016].

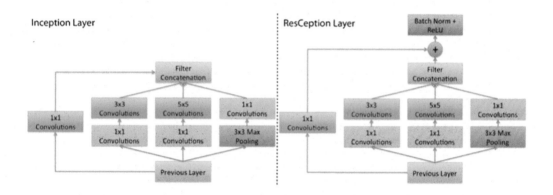

Figure 10.2 A standard Inception layer (left) and A ResCeption layer (right) [Mundhenk et al., 2016].

analysis for identifying the geographical proximity of competitors. [Mundhenk et al., 2016] undertook work to classify cars from satellite images. They began by taking these images and dividing them into grids of size 1024x1024 and assigning testing and training areas.

A new network for classification was devised which they named ResCeption, a combination of Residual Learning with Inception. This network was compared to AlexNet as a baseline as well as GoogLeNet (Inception) on its ability to successfully classify cars from the images.

The training for each network was achieved using Caffe and stochastic gradient descent for 240k iterations with a mini batch size of 64 (Further calibrations can be found in the literature, see [Mundhenk et al., 2016]).

TABLE 10.1 Table from [Mundhenk et al., 2016] shows Inception and ResCeption work noticeably better than AlexNet with ResCeption incrementally surpassing Inception.

Model	Correct
AlexNet	97.62%
Inception	99.12%
ResCeption	99.14%

Industry professionals caution, however, that satellite and other types of aerial surveillance data are best supplemented with other types of data able to provide more detailed estimates of actual foot traffic when it comes to gauging retail sales.

Satellites are also used to track ships, monitor crops and detect activity in ports and oil fields. Collecting socioeconomic data is an important part of economic decision making, be it at governmental level or for small businesses. Governments and large organisations such as the World Bank expend large resources in conducting censuses and surveys to obtain such data [Han et al., 2020]. Use cases include monitoring economic activity that can be captured using aerial coverage, such as agricultural and mineral production and shipments, or the construction of commercial or residential buildings or ships; industrial incidents, such as fires; or car and foot traffic at locations of interest. Related sensor data is contributed by drones that are used in agriculture to monitor crops using infrared light. The use of satellite data to produce probabilistic models of migratory behaviour of animals can be used to measure the impact of pollution or governmental policy decisions by evaluating how these change over time. As well as potentially informing policy decisions, such data from geolocation can also be used to assess land value and its potential for commercial use.

Several challenges often need to be addressed before satellite image data can be reliably used in ML models. In addition to substantial pre-processing, these include accounting for weather conditions such as cloud cover and seasonal effects around holidays. Satellites may also offer only irregular coverage of specific locations that could affect the quality of the predictive signals.

10.3 GEOLOCATION DATA

Geolocation data is another rapidly growing category of alternative data generated by sensors. The recent developments in geolocation technology have meant an increase in the amount of available data since such data no longer requires the use of satellite transmitters or larger global positioning system (GPS) loggers [Merkel et al., 2016]. Geolocation data can be used to determine the physical address of an electronic device. A familiar source is smartphones, with which individuals voluntarily share their geographic location through an application, or from wireless signals such as GPS, CDMA or Wi-Fi that measure foot traffic around places of interest, such as stores, restaurants or event venues. It involves the use of location technologies such as IP address tracking and GPS triangulation. Companies can use geolocation data in sales and marketing, making region-specific offers. Geolocation data can also be used for e-finance, for instance, to determine if unusual or fraudulent requests are coming from a specific location [Curran and Orr, 2011] by using tools such as Google Gears. Social media can also be used to obtain geolocation data, for instance, a user on Twitter can investigate what people are discussing regarding a specific event or business from a particular area. This information could be used by investors or business owners to monitor current affairs and brand perception. A study by EY found that now approximately 28% of hedge fund managers use geolocation data when developing investment strategies. The data on consumer behaviour and the modelling of such data is one such application that is used when making forecasts for investments.

Furthermore, an increasing number of airports, shopping malls and retail stores have installed sensors that track the number and movements of customers. While the original motivation to deploy these sensors was often to measure the impact of marketing activity, the resulting data can also be used to estimate foot traffic or sales. Sensors to capture geolocation data include 3D stereo video and thermal imaging, which lowers privacy concerns but works well with moving objects. There are also sensors attached to ceilings, as well as pressure-sensitive mats. Laser sensors attached to the ceiling, when integrated over time are able to generate a 3D surface, $s(x,y,z)$ of an area with the advantage that such sensors' functionality remains almost unchanged in different lighting conditions [Hernandez-Sosa et al., 2011]. A person crossing the sensor results in a peak of the surface ($\nabla = 0$). Peaks are filtered to determine relevance so that those representing people are kept. The work of [Hernandez-Sosa et al., 2011] demonstrated such technology resulting in 90% of peaks were correctly identified. Some providers use multiple sensors in combination, including vision, audio and cell phone location, for a comprehensive account of the shopper journey, which includes not only the count and duration of visits, but extends to the conversion and measurement of repeat visits.

10.4 WEATHER DATA

Weather data is associated with numbers and facts about the nature of the atmosphere, which may include wind speed, temperature, rain and snow.

Companies can use it to predict their customers' needs. For instance, when meteorologists forecast a hurricane, demand for products such as bottled water, toiletries and bread always increases [Jain and Jain, 2017]. Weather data is collected from sensors. Major weather data is available on meteorological platforms, which publish the forecasts in structured forms.

Weather refers to the atmospheric conditions observed over a short period, while climate reflects the long-term behaviour of the atmosphere. Weather outcomes have been monitored and recorded since around 1850 through a global network of weather stations and satellites. Access to historical weather and climate data, as well as station history information, is available through Climate Data Online (CDO), which offers quality-controlled measurements of temperature, precipitation, wind, degree days, radar data, and 30-year Climate Normals on a daily, monthly, seasonal, and yearly basis [NOAA, 2023].

However, it's important to note that complete weather data records are not always available because certain countries treat their data as proprietary and impose high access fees. This limitation affects data availability, particularly in countries like India. Additionally, the coverage of weather stations varies worldwide, with higher spatial density and longer time series observed in countries with historically higher incomes, such as the U.S. and the EU 15.

There are several business sectors that need weather data to forecast short-term events, reduce immediate damages and look at the long term to see if the current business model is still reasonable. For example, airline companies and freight-forward companies might follow wind conditions and wave conditions to optimize their route, save costs and reduce environmental damage from emissions. Energy supply companies might need data on droughts or heatwaves. If they have information about heatwaves or cold weather, they can assess potential energy demand in the upcoming seasons, from which they adjust their operation and supply. If there is a drought, nuclear power plants might face the risk of insufficient cooling weather, which lead to an emergency shutdown of their operation. Event-based companies like sports, culture, music and so on need to update forecasts of thunderstorms or freak weather conditions, which clearly affect their business activities and incur costs. Finally, insurance companies, agricultural companies and investment or stock broking companies also need information about considerable storms, droughts and floods. Insurance companies might face higher costs for removing damages and adjusting their product prices, agricultural firms can get heavily damaged crops and be charged higher insurance expenses and investment and stock broking companies might recalculate their product prices like weather derivatives or launch new products. Fund managers, investors, traders investing in or trading on these companies can also use these weather data to forecast the companies' operating results and make decisions.

There is a long history of using weather measures as explanatory variables in statistical models. For example, [Fisher, 1925] examined the effects of rainfall on wheat yields and [Virtue, 1929] used weather as an instrumental variable to identify a demand function for oils. [Daglis et al., 2020] also shows that solar weather, quantified by means of Velocity Solar Wind (Vsw) and geomagnetic index (Dst) have forecasting ability over the volatility of NASDAQ's Finance Sector Price Index. Because weather is exogenous and random in most economic applications, it acts like a natural experiment and thus in some settings allows researchers to statistically identify the causal effect of one variable on an economic outcome of interest [Angrist and Krueger, 2001].

In financial models, weather data can be used to calculate risk indicators and define thresholds like average, standard deviation and minimum/maximum values. Analysts calculate the average of weather data like monthly sunshine duration hours, precipitation, soil temperature, wind force and so on to create different indicators in their models. These indicators are then fed into statistical models or machine learning models to calculate expected damages and additional costs, which will be useful in financial evaluation models.

Figure 10.3 shows the process undertaken to take data (including that of the weather) and after cleaning produce a model suitable for making forecasts for crop yield [Schwalbert et al., 2020].

10.5   OTHER SENSORS

Jet tracking offers another source of alternative data, such data can be used to infer meetings between businesses or individuals which could be indicators of possible mergers, new deals being negotiated or acquisitions of commodities. Although in the United States the Federal Aviation Administration (FAA) permits the exclusion of private aeroplanes identities from public display, such information can be extrapolated from the transponder code broadcast from the Automatic Dependent Surveillance–Broadcast (ADS-B) [FlyXO, 2022] allowing the one to track specific flights. Websites such as https://globe.adsbexchange.com/ record such data and present it either in the form of a live view or via an API. Although jet tracking may not be the only method to gain such insights in the future of businesses, the additional content provided can be used to improve modelling performed by quantitative analysts.

10.6   SENSORS DATA PROVIDERS

Raw sensor data is significantly more unstructured and considerably larger than other kinds of alternative data. Sensor data providers use raw data to derive business relevant insights from the satellite images, geolocation, climate

Model pipeline

Model development

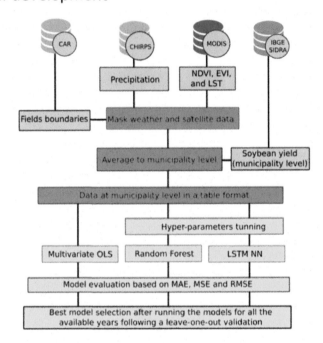

Figure 10.3 Flowchart indicating steps taken by [Schwalbert et al., 2020] for model development using weather data.

and cameras. They help companies solve related business issues across industries. They also help financial participants, including fund managers, investors and traders make better decisions when investing or trading these companies. In Table 1, we list some major providers of satellite, weather and geolocation data. We present the companies name with their official webpage, the business sectors they focus on and salient features of their products.

10.7 DISCUSSION

In conclusion, sensor data is an integral component of alternative data used by business firms to gain a competitive advantage over their business rivals and by finance investors to get an edge in investment results. It is the data gathered using technological devices, including sensors, CCTVs and satellites. Sensor data is collected from the environment, structured and analysed through a series of computerized series.

In this chapter, we have discussed several kinds of sensor data including satellite data, geolocation data, weather data, foot and car traffic data, and others. Satellite data is the information collected by an artificial hovering image tool revolving around the earth. Businesses can use this data to determine

TABLE 10.2 Major sensor data providers

No.	Company Name	Asset Class	Notes
1.	Orbital Insights www.orbitalinsight.com	Agriculture, Oil, Equity	Orbital Insight is a California-based geospatial analytics company. The company uses AI, computer vision and data science to ingest, analyse and transform billions of geospatial data points (including satellite, cell phone "pings", AIS, SAR, aerial, connected devices and so on) into actionable intelligence used to make superior business and policy decisions. Their SaaS technology unlocks global insights reveals hidden infrastructure, supply chain, socio-economic and geopolitical trends.
2.	Windward	Freight	The Windward Mind is a platform that aggregates, analyses and vets all maritime data. It takes the data and creates a unique Vessel Story on each ship worldwide, a complete history of each ship's activities over time. With over 90% of the world's trade transported over the oceans, data on ship activity is critical to decision makers across industries.

(Continued on next page)

TABLE 10.2 (Continued)

S. No.	Company Name	Asset Class	Notes
3.	Terra Bella (Google Skybox) https://www.planet.com/terrabella	Real Estate, Equity, Commodity	Terra Bella provides commercial high-resolution Earth observation satellite imagery, high-definition video and analytics services. Planet has signed an agreement to acquire the Terra Bella business from Google.
S. No.	Company Name	Asset Class	Notes
4.	Placemeter www.placemeter.com	Equity, Real Estate	Placemeter uses advanced computer vision technology to lift data points from video streams and is robust and built to scale. First, the system handles a large amount of video streams. Second, it uses machine learning algorithms to process video and classify objects in a wide range of new contexts. They keep track of retail data like: Store door counts, Pedestrian Traffic in front of store, Street to purchase conversion rate, Impact of black Friday, other sales holidays and seasons.

(Continued on next page)

TABLE 10.2 (Continued)

S. No.	Company Name	Asset Class	Notes
5.	URSA www.ursaspace.com	Oil	URSA provides valuable information from satellite imagery acquired twice a day, anywhere in the world. Ursa's Oil Storage data feed uses radar satellite imagery to measure storage level changes of major oil locations around the world. They focus on critical locations that have little or no data available like China.
6.	Spire www.spire.com	Agriculture, Commodity	Spire uses satellite imagery to provide maritime, weather and aviation data that covers 90% of global trade. They utilize a network of tens or hundreds of CubeSats with radio occultation payloads, which significantly increase the number of opportunities to receive radio occultation and result in better forecasts. The Satellite AIS they use also offers better coverage of remote places than other Terrestrial AIS (more effective in densely packed ports) providers.

(Continued on next page)

TABLE 10.2 (Continued)

S. No.	Company Name	Asset Class	Notes
7.	Understory www.understoryweather.com	Equity, Agriculture, Natural Gas, Commodity	Understory is a weather infrastructure and analytics company that detects rain, hail, wind and other weather events directly at the earth's surface. Understory builds and operates weather networks that are comprised of patented RTi weather stations, which are full-stack weather solutions. The stations measure 50,000 times a second and power a cloud-based artificial intelligence core that stitches the measurements together to provide an understanding of weather events. The company collates observations across distances ranging from kilometres to 100 m resolution.
8.	Black Sky www.blacksky.com	All Asset Classes	Black Sky provides color satellite imagery at a resolution of one meter (1 square meter = 1 image pixel) which makes monitoring economic activities such as observing ships in ports, earthquake damage or herd migration much easier. They combine satellite imagery, social media, news and other data feeds to create timely and relevant insights.

(Continued on next page)

TABLE 10.2 (Continued)

S. No.	Company Name	Asset Class	Notes
9.	Descartes Labs www.descarteslabs.com	Commodity, Agriculture	Descartes labs have full imagery archives (some including data only a few hours old) from hundreds of satellites. The Descartes Platform is built to ingest several kinds of data, including satellite, weather data, commodity price histories, web crawls and sentiment analysis from social media networks. Currently, the Descartes Platform ingests 5 terabytes (TB) of near real-time data per day, roughly equivalent to 5,000 hours of standard video. Their current corpus is over 3 petabytes of data (3,000 TB) with the ability to grow larger. With sensor data growing exponentially, the Descartes Platform is designed to respond elastically to this data explosion and harness it for real time forecasting. Descartes Labs provides forecasts for different commodities (including cotton, rice, soy and wheat) across different growing regions (including US, Brazil, Argentina, Russia and Ukraine).

the population of their target consumers' market and consumer behaviors. Geological data, on the other hand, is mainly applied by the businesses seeking to determine the physical address of the target location. Weather data informs the companies about the weather pattern of the target locations. The data can be applied in determining whether the weather pattern in the target market location is favorable for the business. Finally, foot and car traffic data involve the information of the people and vehicles passing a target location. Business firms can use this data type to determine consumers' population in specific areas like chain stores and restaurants.

At the end of this chapter, we also present a list of different sensor data providers for readers to investigate relevant data sources for their applications.

REFERENCES

Angrist, J. D. and Krueger, A. B. (2001). Instrumental variables and the search for identification: from supply and demand to natural experiments. *Journal of Economic Perspectives*, 15(4):69–85.

Curran, K. and Orr, J. (2011). Integrating geolocation into electronic finance applications for additional security. *International Journal of Electronic Finance*, 5(3):272–285.

Daglis, T., Konstantakis, K. N., Michaelides, P. G., and Papadakis, T. E. (2020). The forecasting ability of solar and space weather data on NASDAQ's finance sector price index volatility. *Research in International Business and Finance*, 52:101–147.

Fisher, R. A. (1925). III. The influence of rainfall on the yield of wheat at Rothamsted. *Philosophical Transactions of the Royal Society of London. Series B, Containing Papers of a Biological Character*, 213(402-410):89–142.

FlyXO (2022). Tracking private jets for investment tips! https://flyxo.com/blog/tracking-private-jets-investment-tips/.

Han, S., Ahn, D., Park, S., Yang, J., Lee, S., Kim, J., Yang, H., Park, S., and Cha, M. (2020). Learning to score economic development from satellite imagery. In *Proceedings of the 26th ACM SIGKDD International Conference on Knowledge Discovery & Data Mining*, pages 2970–2979.

Henderson, J. V., Storeygard, A., and Weil, D. N. (2012). Measuring economic growth from outer space. *American Economic Review*, 102(2):994–1028.

Hernandez-Sosa, D., Castrillón-Santana, M., and Lorenzo-Navarro, J. (2011). Multi-sensor people counting. In *Iberian Conference on Pattern Recognition and Image Analysis*, pages 321–328. Springer.

Jain, H. and Jain, R. (2017). Big data in weather forecasting: Applications and challenges. In *2017 International Conference on Big Data Analytics and Computational Intelligence (ICBDAC)*, pages 138–142. IEEE.

Merkel, B., Phillips, R. A., Descamps, S., Yoccoz, N. G., Moe, B., and Strom, H. (2016). A probabilistic algorithm to process geolocation data. *Movement Ecology*, 4(1):1–11.

Mundhenk, T. N., Konjevod, G., Sakla, W. A., and Boakye, K. (2016). A large contextual dataset for classification, detection and counting of cars with deep learning. In *European Conference on Computer Vision*, pages 785–800. Springer.

National Centers for Environmental Information - National Oceanic and Atmospheric Administration. https://www.ncdc.noaa.gov/cdo-web/ Accessed: 2023-05-27.

Schwalbert, R. A., Amado, T., Corassa, G., Pott, L. P., Prasad, P. V., and Ciampitti, I. A. (2020). Satellite-based soybean yield forecast: integrating machine learning and weather data for improving crop yield prediction in southern brazil. *Agricultural and Forest Meteorology*, 284:107886.

Virtue, G. (1929). The tariff on animal and vegetable oils. JSTOR.

(IV)

ALTERNATIVE DATA USE CASES
IN FINANCE

(IV.A)

Application in Trading and Fund Management
(Finding New Alpha)

CHAPTER 11

Media Sentiment Momentum

Anthony Luciani

MarketPsych Data

Changjie Liu

MarketPsych Data

Richard Peterson

MarketPsych Data

CONTENTS

11.1   INTRODUCTION

> *"As there are so many people who cannot wait to follow the prevailing trend of opinion, I am not surprised that a small group becomes an army. [Most people] think only of doing what the others do and of following their examples...."*
> — Josef de la Vega, 1688, "Confusion de Confusiones" [de la Vega, 1688]

Published in 1688, Confusion de Confusiones is the earliest book written on financial markets. Its author, Josef de la Vega, noted that the transmission and spread of opinion drove price trends in the Amsterdam Exchange. De la Vega's opening quote is the subject of this paper, and its content embodies two

DOI: 10.1201/9781003293644-11

converging lines of academic inquiry: stock price underreaction to information [Barberis et al., 1998] and social information transmission [Hirshleifer, 2020].

In the academic literature on share price patterns, two are broadly described in response to the release of new information: overreaction and underreaction. Both patterns are defined in hindsight. A price movement is termed overreaction if it is more likely than random chance to reverse or mean-revert. Underreaction describes a persistent response (a trend-like component) to information over time. These terms refer to both the security price patterns and investors' behavioural responses to information (share buying and selling) that are hypothesized to drive those patterns.

While it is clear that stock prices rapidly incorporate new information (breaking news), sometimes stock prices respond to republished news, stale information, or information obtained through social networks. Recent research on social transmission of investment ideas is delineating the specific mechanisms, circumstances and characteristics of such transmission.

Using a proprietary commercial dataset of media sentiments for 100,000+ global equities point-in-time from 1998 to 2021, this paper identifies the magnitudes, durations and regional variations of stock price responses to an aggregate of new and stale information derived from aggregated financial news media and investment social media. While the impact of specific news events on stock prices has been studied more precisely, this paper explores the connection between the sentiments derived from aggregations of news over weeks to months in size and subsequent weekly to annual stock returns, an effect called "sentiment momentum".

The rest of this chapter is organized as follows. In Section 11.2 we introduce the Linking of Price Momentum to Underreaction and Social Transmission and in Section 11.3, we describe the Data that we use in our investigation. In Sections 11.4 and 11.5, readers are introduced to the experimental methodology and results of this chapter's analysis of sentiment momentum in global equity markets. We conclude the chapter with a discussion of our investigation; our conclusions are presented in Section 11.6 in a summary form.

11.2 LINKING PRICE MOMENTUM, UNDERREACTION AND SOCIAL TRANSMISSION

Price momentum is the tendency for past stock returns to be indicative of future returns. First described in the academic literature by [Jegadeesh and Titman, 1993], this effect has remained broadly observable since its discovery [Griffin et al., 2005]; [Ehsani and Linnainmaa, 2022], although its magnitude is eroding as daily stock price return autocorrelation has been decreasing since the 1970s [Lo, 2017]. Journalists interfacing with investment industry participants report that the momentum investing space is crowded and the alpha less accessible in recent years [Lee, 2020]. There are two behavioural explanations for price momentum: profit-maximizing investors chasing returns which in turn cause a multiplicative effect [Grinblatt et al., 1995], and investor underreaction (delayed reactions) to new and stale news [Daniel et al., 1998]; [Hong and Stein, 1999]. One of the earliest defined examples of price momentum occurs around corporate earnings announcements. Following corporate earnings surprises, share prices drift in the direction of the surprise for several months [Bernard and Thomas, 1989]. Stocks outperform in the three quarters after extreme positive earnings surprises and underperform in the three quarters following extreme negative earnings surprises [Hirshleifer et al., 2008]. This phenomena, named post-earnings announcement drift (PEAD), has been found globally and has remained consistent for decades [Mendenhall, 2004]; [Chen et al., 2017], researchers have postulated that investors' delay in updating their expectations using relevant information from corporate events, termed investor underreaction, is one of the reasons for the existence of this price pattern. Moving from the synchronous format of earnings announcements to the asynchronous nature of news publishing, researchers found that sentiments quantified in news articles via positive and negative word counts from Wall Street Journal and Dow Jones content were correlated with future returns several days into the future [Tetlock et al., 2008]. The authors postulated that the delayed response following positive and negative news is due to investor underreaction. Similarly, [Li, 2006] quantified "risk sentiment" using negative words and found that high risk sentiment in firms' annual SEC filings predicted both negative future earnings and returns with a yearly forecast window. [Tetlock, 2011] identified that republished "stale" news leads to price momentum in stocks.

Beyond the news content, context matters. Factors such as investor attention and the primary audience of the news (retail investors vs professional analysts) also play a role in price reaction patterns. Attention around earnings announcements predicts the response of stock prices to earnings surprises, with greater underreaction among stocks with less investor attention [Hirshleifer, 2020]. The underreaction effect from firm-specific news stories is stronger for news that focuses on fundamentals [Tetlock et al., 2008]. Media reputation and visibility also affects how investors react to surprises, with celebrity firms experiencing stronger returns on surprises than other companies [Pfarrer et al., 2010]. Furthermore, the medium matters – sentiment expressed in social media has a different impact than that expressed in news media [Beckers, 2018].

Occasionally a firm's share price will experience both under- and overreaction simultaneously, underreacting to routine news while overreacting to dramatic unexpected news. Among pharmaceutical companies, while investors underreact to routine quarterly earnings news, they overreact to dramatic events such as pharmaceutical product announcements [Fischer, 2012].

Sometimes underreaction can be followed by overreaction. When merger news is announced, a price drift characteristic of underreaction occurs, but if the merger deal remains without follow-on media updates after 30 weeks, an over-reaction pattern (price reversal) subsequently dominates [Giglio and Shue, 2014].

While there are high correlations between sentiment and contemporaneous returns, as news often reports on the recent price changes following key news announcements, the previously cited studies suggest there may be a predictive value in the media's reporting on price action itself.

With investment communities increasingly relying on social media platforms to express opinions and establish consensus, there are several academic findings of interest related to investment social networks. Investment ideas have been found to more readily spread through social connections online when the sources of information hold more credibility, with factors like age, education and income all correlating to faster information transmission [Rantala, 2019]. Social sharing is not only limited to retail traders; one-third of institutional investors report gaining investment ideas from social media according to a 2015 survey [Connell, 2015]. Interestingly, the traders most prolifically disseminating investment ideas on social platforms are more likely to make mistakes – the greater the attention paid them, the more susceptible they are to common behavioural biases [Glaser and Risius, 2018].

Investor behaviour and performance are also altered by the medium across which investment ideas are transmitted. Analysis of investor behaviour at the online brokerage Robinhood – with graphical user interfaces (GUIs) designed to facilitate social transmission of investment ideas – demonstrates that account holders on average outperform professional investors [Welch, 2022]. However, when Robinhood investors jump into specific sentiment stocks at the same time following significant positive price movement ("herding events"), those investors tend to underperform [Barber et al., 2021]. The concept of social information contagion is described in detail in the American Finance Association Presidential Address of [Hirshleifer, 2020], reflecting the prominence of such concepts in modern academic thought.

Most pertinent to this study, business journalists drive both social transmission and price underreaction by correlating the tone of their reports with recent market price movements, thus reiterating the tone of price-impacting narratives [Dougal et al., 2012]. The repetition of narratives may be a key driver of price underreaction. Importantly, as identified by [Tetlock et al., 2008], stale news that captures attention can itself be predictive. Based on prior research [Cahan, 2018a], [Cahan, 2018b], recent news sentiment can be predictive of future stock prices globally.

11.3   DATA

In partnership with Refinitiv, MarketPsych Data produces a granular financial sentiment dataset called the Refinitiv MarketPsych Analytics (RMA). The data is based on a combination of dictionary and machine-learning approaches to natural language processing tasks. The accuracy of the Natural Language Processing engine is calibrated using human raters, not trained with market price data. For equities, sentiments are derived from significant events such as earnings reports, analyst ratings changes, drugs trial outcomes, as well as emotionally charged statements, themes and opinions. The sentiment dictionaries are built upon a modular framework of customized tools that employ domain-specific language sets to allow the balance between positive and negative events transpiring in the media to shape the final quantification of sentiment around those topics. More information on the RMA is available in [Peterson, 2013].

The RMA scores used in this paper consist of streaming sentiments and themes for 100,000+ global companies. The RMA data represent aggregate scores from 2 million financial articles daily downloaded in real-time from tens of thousands of news feeds, social media hubs, blogs and chat rooms. Non-English articles in 13 languages are translated into English and then analysed for sentiment weightings. MarketPsych reports the dataset has clients in 25+ countries and 90+ academic papers have been written using this dataset.

In particular, the following formula summarizes the daily calculation for the sentiment index for asset a as:

$$Sentiment_{t,a} = \frac{Positive\ References_{t,a} - Negative\ References_{t,a}}{Buzz_{t,a}} \quad (11.1)$$

where Positive (Negative) References are the sum of all positive (negative) terms and phrases about asset a captured in traditional news and social media across time period t. For example, while gloom, fear and anger are negative subsets of the sentiment index for a, joy and optimism are the positive subsets of the sentiment index for a. $Buzz_{t,a}$ captures the total number of references about the asset a across all topics, events and themes. Higher buzz implies higher discussion.

Preliminary industry research at Empirical Research Partners, using a mix of news, social media and earnings-specific sentiment from the RMA data, found sentiment momentum in prices across both U.S. and global stocks [Cahan, 2018a], [Cahan, 2018b] after controlling for traditional factors including value and momentum.

The primary group of equities tested in this research are Russell 3000 index constituents. The sample period was selected to begin in January 1, 2006, which is the first full year with Russell 3000 index constituents available to MarketPsych researchers. The sample period concludes at the end of 2020. Stock price data for Russell 3000 index constituents is retrieved from DataStream via Refinitiv's QAD platform and is adjusted to include dividends and capital change events.

As news is often published as reactive commentary to recent price changes, a high contemporaneous correlation between company sentiments and stock prices is well established in existing research and is not the subject of this paper.

11.4 METHODOLOGY

Price- and Sentiment-Momentum anomalies are established across large windows from weeks to quarters in the existing research, this forms the basis for the window size of the variable used to represent the prevailing investor sentiments for the Russell 3000 constituents. A feature based on news sentiment is first constructed over a 30-day window. We observe how quickly (or slowly) the information contained within the media from that month becomes factored into stock prices over the following weeks and months. From the available sentiment metrics in the data, the general and overarching "Sentiment" score is utilized along with the "Buzz" score. From these two dimensions, a feature X_t is constructed for each asset a, which allows for investor attention (the 'Buzz' variable below) to weight the importance of daily sentiments (the 'Sentiment' variable) across a 30-day window. This variable represents attention-weighted average sentiment over that 30-day period. Using the formula below, in the generation of X_t a daily sentiment score based on 5000 media mentions is weighted 5x more than a daily score derived from only 1000 mentions.

$$ X_{a,t} = \frac{\sum_{i=t-30}^{t}(buzz_{a,i} * sentiment_{a,i})}{\sum_{i=t-30}^{t} buzz_{a,i}} \tag{11.2} $$

This value is determined for each equity, for each Sunday of the 15-year sample period, using an equally weighted look-back window of 30 days. The Russell 3000 constituents will often be ranked and then grouped into quantiles based on this variable.

The time windows over which stock returns are scrutinized range from one week to one year long, stock price changes are observed over the close-to-close starting from the Monday following the Sunday, for the subsequent 7, 30, 90 and 360 days (interpolating across weekends). Across each returns observation window, the cross-sectional mean of returns from all equities is deducted so that the excess returns for each equity (and quantile) relative to the sentiment basket can be viewed more easily.

In order to graphically view the subsequent changes in returns of baskets of assets ranked by the sentiment variable, assets are grouped into equally sized

TABLE 11.1 Spearman correlation values between sentiment ranking and returns

Returns Window Sentiment Window	7-day	30-day	90-day	360-day
7-day	0.005**	0.002**	0.001	−0.001
30-day	0.006**	0.003**	0.004**	0.001
90-day	0.006**	0.007**	0.006**	0.003**

quantiles according to their sentiment ranking X_t, deciles are primarily used in this report in order to visualize both the degree of correlation and nonlinearity of effect. Decile 1 (the lowest 10%) is the group of equities portrayed most negatively in the news media over the past 30 days, while decile 10 is the most positive group. This approach mirrors the style used by [Tetlock et al., 2008], a canonical paper in this field.

A comparison of sentiment against the performance of traditional fundamental factors, a check for consistency across the sample period and an observation of the signal decay across the weeks since the initial day of ranking is performed subsequently.

11.5 RESULTS

A significant and positive correlation appears between 7-day, 30-day and 90-day company sentiment averages and stock returns over the full sample period, with relations appearing stronger for equities that maintain their standing over longer periods of time. Table 11.1 shows the Spearman correlation values between sentiment ranking and returns. Results significant at the 95% confidence level are indicated with a (**).

The 30-day ranking of sentiment appears to lose significance somewhere between the 90 and 360-day returns windows. In the visual plots, this decay of signal appears somewhere in the second quarter following the sentiment ranking.

Figure 11.1 plots the average cumulative returns of each group of sentiment-ranked equities over the subsequent 90 days since their ranking. The correlation between 30-day sentiment ranking and returns is depicted by the connection between the colour gradient and the subsequent returns of groups of stocks relative to each other. The sign of this relationship is also intuitive; positive events and commentary about a company are expected to manifest into positive returns, even though this research highlights a significant delay in investors' ability to incorporate stock-specific information into prices.

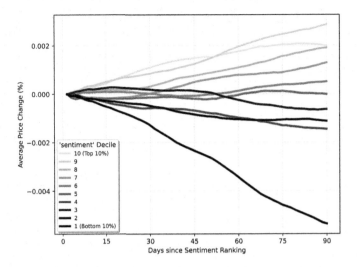

Figure 11.1 90-Day Price Change of Russell 3000 Index Constituents Grouped by their Sentiment Deciles – Sentiment Momentum

Also note the asymmetric price response to sentiment: negative sentiments appear to have a stronger and longer-lasting effect on negative returns than positive sentiments have on positive returns. The relatively larger price drift from negative media is consistent with both the reluctance of social networks and firms to share negative information [Tesser and Rosen, 1975] and [Lim et al., 2020] leading to underreaction, as well as the predisposing bias towards overweighting negative information when it is eventually made public [Rozin and Royzman, 2001]. This underreaction-type price drift can be achieved across varying sizes of look-back and look-forward windows.

11.5.1 Global Equities

Expanding the methodology employed on the Russell 3000 equities to those major indices or broad coverages of the RMA dataset in other regions (APAC, EMEA, Australia's All Ordinaries index and the UK's FTSE-250 index), Figure 11.2 demonstrates correlations between sentiment-ranked groups of equities and subsequent returns across a 90-day window in every region, using the same methodology.

These results suggest that an exercisable relationship exists between financial news-media based stock sentiment and stock returns.

One important step toward verifying the existence of this relationship is to establish how independent it is of traditional factors, especially price momentum given their close conceptual connection.

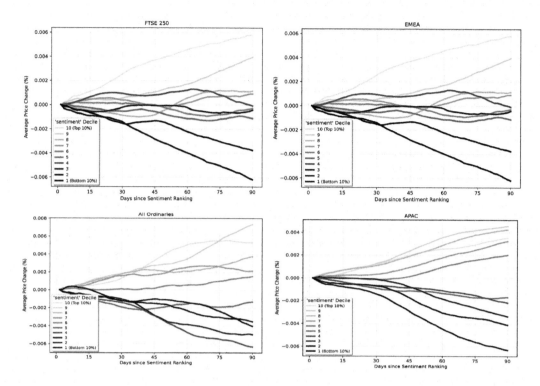

Figure 11.2   The Sentiment Momentum effect is observed across global stock markets

11.5.2   Correlations with Existing Factors

In order to determine whether the above sentiment momentum effect is related to Fama-French factors [Fama and French, 1996], a monthly rebalanced portfolio is formulated in which sentiment is represented as a factor. In construction similar to the Fama-French and momentum [Jegadeesh and Titman, 1993] factors, the portfolio takes from the monthly sentiment rankings the top 30% as the long component and the bottom 30% as the short component, no averaging across Size terciles is performed. The Pearson correlation of returns of these portfolios across the 2006–2020 period is displayed in the table below, all results are significant at the 95% confidence interval.

Observe in Table 11.2 correlations similar in magnitude between the sentiment and the Fama-French factors, and a high degree of correlation (0.73) with momentum, which was expected given the feedback loops between the two variables. To have such a high contemporaneous correlation between Sentiment and Momentum suggests that the financial news media's role as a catalyst for the returns of the Momentum strategy should be re-evaluated.

TABLE 11.2 Factor return correlations with sentiment across the 2006–2020 period

	Market	Size	Value	Momentum	Sentiment
Market	1				
Size	0.36	1			
Value	0.28	0.18	1		
Momentum	−0.36	−0.18	−0.47	1	
Sentiment	−0.39	−0.19	−0.45	0.73	1

TABLE 11.3 p-values derived from Simple Linear Regression of lagged values of feature X against values of feature Y

Lag (Months)	1	2	3	4	5	6
X=Momentum, Y=Sentiment	0.542	0.485	0.568	0.312	0.943	0.120
X=Sentiment, Y=Momentum	0.003	0.054	0.102	0.004	0.456	0.268

11.5.3 Granular Comparsion Against Price Momentum

Attempting to further untangle the link between Sentiment and Momentum, the p-values derived from simple linear regressions of lagged values from each portfolio versus the other over lags of one to six months ahead. This test is intended to ascertain whether sentiment as a factor holds precedence over (Granger-causes) returns from price momentum, or vice versa.

The results of the Granger causality analysis indicate that in the early months (1 through 4, approximately 120 days), the sentiment portfolio holds significant influence over future values of the momentum portfolio, while the momentum portfolio has no significant effect on sentiment.

In a comparison of Sentiment against Momentum for individual equities, a grouping of Russell 3000 equities is performed based on their momentum values using the traditional factor construction, and the prior 30-day sentiment ranking is also used. The subsequent price changes over the coming 90 days are plotted below in figure 11.3, with the original finding from figure 11.1 rescaled. The existing work by [Jegadeesh and Titman, 1993] for components of the Russell 3000 index is validated and a similar response pattern exists in magnitude and degree of correlation for the sentiment ranking. The lowest ranked equities by both factors hold the strongest (decreasing) price response.

By constructing portfolios which combine together price- and sentiment-based momentum, the interactions between these two closely-linked concepts can be further examined.

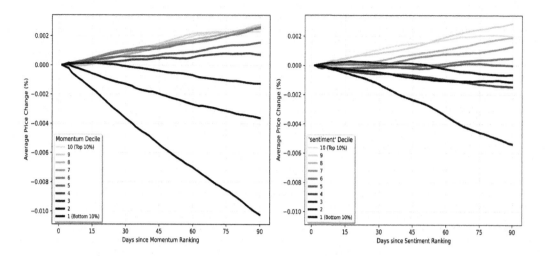

Figure 11.3 Price Momentum (left) versus Sentiment Momentum (right) 90-day price drift graphs

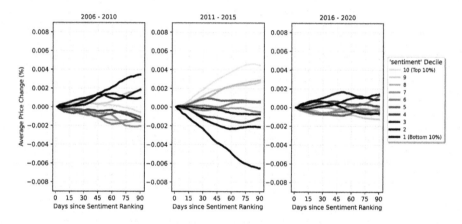

Figure 11.4 The sample period is split into terciles

11.5.4 Time Consistency

Each of the Fama-French factor portfolios do not yield consistent returns over time. Investigating the consistency of the sentiment momentum factor, the dataset is subdivided into 5-year terciles (2006 to 2010, 2011 to 2015 and 2016 to 2020). Using the same sentiment decile ranking methodology, price-drift graphs are constructed in figure 11.4 below.

The 2011–2015 window echoes the relationship observed across the full time period in figure 11.1, but it is not immediately evident that this exists in the other two subsets. However, a common trait of price momentum strategies is their poor performance following a large market drawdown during the recovery period [Daniel, 2011]; the worst stock performers during the drawdown are the most favourable to short based on their recent price history, but they

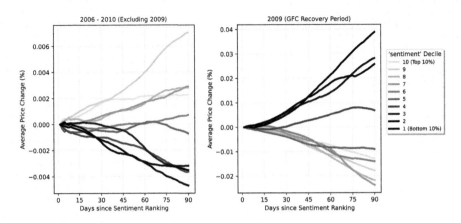

Figure 11.5 Excluding 2009 from the 2006–2010 period removes the strong reversal in momentum performance that occurred during the 2009 market rebound

are heavily undervalued and are the fastest to recover. Given the similarity between characteristics of price- and sentiment-based momentum effects so far, when the recovery periods of the Russell 3000 are manually labelled, it's possible that an inversion of the sentiment-momentum effect can be identified. Isolating the year 2009 – which is primarily comprised of the recovery period following the global financial crisis – from the first tercile, the inversion of the relationship is confirmed. Figure 11.5 below also reveals that excluding 2009 from the first tercile yields the original sentiment momentum effect.

Likewise, in the recoveries following the 2015–2016 global stock market selloff and the March 2020 COVID-19 crash, the Sentiment Momentum effect reversed while the markets recovered. The Price Momentum factor also broke down across these dates. Excluding these periods from the third tercile reveals a less prominent sentiment momentum effect (figure 11.6), which may also be the product of alpha decay.

Early detection of a market recovery is key when moderating the sentiment momentum signal. The timing of such conditions should occur after market declines, as these typically precede recoveries. During the recovery, sentiment momentum (and price momentum) may provide a strong contrarian signal. A key concern with published quantitative findings is alpha decay (loss of factor strength). Commercial media sentiment-based datasets have been marketed since at least 2006. For the 2006 – 2010 period, sentiment momentum appears actionable beyond 90 days as depicted by the steep slopes of the top and bottom two deciles at 90 days, whereas the gradients for the extreme deciles of 2011–2015 appear flatter, potentially indicating a faster incorporation of information into prices (information decay) over a five year period. The third tercile in figure 11.4 also contains smaller market drawdowns and recoveries,

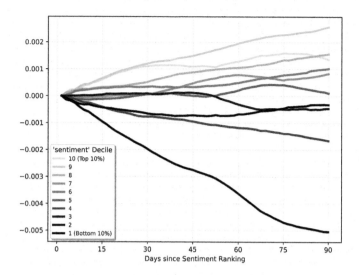

Figure 11.6   Sentiment Momentum in the Third Tercile after Excluding Market Recovery Periods

TABLE 11.4   Excess Weekly Returns of Top-Decile (TD) and Bottom-Decile (BD) equities ranked by sentiment feature X_t

Returns Week	+1	+2	+3	+4	+5	+6	+7	+8	+9	+10	+11
TD, t-Stat	1.19	2.25	1.93	2.97	1.74	1.85	1.76	1.74	0.79	0.83	0.66
TD, p-Value	0.23	0.03	0.05	0	0.08	0.06	0.08	0.08	0.43	0.41	0.51
BD, t-Stat	−2.06	−2.32	−2.03	−1.15	−1.78	−1.22	−0.93	−1.53	−2.05	−1.17	−0.78
BD, p-Value	0.04	0.02	0.04	0.25	0.08	0.22	0.35	0.13	0.04	0.24	0.44

but across this window it's also plausible that more information decay has occurred. Finally, excluding 2009 from the full sample period, the weekly excess returns for decile extremes of Russell 3000 equities (ranked by sentiment) are observed over weeks +1 to +52 after the day of ranking. Two-sided one-sample t-tests are used to examine whether the mean of the distribution of these excess returns deviates from 0. The below table contains these results up to a suitable limit (week +10).

The t-statistics up to week +10 show signs in line with the expected excess returns direction for the given sentiment decile. All are elevated in size compared to the later weeks, with the p-values fluctuating but remaining lower compared to week +10 onwards. This result further validates the existing PEAD and price momentum research and contributes to the notion that investors do underreact to the valuable information contained in news content.

11.6   DISCUSSION

This paper outlines a strong positive connection (termed Sentiment Momentum) between financial news-derived sentiments and subsequent returns for Russell 3000 index components. We suggest that investor "Underreaction" contributes to these trend-like patterns in share prices following periods of excessively positive (or negative) sentiments, with investors eventually pricing assets relative to their perceived value as determined by both stale and novel news events. Broadly speaking, these results reflect the oldest written stock investment advice, De La Vega's suggestion to "follow the prevailing trend of opinion" [de la Vega, 1688].

High correlations between sentiment- and price-momentum-based factor returns across monthly time horizons are established. Sentiment correlates with future values of price momentum while the reverse was not apparent in this framework, indicating that extraneous information, rather than price movement itself, is the primary instigator of price momentum. It is observed that during a market recovery period, the sentiment momentum effect negates itself as the most negatively perceived stocks during the market crash are the fastest to recover, outperforming the rest of the basket in this process. Timing when to ignore (or negate) a momentum strategy will prevent the investor from performing poorly in a recovery period, the removal of such regimes from the 2015 to 2020 tercile reveals that the sentiment momentum effect exists, albeit with a diminished effect size and duration.

In practice, adapting the sentiment momentum effect for a different investment time horizon or to satisfy turnover constraints means tailoring the sentiment feature X_t to best represent a signal suitable for these constraints. If the investment horizon is annual, quarterly windows of data will provide more insight than monthly windows. There are many other ways to transform the underlying data, including exponential windows, absolute or relative changes in sentiment, or the use of more nuanced topical indices such as fear or sentiment around earnings forecasts. More complex feature engineering may yield less frequent but more reliable price response patterns.

The MarketPsych research team, in partnership with StarMine (a LSEG/Refinitiv company) developed a stock relative return predictive model based on the sentiment momentum and other media findings [Liu, 2020]. That model was finalized in August 2019 and, as designed, the top vs bottom decile spread has demonstrated significant alpha since launch. Thus we believe there is solid forward-tested evidence that media-based price patterns, such as the sentiment momentum effect discussed in this paper, are persistent across equity markets.

One important interaction highlighted in this report is that between investor attention and sentiment. Further testing the significance of sentiment for different brackets of investor attention (e.g. high buzz, low buzz) may help narrow down a smaller group for which the sentiment momentum effect is more significant. The multiplicative effect of reactive sentiments, as well as the effects of sentiments associated with novel events, should also be thoroughly investigated, as each of these effects are likely to affect asset behaviour differently.

REFERENCES

Barber, B. M., Huang, X., Odean, T., and Schwarz, C. (2021). Attention induced trading and returns: evidence from RobinHood users. *Available at SSRN:3715077.*

Barberis, N., Shleifer, A., and Vishny, R. (1998). A model of investor sentiment. *Journal of Financial Economics*, 49(3):307–343.

Beckers, S. (2018). Do social media trump news? The relative importance of social media and news based sentiment for market timing. *The Journal of Portfolio Management*, 45(2):58–67.

Bernard, V. L. and Thomas, J. K. (1989). Post-earnings-announcement drift: delayed price response or risk premium? *Journal of Accounting Research*, 27:1–36.

Cahan, R., Bai, Y., and Yang, S. (2018a). Big Data: Harnessing News and Social Media to Improve Our Timing. *Empirical Research Partners.*

Cahan, R., Bai, Y., and Yang, S. (2018b). Big Data: Media Sentiment: Useful Around the World. *Empirical Research Partners.*

Chen, J. Z., Lobo, G. J., and Zhang, J. H. (2017). Accounting quality, liquidity risk, and post-earnings-announcement drift. *Contemporary Accounting Research*, 34(3):1649–1680.

Connell, D. (2015). Institutional investing: how social media informs and shapes the investing process. https://www.greenwich.com/asset-management/institutional-investing-how-social-mediainforms-and-shapes-investing-process.

Daniel, K. D. (2011). Momentum crashes. *Columbia Business School Research Paper. Available at SSRN: 1914673*, pages 11–03.

Daniel, K. D., Hirshleifer, D. A., and Subrahmanyam, A. (1998). A theory of overconfidence, self-attribution, and security market under-and over-reactions. *Journal of Finance*, 53(6):1839–1885.

de la Vega, J. (1688). Confusion de confusiones. *Colchis Books*, 13:289–293.

Dougal, C., Engelberg, J., Garcia, D., and Parsons, C. A. (2012). Journalists and the stock market. *The Review of Financial Studies*, 25(3):639–679.

Ehsani, S. and Linnainmaa, J. T. (2022). Factor momentum and the momentum factor. *The Journal of Finance*, 77(3):1877–1919.

Fama, E. F. and French, K. R. (1996). Multifactor explanations of asset pricing anomalies. *The Journal of Finance*, 51(1):55–84.

Fischer, D. (2012). Investor underreaction to earnings surprises and overreaction to product news in the drug industry. *The Journal of Business and Economic Studies*, 18(2):82.

Giglio, S. and Shue, K. (2014). No news is news: do markets underreact to nothing? *The Review of Financial Studies*, 27(12):3389–3440.

Glaser, F. and Risius, M. (2018). Effects of transparency: analyzing social biases on trader performance in social trading. *Journal of Information Technology*, 33(1):19–30.

Griffin, J. M., Ji, X., and Martin, J. S. (2005). Global momentum strategies. *The Journal of Portfolio Management*, 31(2):23–39.

Grinblatt, M., Titman, S., and Wermers, R. (1995). Momentum investment strategies, portfolio performance, and herding: a study of mutual fund behavior. *The American Economic Review*, 85(5):1088–1105.

Hirshleifer, D. (2020). Presidential address: social transmission bias in economics and finance. *The Journal of Finance*, 75(4):1779–1831.

Hirshleifer, D. A., Myers, J. N., Myers, L. A., and Teoh, S. H. (2008). Do individual investors cause post-earnings announcement drift? Direct evidence from personal trades. *The Accounting Review*, 83(6):1521–1550.

Hong, H. and Stein, J. C. (1999). A unified theory of underreaction, momentum trading, and overreaction in asset markets. *The Journal of Finance*, 54(6):2143–2184.

Jegadeesh, N. and Titman, S. (1993). Returns to buying winners and selling losers: implications for stock market efficiency. *The Journal of Finance*, 48(1):65–91.

Lee, J. (2020). Momentum trade plunges the most on record in rotation frenzy. https://www.bloomberg.com/news/articles/2020-11-09/ momentum-trade-plunges-the-most-on-record-in-rotation-frenzy.

Li, F. (2006). Do stock market investors understand the risk sentiment of corporate annual reports? *Available at SSRN 898181.*

Lim, S., Lane, J., and Uzzi, B. (2020). Biased information transmission in investor social networks: evidence from professional traders. In *Academy of Management Proceedings*, volume 2020. Academy of Management Briarcliff Manor, NY 10510.

Liu, C., L. A. . P. R. (2020). StarMine MarketPsych Media Sentiment Model Product Whitepaper. *Refinitiv.*

Lo, A. W. (2017). Adaptive markets: financial evolution at the speed of thought. In *Adaptive Markets*, pages 281–282. Princeton University Press.

Mendenhall, R. R. (2004). Arbitrage risk and post-earnings-announcement drift. *The Journal of Business*, 77(4):875–894.

Peterson, R. & Fischkin, E. (2013). Thomson Reuters MarketPsych Analytics (TRMA) White Paper. *Thomson Reuters.*

Pfarrer, M. D., Pollock, T. G., and Rindova, V. P. (2010). A tale of two assets: The effects of firm reputation and celebrity on earnings surprises and investors' reactions. *Academy of Management Journal*, 53(5):1131–1152.

Rantala, V. (2019). How do investment ideas spread through social interaction? evidence from a ponzi scheme. *The Journal of Finance*, 74(5):2349–2389.

Rozin, P. and Royzman, E. B. (2001). Negativity bias, negativity dominance, and contagion. *Personality and Social Psychology Review*, 5(4):296–320.

Tesser, A. and Rosen, S. (1975). The reluctance to transmit bad news. In *Advances in Experimental Social Psychology*, volume 8, pages 193–232. Elsevier.

Tetlock, P. C. (2011). All the news that's fit to reprint: Do investors react to stale information? *The Review of Financial Studies*, 24(5):1481–1512.

Tetlock, P. C., Saar-Tsechansky, M., and Macskassy, S. (2008). More than words: quantifying language to measure firms' fundamentals. *The Journal of Finance*, 63(3):1437–1467.

Welch, I. (2022). The wisdom of the robinhood crowd. *The Journal of Finance*, 77(3):1489–1527.

CHAPTER 12

Defining Market States with Media Sentiment

Tiago Quevedo Teodoro

MarketPsych Data

Joshua Clark-Bell

MarketPsych Data

Richard L. Peterson

MarketPsych Data

CONTENTS

12.1 INTRODUCTION

There is evidence to suggest a regime-switching character of equity markets [Schaller and Norden, 1997]. Formally, this implies that returns are not drawn from a single (Gaussian) distribution. For example, if we assume a binary state S and allow for different mean μ and variance σ^2, the returns of a market index during period t can be defined as:

$$r_t = \mu_0(1 - S_t) + \mu_1 S_t [\sigma_0^2(1 - S_t) + \sigma_1^2 S_t]\varepsilon_t \tag{12.1}$$

A regime-switching description appears to match the tendency of financial markets to change their behaviour abruptly and for such new behavior to persist for several subsequent periods [Ang and Timmermann, 2012].

To demonstrate this we use a Gaussian Mixture Model (GMM) [Pedregosa et al., 2011]. A GMM assumes that the data points are generated from a mixture of a given number of Gaussian distributions with unknown parameters.

DOI: 10.1201/9781003293644-12

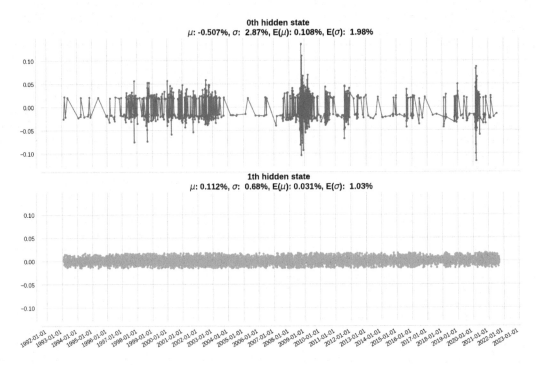

Figure 12.1 A GMM distribution of the SPY log-returns into two states.

A GMM is a parallel to "generalizing k-means clustering to incorporate information about the covariance structure of the data as well as the centers of the latent Gaussians"[Pedregosa et al., 2011]. In this example we use a GMM to define two market regimes from the time-series of historical SPY log-returns (adjusted by distributions) as shown in Figure 12.1.

Note how the 0th state includes the extreme price changes clustered during the three major market drawdowns (dot com bubble, global financial crisis and the COVID-19 pandemic). The leverage effect [Black, 1976; Christie, 1982] is also observed here given that the 0th state has a lower mean and higher volatility than the 1th state. In Figure 12.2 we display the QQ-plots of the overall returns, and the returns of each transformation state. Both states display a kurtosis value closer to zero than the distribution of all returns.

Such observations have considerable implications for the optimal portfolio choice. For example, instead of a traditional fixed 60/40 (equities/bonds) portfolio, one could adjust the allocation according to the properties of each regime.

In this study we attempted to use the framework of switching regimes to construct a simple portfolio that moves from risk-on (an equity index tracker, the SPY ETF) to risk-off (a bond index tracker, the AGG ETF). However, in comparison with a typical application of only price data, we also incorporated information about market sentiment. The purpose was to include a more

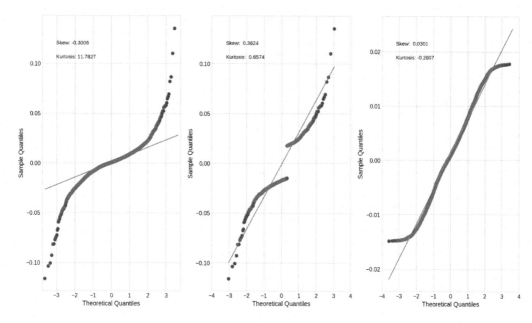

Figure 12.2 QQ-plot of the overall SPY log-returns (left), in state 0 (center) and state 1 (right).

explicit behavioral dimension [Shiller, 2003] to the analysis. Additionally, market sentiment has been shown to be an important explanatory variable of future market returns [Baker and Wurgler, 2006; Jiang et al., 2019]. *Sentiment* in this analysis is represented by the Refinitiv MarketPsych Sentiment score for the S&P500.

The rest of this chapter is organised in the following way. In Section 12.2 we present and explain Refinitiv MarketPsych Analytics. In Sections 12.3 and 12.4 we present the Methodology, consider the Results and discuss these in detail. The chapter is concluded with a summary presentation of our conclusions in Section 12.5.

12.2 REFINITIV MARKETPSYCH ANALYTICS

The Refinitiv MarketPsych Analytics (RMA) provides media (news and social) sentiment scores for more than 110,000 assets, including stock indices such as the S&P500. Millions of articles and social media commentaries in 12 languages are analysed each day through natural language processing (NLP) and converted into structured data, including the textual sentiment data applied in this study. The feeds are updated every 60-seconds[1] and the data goes back as far as January 1998. In order to compute this sentiment score for a given time range t and asset a, firstly, MarketPsych computes the sum of the

---

[1]In this study we use 24-hour aggregations as provided out of the box.

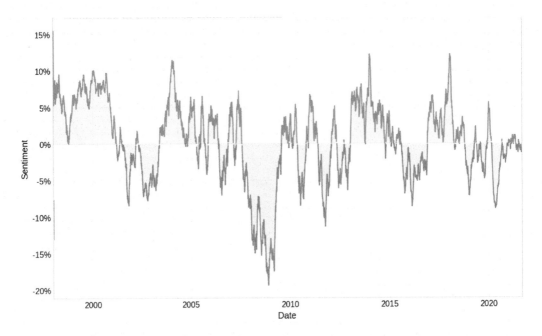

Figure 12.3 A EWM(90) sentiment time-series computed from news and social media commentary around the S&P500 and related ETFs (e.g., SPY) as available in the Refinitiv MarketPsych Analytics data feeds.

absolute values v of all indexed meanings u in documents d:

$$b_{a,t} = \sum_{d \in t} \sum_{u \in d} |v_{a,d,u}| \qquad (12.2)$$

In simple terms, $b_{a,t}$, the so-called "buzz", is proportional to the number of textual references to an asset. In this study, t refers to the 24-hour window between 15:31 ET of day $t-1$ and 15:30 ET of day t; and a is the S&P500 and associated ETFs (SPY, IVV, VOO, etc.). Finally, sentiment, $s \in [-1, 1]$, is then computed as:

$$s_{a,t} = \frac{{}^{+}b_{a,t} - {}^{-}b_{a,t}}{b_{a,t}} \qquad (12.3)$$

where ${}^{+}b$ and ${}^{-}b$ are the subsets of positive and negative references, respectively.

As shown in Figure 12.3, cycles of boom and bust clearly have an effect in the phraseology used by the media. In fact, the contemporaneous correlation between the daily sentiment and daily returns is 18%. On the other hand, the overarching themes tracked by the media tend to be stable over time as the overall language likely shifts at a much slower pace than, e.g., market policies. We will then apply media sentiment as an extra layer to define the market regime and act upon it in a point-in-time manner as to simulate a trading strategy.

12.3 METHODOLOGY AND DISCUSSION

Firstly, we considered the following portfolio:

$$r_{d+1} = \begin{cases} ^{\mathrm{SPY}}r_{d+1}, & \text{if } E[^{\mathrm{SPY}}r|d] \geq 0 \\ ^{\mathrm{AGG}}r_{d+1}, & \text{if } E[^{\mathrm{SPY}}r|d] < 0 \end{cases} \qquad (12.4)$$

Thus, we wanted to construct a strategy with daily turnover that rotates from full allocation (100%) long SPY (risk-on) to 100% long AGG (risk-off) when the expectation for the 1-day forward SPY price change is negative. Although the choice for bonds as risk-off portfolio is questionable [Andersen et al., 2007], there is evidence for its use as hedge against equities over long periods [Ciner et al., 2013]. The use of long only portfolios is a choice of convenience as the implementation of optimal allocations is not straightforward [Chaves et al., 2011] and thus beyond the scope of this study. Also for means of simplicity, the portfolio choice is based on a threshold of 0 for the expected returns. This choice likely introduces some bias towards the risk-on position which is reasonable even without considering the last decade's bull market [Bali et al., 2009]. Finally, another matter of convenience is the choice of daily returns. The high turnover from the use of daily data is taken into account later on by simulating the costs of transaction.

In order to have a proper verification set up, we divided the series into three periods, with the divisions between them defined to account for at least one period of boom and bust in the US equity markets:

- *In-sample*: Jan/1998 to Dec/2005

- *Validation*: Jan/2006 to Dec/2015

- *Out-of-sample*: Jan/2016 to Sep/2021

An overview of the methodology is displayed as a pseudo-code below (some choices will be discussed in the next section).

```
 1: for d = 31/Dec/98, ..., 07/Sep/21 do
 2:     Get SPY [d − 252, ..., d] log-returns
 3:     Distribute returns into 3 GMM states
 4:     Compute E_{d+1}[μ] for the state S_d
 5:     Get sentiment s_d
 6:     if E_{d+1}[μ] < 0 AND s_d < 0 then
 7:         Long AGG
 8:     else
 9:         Long SPY
10:     end if
11: end for
```

The first three lines in the pseudo-code show that every day we distribute the returns of the previous 252 trading days into three GMM states. The number of states were chosen as the lowest average Akaike Information Criterion (AIC) [Akaike, 1998] and Bayesian Information Criterion (BIC) [Neath and Cavanaugh, 2012] that could be obtained in the in-sample period. The choice of 252 days (1 year as only business days are considered) is arbitrary. In the fourth row, we compute the expected return by averaging the 1-day forward returns of days following the last computed state. Table 12.1 summarizes the results found with this methodology in the in-sample period.

The in-sample period (excluding the first batch of data until December 1998) includes 1760 data points. In 911 of those, the most likely state S_d had a positive expected return. The actual average return of the days following these states was 2.93 basis points, which compares to the -1.82 average for days that had a negative expected return. Furthermore we tried breaking down these results with another layer that includes the sentiment at $t-1$, thus leading to four possibilities: days with positive/negative expected returns intersected with days of positive/negative sentiment. The results indicate that the positive returns of the E^+ states are actually concentrated in days that additionally had a negative sentiment (thus E^+s^-). On the other hand, days following the E^-s^- state had negative returns. The third to last column in Table 12.1 shows the AGG average return in the same period (already excluding management fees). The average daily return is about 1.3 basis points. A sensible strategy would then be to move from the SPY and into AGG when in states E^+s^+ and E^-s^-. However, due to the sentiment volatility, the turnover would be 42%. To account for the turnover we simulate trading costs as:

- Spread: 4 cents

- Fees: 0.5 basis points

The resulting portfolio after taking the above costs into account results in the properties shown under the P1 header. Due to the turnover, the average return is lower than that of the AGG buy & hold. To reduce the turnover and also to have bias towards equities [Bali et al., 2009], we tried a second portfolio that only moves into bonds when in state E^-s^-. The turnover in such case decreases to 23%. The properties of this portfolio are shown under the P2 header. Despite now including the E^+s^+ state in the risk-on portfolio, the average return almost doubled due to the reduction in friction costs. Thus, the final model adds a sentiment layer as described in lines 5 to 10 in the pseudo-code. The portfolio returns are then described as:

$$
r_{d+1} = \begin{cases} {}^{*\mathrm{AGG}}r_{d+1}, & \text{if } E[{}^{\mathrm{SPY}}r|d] < 0, s_d < 0 \\ {}^{*\mathrm{SPY}}r_{d+1}, & \text{else} \end{cases} \tag{12.5}
$$

TABLE 12.1 Statistics about the in-sample log-returns of the SPY and of the different point-in-time market regimes as defined in this study. $E^{+/-}$ represents the sign of the expected return and $s^{+/-}$ the sign of sentiment. AGG refers to the log-returns of the ETF of same name. P1 refers to a strategy that rotates from the SPY into AGG when in state E^+s^+ and E^-s^-. P2 shows the results of a strategy that rotates from the SPY into AGG when in state E^-s^-. The Sharpe Ratio was calculated assuming the risk free rate is the 13 week treasury bill (IRX).

	SPY	E^+	E^-	E^+s^+	E^+s^-	E^-s^+	E^-s^-	AGG	P1	P2
count	1760	911	849	567	344	466	383	1760	1760	1760
mean (bp)	0.64	2.93	−1.82	−0.36	8.35	3.72	−8.57	1.33	1.02	1.81
std	1.21%	1.08%	1.33%	1.00%	1.21%	1.16%	1.52%	0.14%	0.98%	0.98%
min	−5.89%	−5.89%	−5.36%	−5.34%	−5.89%	−3.99%	−5.37%	−0.92%	−5.93%	−5.89%
median (bp)	6.51	7.50	4.48	4.54	10.01	7.41	−1.75	1.50	0.70	0.60
max	5.79%	4.76%	5.79%	2.88%	4.76%	5.64%	5.79%	0.85%	5.60%	5.64%
Sharpe Ratio	−0.52							−2.13	−0.30	

Figure 12.4 Horse race between the 4 states defined in the in-sample period (top subplot); and between 3 different strategies (bottom subplot): namely, B&H SPY, B&H AGG and the rotation from SPY into AGG when in state $E^- s^-$.

The asterisks are to denote the deduction of trading costs when turning over the portfolio. With this choice of model, we then try this choice of portfolio allocation in two following periods, which we defined above as *validation* and *out-of-sample*. We call the former *validation* as its intent was to validate the in-sample results in an actionable portfolio framework. The *out-of-sample* period would only be tested once and only if positive results were found in the *validation* period.

12.4 RESULTS

The results in the validation period are displayed in Table 12.2. The overall picture is similar to what was observed in the in-sample period. The SPY performs better on days with a positive expected return (E^+) than in days with a negative expected return (E^-). Once again, the $E^+ s^+$ state performs worse than $E^+ s^-$. Also, $E^- s^-$ is again the worst performer. To also demonstrate the evolution of the performance of each of these four states along the years, we display the cumulative performance of each during the validation period in Figure 12.4. The days in state $E^- s^-$ are particularly concentrated in the downturn of the global financial crisis. As the sub-plot in the bottom of Fig. 12.4 shows, this would have been important for avoiding the sharp downturn

TABLE 12.2 Statistics about the validation and out-of-sample log-returns of the SPY and of the different point-in-time market regimes as defined in this study. $E^{+/-}$ represents the sign of the expected return and $s^{+/-}$ the sign of sentiment. AGG refers to the log-returns of the ETF of same name. Portfolio refers to a strategy that rotates from the SPY into AGG when in state in state $E^- s^-$

		SPY	E^+	E^-	$E^+ s^+$	$E^+ s^-$	$E^- s^+$	$E^- s^-$	AGG	Portfolio
Validation	count	2517	1704	813	824	880	304	509	2517	2517
	mean (bp)	2.77	3.58	1.09	2.45	4.64	7.06	−2.48	1.68	3.38
	std	1.30%	1.05%	1.71%	0.70%	1.29%	1.11%	1.98 %	0.33%	0.98%
	min	−10.36%	−7.71%	−10.36%	−3.31%	−7.71%	−4.28%	−10.36%	−7.08%	−7.76%
	median (bp)	6.69	7.16	6.04	6.13	9.70	6.33	5.88	2.03	4.43
	max	13.56%	11.05%	13.56%	3.29%	11.05%	4.03%	13.56%	3.80%	11.05%
Sharpe Ratio	0.13							0.00	0.27	
Out-of-sample	count	1515	1332	183	579	753	59	124	1515	1515
	mean (bp)	6.28	7.53	−2.88	7.63	7.46	18.47	−13.03	1.21	7.05
	std	1.12%	0.93%	2.12%	0.63%	1.11%	0.76 %	2.52 %	0.28%	0.90%
	min	−11.59%	−5.94%	−11.59%	−3.50%	−5.94%	−2.25%	−11.56%	−4.08%	−5.94%
	median (bp)	7.87	7.50	10.19	4.83	12.68	14.18	8.68	1.82	0.60
	max	8.67%	6.50%	8.67%	2.32%	6.50%	2.28%	8.67%	2.34%	6.47%
Sharpe Ratio	0.67							−0.11	0.98	

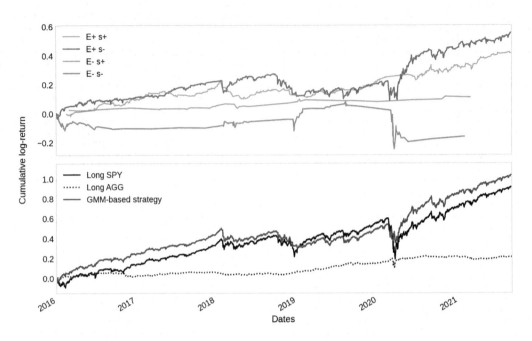

Figure 12.5 Horse race between the four states defined in the in-sample period (top subplot); and between 3 different strategies (bottom subplot): namely B&H SPY, B&H AGG and the rotation from SPY into AGG when in state E^-s^-.

in the GFC months. However, the strategy performs relatively worse in the following years with the gap between the rotation and the naive buy & hold closing in at the end of the period. On the other hand, given that in the overall period the strategy still performed considerably better on a risk-adjusted basis than the naive SPY B&H, we proceeded with the analysis in the out-of-sample period.

Results for the out-of-sample period are also displayed in Table 12.2. Equity curves derived from the several portfolios are displayed in Figure 12.5. The main points drawn in validation period appeared again in the out-of-sample period. Likely due to the raging bull market of the last 6 years, only about 12% of the points showed a negative expected return. And only 8% were in state E^-s^-. Naturally, the results of the strategy followed the SPY returns closely. However, as it was the case in 2008, the rotation between the portfolios helped with avoiding the Coronavirus pandemic draw-down to a large extent (the maximum draw-down of the strategy was 15.2% vs 32.5% in the SPY.

12.5 CONCLUSION

By applying a GMM model on the SPY log-returns, we were able to identify 1-day forward periods with negative/positive expected returns. When each of these two states were decomposed according to the sign of sentiment, often the subsets presented quite different properties. Most notably, one-day periods following days when the SPY could be characterised by a market regime with a negative expected return and negative sentiment, namely E^-s^-, were found to considerably under-perform periods following the sibling state E^-s^+ (negative expected return but positive sentiment). Based on this information, which was initially drawn from the in-sample period (1998–2005), we proposed a portfolio that rotates out of the SPY and into a bond-index ETF (AGG) when the SPY is in state E^-s^-. The strategy was then tested in validation and out-of-sample periods which addressed the 2006–2021 years. Results were consistent in both periods. This simple rotation between SPY and AGG ETFs demonstrates how the regime switching character of financial markets can be applied in a framework of risk-on/risk-off portfolios. It also demonstrates how sentiment can be used as an additional dimension to more accurately determine the market regime.

REFERENCES

Akaike, H. (1998). Information theory and an extension of the maximum likelihood principle. In *Selected Papers of Hirotugu Akaike*, pages 199–213. Springer.

Andersen, T. G., Bollerslev, T., Diebold, F. X., and Vega, C. (2007). Real-time price discovery in global stock, bond and foreign exchange markets. *Journal of International Economics*, 73(2):251–277.

Ang, A. and Timmermann, A. (2012). Regime changes and financial markets. *Annual Review of Financial Economics*, 4(1):313–337.

Baker, M. and Wurgler, J. (2006). Investor sentiment and the cross-section of stock returns. *The Journal of Finance*, 61(4):1645–1680.

Bali, T. G., Demirtas, K. O., Levy, H., and Wolf, A. (2009). Bonds versus stocks: investors' age and risk taking. *Journal of Monetary Economics*, 56(6):817–830.

Black, F. (1976). Studies of stock market volatility changes. *1976 Proceedings of the American Statistical Association Business and Economic Statistics section*. American Statistical Association.

Chaves, D., Hsu, J., Li, F., and Shakernia, O. (2011). Risk parity portfolio vs. other asset allocation heuristic portfolios. *The Journal of Investing*, 20(1):108–118.

Christie, A. A. (1982). The stochastic behavior of common stock variances: value, leverage and interest rate effects. *Journal of Financial Economics*, 10(4):407–432.

Ciner, C., Gurdgiev, C., and Lucey, B. M. (2013). Hedges and safe havens: an examination of stocks, bonds, gold, oil and exchange rates. *International Review of Financial Analysis*, 29:202–211.

Jiang, F., Lee, J., Martin, X., and Zhou, G. (2019). Manager sentiment and stock returns. *Journal of Financial Economics*, 132(1):126–149.

Neath, A. A. and Cavanaugh, J. E. (2012). The bayesian information criterion: background, derivation, and applications. *Wiley Interdisciplinary Reviews: Computational Statistics*, 4(2):199–203.

Pedregosa, F., Varoquaux, G., Gramfort, A., Michel, V., Thirion, B., Grisel, O., Blondel, M., Prettenhofer, P., Weiss, R., Dubourg, V., Vanderplas, J., Passos, A., Cournapeau, D., Brucher, M., Perrot, M., and Duchesnay, E. (2011). Scikit-learn: machine learning in Python. *Journal of Machine Learning Research*, 12:2825–2830.

Schaller, H. and Norden, S. V. (1997). Regime switching in stock market returns. *Applied Financial Economics*, 7(2):177–191.

Shiller, R. J. (2003). From efficient markets theory to behavioral finance. *Journal of Economic Perspectives*, 17(1):83–104.

(IV.B)

Application in Risk Control

CHAPTER 13

A Quantitative Metric for Corporate Sustainability

Dan diBartolomeo

Northfield Information Services

William Zieff

Northfield Information Services

CONTENTS

13.1 INTRODUCTION AND BACKGROUND

13.1.1 Status Review

"Sustainability" has become a buzzword for a set of related and increasingly popular investment themes. The common thread through all these ideas is that superior investment results will arise from long-lived enterprises. Enterprises may be long lived for several reasons. They may have activities that make ongoing contributions to societal well-being (health care, renewable

DOI: 10.1201/9781003293644-13

energy). Another cohort of sustainable companies may be those that that are better managed through superior internal governance. Yet another reason is that their business activities involve high barriers to competition, or technical obsolescence. In this study we will consider multiple aspects of the question of investing with strategic attention on sustainability.

The bulk of this study will be based on a quantitative model for forming expectations of firm sustainability that was first introduced in [DiBartolomeo, 2010]. Our primary purpose is to differentiate sustainability as purely the concept of investor expectations of firm of how long a firm will continue as an economically viable business enterprise. This is quite a different than the typical investor perspective for making an inference about the sustainability of a firm. Such inferences are typically based on one of two general categories of information.

The first popular information set is a broad range of metrics associated with the so-called ESG (environmental, social and governance) attributes of a firm. Globally embodied in the United Nations Principles on Responsible Investing (2006), the ESG perspective on this issue has created a small industry of competing reporting standards and commercial providers of the purportedly relevant information.

Another second line of recent research on sustainability has focused on the concept of barriers to entry or "corporate moats." A classic example of the corporate moat concept is the 1998 settlement of the lawsuit between a number of US States and tobacco companies. Ultimately, the tobacco firms paid $20 Billion, but in return obtained an oligopoly position in the cigarette business as any new tobacco companies who were not part of the settlement would be subject to massive future legal costs.

Our basic conclusion is that there is an extremely strong relationship between higher sustainability and lower risk for equities that is statistically significant for all periods studied. There is positive relationship between sustainability and equity returns, but the relationship is not statistically significant for all types of stocks for all periods.

13.1.2   Selected Research on Sustainable Investing

On September 13th, 1970 prominent economist Milton Friedman was famously quoted in the New York Times1 as saying "the only corporate social responsibility a company has is to maximize it's profits" [Friedman, 1970]. However, Friedman did not specify the time horizon over which profitability was to be measured. Does his comment refer to short term considerations (e.g. next quarter profits), or does it relate to the long-term profitability of an enterprise? Essentially every capital budgeting decision made by a company involves short-term expenditure to promote long-term profitability so there may agreement rather than conflict in Friedman's view and the conception of sustainable investing.

The earliest academic work on sustainable investing was a paper called "Is Pollution Profitable?" by [Bragdon and Marlin, 1972]. Since then, there has a been a massive academic literature in the closely related concepts of socially responsible investing and investing based on ESG attributes.There is even a prominent award for the best relevant research known as the Moskowitz Prize. Despite hundreds of studies, it would be fair to assert that no consensus has been reached as to whether a persistent and statistically significant relationship exists between SRI/ESG attributes of firms and their economic returns to shareholders.

Bragdon went on to write several books that framed the definition of sustainable investment as being centered in enterprises that are long lived. As an ongoing collateral activity, he created a 60 company stock index called LAMP that seeks to track performance of long-lived companies. Index data is available from 1995 to date. LAMP index relative has been very positive compared to tradition equity indices.

[Kurtz and DiBartolomeo, 1996] and [Kurtz et al., 2011] found that traditional financial variables explain differences in stock performance between broad market indices and SRI/ESG focused indices, suggesting little influence of SRI/ESG to those time points. [Kantos and DiBartolomeo, 2020] revises traditional CAPM asset pricing to account for investor sustainability concerns over large events (war, pandemic, climate change) and found that assuming investors incorporate expectations about large events resolved several purported anomalies in equity performa. diBartolomeo (2020, Journal of Performance Measurement) summarizes empirical findings of several studies of large events aptly named the "Four Horsemen of the Financial Apocalypse."

Another key question for investors is whether sustainability is closely associated with the quality of corporate governance. The "G" part of the ESG literature is a vast spectrum. Most of the early research appearing in legal and business journals are papers that assert "CEOs are greedy and boards of directors are stupid," so strict governance is required to protect shareholders from unfair expropriation of economic benefit. More recent research indicates that the issues of what is good governance are much more nuanced, and way be very different across sectors (tech, cyclical, finance). Evidence suggests that how a business is positioned within the macro-economy influences the relationships between stakeholders, as described in [Kurtz, 2012], http://www.northinfo.com/documents/523.pdf.

Particular sectors of the economy (e.g. finance) have also been subject of various regulatory efforts to explicitly improve corporate sustainability. In effect, governments have tried to install regulatory barriers to management stupidity. In the wake of the Global Financial Crisis (2007-2009), lots of new financial regulation has come into effect to prevent systemic instability. In the USA during the GFC, a single issue led to the failure of numerous large financial institutions. During the period before and during the GFC, residential mortgage foreclosures spiked sharply upward across the USA. A fact that was largely overlooked was that a few US states (e.g. Vermont) had no increase in foreclosures. Vermont has a high capital gain tax on profits from real estate sales unless the property has been held a material period, effectively eliminating "house flipping." The downside for property owners this state has experienced low rates of property appreciation compared to many other states. In effect, stability had come at the cost of slow growth in property values, so it is initially unclear as to the preferred economic course for investors.

Another line of recent research on sustainability has focused on the concept of barriers to entry or "corporate moats." The investment information firm Morningstar includes "moats" as an important aspect of their company research. [Kanuri and McLeod, 2016] summarizes the concept and its relationship to ESG and sustainability, http://www.northinfo.com/documents/730.pdf.

It should be noted that clear evidence of "moat effects" can be demonstrated quantitatively. One recent example is what we have chosen to call the Amazon Effect. Shortly before the announcement of the 2017 acquisition of Whole Foods by Amazon, our analytical models provided clear evidence of how pervasively Amazon was impacting the retail sector. As of May 31, 2017 our data showed an interesting anomaly which was a statistically detectable, but transient component to equity return correlations. The highest exposure to transient risk factor Amazon with a "beta" of 4.2. Many traditional retailers were at other end of the spectrum for this factor, with high magnitude negative exposure values. Macys, Staples, Dollar Tree, Dollar General, Foot Locker, AutoZone, Advanced Auto Parts are among the firms that the model evidenced being impacted by Amazon.

13.1.3 Organization of the chapter: A Guided Tour

Rest of this chapter is organized in the following way. In Section 13.2 we introduce our Test of Intentional Survivorship Bias ; and in Section 13.3 we discuss Differential Effects of Sector Definitions of Sustainability. In Section 13.4 we describe the The Sustainability Model. In Section 13.5 we set out a historical Study of the Returns using Quantitative Sustainability Metric. This section is presented in two subsections describing first the study and then Sustainability and the Relationship to Returns and Risks. The study is illustrated using altogether six (6) tabulated information/data. Conclusions are summarized in Section 13.6 which are followed by Acknowledgements and References.

13.2   TEST OF INTENTIONAL SURVIVORSHIP BIAS

We begin our empirical analysis with a "perfect foresight" test as if we could have known more than 25 years ago which firms would survive. In our first test, we isolated the set of 445 companies that existed in our US universe that "survived" from January 31, 1989 to September 30, 2017. Due to the impact of multiple mergers and acquisitions through time, we only included firms whole basic identifying characteristics such as sector participation were unchanged. At the start of the test, the universe had 3398 members of which 445 survivors represent 13%. The implied dropout rate is 6.8% per annum over the nearly 25 years of sample period. If we assume half of the dropouts were favorable (e.g. going private or acquired) and half were unfavorable, we get 3.4% per annum, consistent with expectations of the corporate default model presented in the aforementioned [DiBartolomeo, 2010].

If we believe in sustainability as an investment thesis (e.g. the idea behind LAMP), we would expect material outperformance of the survivor portfolio compared to traditional benchmarks. We would also expect that equally weighted portfolios of survivors would substantially outperform since rebalancing would put more money in temporarily depressed firms that we know a priori will survive.

The alternative hypothesis is that survivors operate in a pathologically conservative fashion which promotes survival but reduces investor returns (like Vermont housing). Over the sample period the universe of US equities the capitalization weighted survivor portfolio produced a cumulative return of 2808%, compared to 1585% for the S&P 500. Mean monthly return was 1.07 for the survivors versus .91 for the widely known S&P 500 index. Monthly volatility was essentially identical at 4.12 versus 4.10. Annual alpha was 2% with a tracking error of 2.98%. The T stat on outperformance was 3.17 (highly statistically significant). Based on the risk factors a commercially available endogenous model of equity risk, one-third of alpha was explained by model factors but was borderline on statistical significance. However, about two-thirds of the alpha appeared to be *security specific and highly significant (T = 2.87). The results support the thesis that there is something uniquely important about sustainability.*

Given the nature of the perfect foresight test, the portfolio weighting scheme is obviously influential. To test the direct influence of the degree of survivorship bias associated with equal weighting versus capitalization weighting, we ran the two survivor portfolios against one another. As previously described the alpha was around 30 basis points per month (T = 5.04) in favour of the equal weighted portfolio. The annual tracking error was 8.58, *which is higher than either portfolio against the S&P 500.* The T statistic on relative performance was 2.21 which is lower than either portfolio against the S&P

500. The alpha contribution of factor bets was not statistically significant. All the outperformance shows up as outside the scope of the factor model which we knew in advance in this case. The outperformance arises from the "rebalancing" gains associated with buying securities with depressed prices with foreknowledge that the firms would survive.

Accumulated over near 30 years this effect is massive. Over the sample period, the equally weighted survivor portfolio produced a cumulative return of 7695%, compared to 1585% for the S&P 500. Mean monthly return was 1.38 for the survivors versus .91 for the S&P. Portfolio monthly volatility was essentially identical at 4.60 versus 4.10. Annual alpha was 4.9% with a tracking error of 8. The T stat on outperformance was 3.58 (highly statistically significant). Using the same analytical models as the published studies by diBartolomeo and Kurtz, about 80% of the alpha appeared to be security specific and highly significant (T = 10.18). Twenty percent of alpha was explained by model factors but was not statistically significant. *The results clearly support the investment thesis of sustainability as a superior investment strategy if we can predict which firms would survive in the future.*

The analysis so far indicates that if we had perfect foresight, survivor portfolios did generally provide good performance, with the equal weighted portfolios holding a very large advantage, all consistent with expectations. There were no pervasive returns to factor exposures either at the style factor or sector level that would account for improved performance relative to traditional benchmarks. Most of the return differences were explained by statistically significant differences in "within sector" performance

13.3 DIFFERENTIAL EFFECTS OF SECTOR DEFINITIONS OF SUSTAINABILITY

In the foregoing discussion, various aspects of the relationship between sustainability and investment performance hinged on the differences in sustainability across sectors and across firms within a sector. It therefore follows that changing the taxonomy of companies into sectors and subsidiary industry groups might have a material influence on whether sustainability should be an important component of investment strategies. The next aspect of our study examines this issue by comparing the performance of a broad universe of US stocks over ten-year sample period using two different schemes of sector definition. The conventional sector structure was the GICS classification scheme which is jointly operated by S&P and MSCI. Our second sector structure (which we will abbreviate as SICS) was that created by the Sustainable Accounting Standards Board, a non-profit organization that seeks to assist investors in undertaking sustainable investment practices. The period studied was the 120 months starting in October 2007 and ending in September of 2017.

The universe of securities to studied was approximately the largest one thousand companies traded on US markets at each moment in time. This universe definition is similar to the Russell 1000, a popular market index intended to reflect performance of large capitalization US stocks. However, our universe is inclusive of American Depository Receipts (ADRs) which a mechanism to trade shares of foreign firms on US stock exchanges.

Each of the classification systems divided the universe of securities into eleven sectors. For some sectors (e.g. Financial, Health Care), the two systems were quite similar. Some of the sectors were defined quite different by intent. For example, the GICS sectors look at the energy industry in what might be considered under a traditional definition. The SASB sector definitions involve a separate sector for "Renewable/Alternative Energy" while putting conventional oil extraction into a "Extractive Industries" sector that also includes mining firms. The intuition behind the SASB structure should be clear. At some point in the future, an oil well or a copper mine must be exhausted so the basic business operation must cease. An extensive economic framework for finite resource enterprises is provided in [Hotelling, 1931].

If investors are interested in "sustainable" companies, it follows logically that they would want to differentiate firms on this basis. However, there are effects that may blur these kinds of distinctions. The first is that as technology improves the life of a resource deposits may be extended (e.g. oil well "fracturing"). In addition, the time scale of an extractive enterprise, while finite may be quite long relative to the market implied expected life of firms so that the concept of sustainability may have more to do with the impact on society and the environment, while saying less about the explicit probability that firm will cease to operate within a time horizon of practical interest to investors.

In framing the discussion of the study results, we must consider what may be reasonable expectations of such an effort. Clearly, a ten-year sample period will not be sufficient to empirically test hypotheses about "sustainable investing" that span time horizons of many decades rather than many months. This insufficiency of data also implies that statistical analyses comparing the two classification schemes may be extremely limited in our ability to assert statistical significance to the observed results.

To address these concerns, we choose to consider the matter in a somewhat unconventional way. We will consider the GICS scheme as an "incumbent" which is in existing wide used in the investment industry. We will consider the SASB classification scheme as an "alternative" which is of current interest to some subset of investors. The study focused on whether formulating investment strategies that encompass the SASB classifications rather than the GICS classifications would represent a material disadvantage to investors in some way. In essence, we are allowing ourselves the luxury of reporting inconclusive results rather than limiting our comments to hypotheses that can be demonstrated to a statistically significant degree.

Our first empirical question is if the two schemes identify clusters of companies in ways that offered investors a greater degree of discrimination. To test this question, we considered whether the stock market returns over the 130 months were more (less) correlated across the sectors defined under the different classification schemes. We calculated the returns for the each of the 22 sector definitions (11 each) on both an equal weighted and capitalization weighted basis. The average cross-sector correlation for the equal weighted GICS sectors was .893 while for the equal weighted SICS sectors the comparable value was .923. For market capitalization, weighting the GICS value was .793 with the SICS value at .813. In both cases the GICS sectors were slightly less correlated to one another; however, the differences were not statistically significant under "a difference in means" test. Given that the security universes are identical, the fact that the GICS classification monthly returns were slightly less correlated across sectors then requires that the average correlation of the securities grouped within each sector be slightly higher. The GICS sectors represent slightly more homogenous behavior within sector and slightly more heterogeneous behavior across sectors. Investors basing their investment choices on the work of fundamental analysts might find selecting stocks from within more homogeneous groupings to be easier to accomplish effectively.

Another way to think about the discriminating power of the classification scheme was to consider the degree of cross-sectional dispersion in the monthly sector returns. If the sector definitions are going to be important to investors some sectors must provide different returns than other sectors during each monthly observation. If all sectors had similar returns, investors would presumably not care about which sectors in which they invested or how the sectors were defined. The average cross-sectional dispersion of monthly returns for the equal weighted SICS sectors was 1.94%, while for the (equal weighted) GICS sectors we obtain a very similar 1.98%. With capitalization weighting applied, the average cross-sectional dispersion of the SICS sectors was 2.50% with the GICS sectors coming in at 2.47%. Again, none of these differences are statistically significant. In this case, our results are mixed with both systems having very modest advantages in two of the four comparisons.

One might also make the argument that the concept of sectors based on sustainability would require that we consider the usefulness of the classifications on a long-term basis. To consider this, we calculated the sample period cumulative return to each of the twenty-two defined sectors. If the sector definitions were meaningful in distinguishing between the long-term economic returns of the sectors, we would expect that the cross-sectional dispersion of the cumulative returns would be larger. For the equal weighted GICS sectors, the dispersion was 44.4% while for the SICS sectors the comparable value was 50.4%. For the capitalization weighted returns, the dispersion of the cumulative returns was 52.4% while the comparable figure for the SICS sectors was 49.0%. Again, the results are similar. The advantage for the SICS sectors in the equal weighted case may arise from the idea that many of the firms involved in sectors where SICS is materially different from GICS (e.g. Renewable/Alternative Energy) are largely populated by younger, smaller firms so their results "stand out from the crowd" more than if the all sectors were dominated by a larger, mature firms.

13.4   THE SUSTAINABILITY MODEL

A practical question is whether we can obtain superior performance with imperfect predictions of which firms will survive. [DiBartolomeo, 2010] provides a predictive model of firm survival based on a variation of the Merton "contingent claims" framework. [Merton, 1974] poses the equity of a firm as a European call option on the firm's assets, with a strike price equal to the face value of the firm's debt Alternatively, lenders are short a put on the firm assets. Default can occur only at debt maturity. The subsequent paper by [Black and Cox, 1976] provides model similar to an American option in which default can occur before debt maturity. Firm extinction is assumed if asset values hit a boundary value (i.e. specified by bond covenants). Another refinement to this line of research was [Leland, 1994] and [Leland and Toft, 1996] which account for the tax deductibility of interest payments and costs of bankruptcy.

The Merton approach was developed for the analysis of credit risk of firms. The obvious parallel between credit analysis and sustainable investing is that a firm must continue to exist long enough to pay off it's debts when they come due. There are three distinctions between the method put forward in diBartolomeo and a classic Merton style analysis of more than three decades earlier. The first distinction is that in the interim period sophisticated models of ex-ante equity security risk (commercially available) have become a standard part of institutional investing practices. The second is the newer model makes provision for firms that currently have no debt but may choose to borrow in the future. The third is that the resultant metrics are put in annual time units for investor expectations of corporate survival rather than as default probabilities.

Framing the sustainability model as being similar to credit risk provides an important discriminating property compared to conventional empirical studies. For equity investors, the question of time horizon makes the benefit the relationship between typical ESG criteria and corporate performance difficult to analyse. For example, an oil company that is less careful about environmental safeguards is apt to save expense in the short run but run greater (although still small) long-term risk of liability for a massive oil spill. Similarly, firms that operate with unfair labor practices will save costs in the short term but run greater risk of labor strikes or regulatory penalties.

In making decisions, equity investors must weigh near term benefits with high certainty against long-term benefits of reducing the likelihood of a future disaster. However, for creditors of the firm there is almost no economic benefit to increasing a firm's short-term profitability. They will be repaid just as they previously expected to be. On the other hand, lenders should be very averse to having a borrower increase the potential risks of a future extremely negative event that would trigger a default of repayment. As such, it is natural that a credit risk related measure would provide greater clarity in our understanding of this matter.

Direct studies on the relationship of credit risk and ESG sustainability criteria such as [Amiraslani et al., 2018] confirm that firms with higher ESG scores are perceived as more creditworthy by lenders and enjoy statistically significantly lower borrowing costs. [Henke, 2016] finds that fixed income investment funds using ESG as a strategic input have enjoyed better performance.

As applied in the 2010 study, the Merton process requires several sophisticated inputs. The "underlying of the Merton option" is the firm's assets, with asset volatility determined from a separate model of equity volatility. Essentially, the asset volatility is the what the volatility of the traded equity would be if the firm had no debt. We can use this construct to solve numerically for the "implied expiration date" of the option that equates the option value to the stock price "market implied expected life of the firm."

For details on the computation of a perpetual Americans call option see [Yaksick, 1995]. The process also included a term structure of interest rates so that as the implied expiration date changes the embedded interest rate changes appropriately. One could simplify the computational process by choosing Black-Scholes as your option model. You can then solve BS for the implied time to expiration using a Taylor series approximation, but the fixed interest rate nature of Black-Scholes is difficult to justify in a long- term context. More complex option models allow for realistic representation of stochastic interest rates.

There were two key empirical findings in this part of the study. The first was that over the entire sample period of 1992 to 2010, the shares of US publicly traded companies were priced to imply a "half-life" of around 20 years, or an extinction rate of around 3%. It should be noted that this figure is materially higher than typically observed levels of defaults of debt issues. The second finding was that while most sectors of the market had an intuitive positive relationship between firm size and expected life, the financial services sector exhibited an inverse relationship when size was defined by revenues rather than market capitalization. This result suggests that very large financial services firms are well aware of the concept of "too big to fail" and take more business risks than smaller firms. These "mega" firms consistently operated in the expectation of government bailouts in the event of problems.

This quantitative measure of sustainability has also shown to be materially related to ESG concepts. There was a strong correlation between predicted firm life and inclusion in ESG sensitive indices (e.g. the MSCI DSI 400). The persistence of this relationship was confirmed through 2017 in [Dyer and Zieff, 2017]. They calculated the performance of members of the DSI who were not in the S&P 500 against the performance of members of the S&P 500 that were not in the DSI. In their sample period, the mean number of holdings in the DSI but not the S&P 500 was 156. The mean number of holdings in the S&P 500 but not the DSI was 265. Using capitalization weighting for the period from February 1992 through February 2016, the "DSI only" portfolio outperformed the "S&P only" portfolio by an average .24% per month with a standard deviation of 2.27% (T=1.8). This outperformance is about 3% per annum, consistent with other studies showing a positive relationship between a positive relationship between some conception of sustainability and equity returns such as [Khan et al., 2016] and [Kanuri and McLeod, 2016].

13.5 A HISTORICAL STUDY OF THE RETURNS USING QUANTITATIVE SUSTAINABILITY METRIC

13.5.1 The study

The third and largest part of our empirical study is was to create a new 30-year dataset for our quantitative metric for the expected life of firms. Our experimental design was consistent with the previous study. We used a simple Black Scholes European option model to compute the Merton contingent claim analysis.

The key input to such a process is the volatility of the firm's assets. This is logically equivalent to asking "How volatile would a firm's equity be if the firm had no debt?" The forward-looking estimates volatility of equity were taken from the Northfield US Fundamental Model (commercially available) and then were adjusted for corporate balance sheet leverage. The horizon for the risk forecast was one year.

The sample period of the study was from December 31, 1991 through December 31, 2021. The full period was broken into two sub-periods, the first being from 1992 through February 2010 (the period of previously published 2010 study) and the other being the remainder from March 2010 through the end of 2021. Membership of firms into the five quintiles was updated monthly.

Results were analysed for two overlapping universes. The first was the full set of companies covered by the risk model, and the second was non-financial companies only to remove the impact of government bailouts to financial firms during the Global Financial Crisis period of 2007–2009.

Over the 30-year sample period, the distribution of expected lives varied over time. The key drivers of that variation were changes in stock market valuations (firms can sell shares to raise cash to pay off debt) and the expected

TABLE 13.1   Full Period: 01/1992–12/2021 (30 Years), All Firms

Entire Period: 01/1992–12/2021 (30 Years), All Firms					
Portfolio	Mean Monthly Return	Monthly Standard Deviation	Sharpe Ratio	Cumulative Returns	Annual Compound Return
Q5 EqualWt (High)	0.99	3.57	0.28	2689.85	11.73
Q4 EqualWt	1.21	4.75	0.25	4913.92	13.94
Q3 EqualWt	1.15	5.63	0.20	3379.98	12.56
Q2 EqualWt	1.29	6.68	0.19	4462.57	13.58
Q1 EqualWt (Low)	1.12	8.72	0.13	1382.59	9.40
Q5 CapWt (High)	0.89	3.73	0.24	1788.49	10.29
Q4 CapWt	1.00	4.81	0.21	2291.53	11.16
Q3 CapWt	0.88	4.83	0.18	1448.29	9.56
Q2 CapWt	1.03	5.41	0.19	2271.39	11.13
Q1 CapWt (Low)	0.81	6.37	0.13	767.64	7.47
S&P 500	0.93	4.16	0.22	1982.17	10.65

volatilities of corporate assets as derived from the model. A third source of variation was the changes in the relative frequency of large and small firms over time. The minimum number of firms in the sample at a moment in time was 4660 while the largest number was 8309.

Our empirical results are presented in Tables 13.1 through 13.6. The broadest data is in Table 13.1 which contains the full sample of companies for the full period of thirty years. Tables 13.2 through 13.6 provide various subsets of the full data either by the extent of firms included, or by sub-period. Returns, volatility and other measures are presented for quintiles of universe with the lowest quintile (Q1) representing firms with the shortest survival time expectation and the highest quantile (Q5) containing firms with the longest expectations of expected life. Results for each quintile is presented on both an equal-weighted and market value weighted basis.

13.5.2   Sustainability and the Relationship to Returns and Risks

The quintile (Q1) of stocks with the lowest expectations of survival times produce materially lower returns over full thirty-year sample. This is also true for the full sample with financial firms excluded for both equal weighted and value weighted samples.

Over the first part of the sample period (1/1992 to 2/2010) the Q1 stocks produce materially lower returns on a capitalization weighted basis (focused on large firms). The result was the same with financial stocks both included and excluded. Q1 stocks produce slightly lower returns on an equal weighted basis, but the differences were smaller.

Over the second part of the sample period (3/2010 to 12/2021), the Q1 stocks produce dramatically lower returns than the first period when financial firms are included. Without financial firms, results are inconclusive when capitalization weighted.

TABLE 13.2 Full Period: 01/1992–12/2021 (30 Years), All Firms ex-Financial

Portfolio	Mean Monthly Return	Monthly Standard Deviation	Sharpe Ratio	Cumulative Returns	Annual Compound Return
Entire Period: 01/1992–12/2021 (30 Years), All Firms ex-Financial					
Q5 EqualWt (High)	1.04	4.23	0.25	2913.25	12.02
Q4 EqualWt	1.18	5.39	0.22	3922.48	13.11
Q3 EqualWt	1.18	6.20	0.19	3319.74	12.49
Q2 EqualWt	1.25	7.28	0.17	3297.32	12.47
Q1 EqualWt (Low)	1.22	9.78	0.13	1462.81	9.60
Q5 CapWt (High)	0.80	3.81	0.21	1242.02	9.04
Q4 CapWt	1.09	5.33	0.20	2963.40	12.08
Q3 CapWt	0.92	5.17	0.18	1565.63	9.83
Q2 CapWt	1.04	5.55	0.19	2295.74	11.17
Q1 CapWt (Low)	1.00	6.97	0.14	1414.00	9.48
S&P 500	0.93	4.16	0.22	1982.17	10.65

TABLE 13.3 First Period: 01/1992–02/2010, All Firms

Portfolio	Mean Monthly Return	Monthly Standard Deviation	Sharpe Ratio	Cumulative Returns	Annual Compound Return
First Period: 01/1992–02/2010, All Firms					
Q5 EqualWt (High)	0.99	3.56	0.28	643.85	11.62
Q4 EqualWt	1.25	4.79	0.26	1080.76	14.49
Q3 EqualWt	1.19	5.71	0.21	822.60	12.95
Q2 EqualWt	1.35	6.98	0.19	1011.32	14.11
Q1 EqualWt (Low)	1.40	9.22	0.15	764.39	12.55
Q5 CapWt (High)	0.83	3.74	0.22	424.90	9.51
Q4 CapWt	0.97	5.15	0.19	522.37	10.54
Q3 CapWt	0.68	5.05	0.14	234.93	6.85
Q2 CapWt	0.96	5.88	0.16	454.69	9.84
Q1 CapWt (Low)	0.78	6.59	0.12	238.46	6.91
S&P 500	0.73	4.27	0.17	303.74	7.95

TABLE 13.4 First Period: 01/1992–02/2010, All Firms

Portfolio	Mean Monthly Return	Monthly Standard Deviation	Sharpe Ratio	Cumulative Returns	Annual Compound Return
First Period: 01/19920–02/2010, All Firms					
Q5 EqualWt (High)	1.01	4.28	0.24	635.04	11.55
Q4 EqualWt	1.23	5.41	0.23	959.74	13.81
Q3 EqualWt	1.21	6.35	0.19	797.01	12.77
Q2 EqualWt	1.35	7.70	0.18	891.20	13.39
Q1 EqualWt (Low)	1.58	10.58	0.15	881.38	13.33
Q5 CapWt (High)	0.68	3.81	0.18	276.62	7.54
Q4 CapWt	1.06	5.90	0.18	621.12	11.43
Q3 CapWt	0.69	5.55	0.12	217.71	6.54
Q2 CapWt	1.02	6.12	0.17	514.87	10.46
Q1 CapWt (Low)	0.88	7.11	0.12	291.83	7.77
S&P 500	0.73	4.27	0.17	303.74	7.95

Much more dramatic was the relationship and return volatility for the quintile portfolios. The portfolio level relationship between *expected sustainability* and portfolio volatility is almost perfectly monotonic. Across the six tables, the risk levels line up correctly in 46 of 48 possible comparisons between the risk of a quintile and the adjacent quintiles. Lower volatility implies

TABLE 13.5 Second Period: 03/2010–12/2021, All Firms

Second Period: 03/2010 - 12/2021, All Firms					
Portfolio	Mean Monthly Return	Monthly Standard Deviation	Sharpe Ratio	Cumulative Returns	Annual Compound Return
Q5 EqualWt (High)	1.01	3.61	0.28	275.06	11.91
Q4 EqualWt	1.14	4.71	0.24	324.63	13.10
Q3 EqualWt	1.10	5.53	0.20	277.19	11.96
Q2 EqualWt	1.20	6.21	0.19	310.56	12.77
Q1 EqualWt (Low)	0.69	7.90	0.09	71.52	4.70
Q5 CapWt (High)	0.98	3.73	0.26	259.78	11.51
Q4 CapWt	1.05	4.25	0.25	284.26	12.14
Q3 CapWt	1.19	4.47	0.27	362.28	13.92
Q2 CapWt	1.14	4.61	0.25	327.52	13.16
Q1 CapWt (Low)	0.85	6.03	0.14	156.35	8.34
S&P 500	1.25	3.99	0.31	415.71	14.98

TABLE 13.6 Second Period: 03/2010–12/2021, All Firms ex-Financial

Second Period: 03/2010 - 12/2021, All Firms ex-Financial					
Portfolio	Mean Monthly Return	Monthly Standard Deviation	Sharpe Ratio	Cumulative Returns	Annual Compound Return
Q5 EqualWt (High)	1.09	4.15	0.26	309.95	12.76
Q4 EqualWt	1.09	5.37	0.20	279.57	12.02
Q3 EqualWt	1.13	5.99	0.19	281.24	12.06
Q2 EqualWt	1.09	6.60	0.17	242.75	11.05
Q1 EqualWt (Low)	0.67	8.40	0.08	59.25	4.04
Q5 CapWt (High)	0.98	3.81	0.26	256.34	11.42
Q4 CapWt	1.12	4.31	0.26	324.81	13.10
Q3 CapWt	1.28	4.52	0.28	424.26	15.14
Q2 CapWt	1.07	4.54	0.24	289.63	12.27
Q1 CapWt (Low)	1.18	6.77	0.18	286.39	12.19
S&P 500	1.25	3.99	0.31	415.71	14.98

much lower variances for the high sustainability portfolios. As the difference between arithmetic mean returns and geometric mean returns (i.e. with compounding) is an algebraic function of the variance, lower volatility provides a material economic benefit in terms of.

The difference in volatility magnitude between Quintile 1 and Quintile 5 is statistically significantly different in every case. The tables also show return/risk ratios (i.e. Return/StdDev). These are *approximate* Sharpe ratios, omitting the risk-free rate, which were of low magnitude throughout the sample period. Differences in Q1 and Q5 Sharpe ratios are statistically significant in every case. Calculating T statistics and P values on Sharpe ratios is algebraically complex. Methods have been put forward in [Jobson and Korkie, 1981], [Memmel, 2003] and [Ledoit and Wolf, 2008].

To provide a more intuitive way to consider return/risk tradeoffs, we can take the volatility value for the highest risk quintile as a base value. We can then estimate returns for the lower risk quantiles if they had been leveraged with cash to the same volatility as the highest risk quintile. You can then compare the returns directly as the risk are the same by construction. This perspective is inherent in the economic nature of Sharpe Ratios as described in [Plantinga and de Groot, 2001].

In summary, we can say that while more sustainable firms generally produced higher returns than less sustainable firms, the relationship to returns measured by quintiles was not monotonic, nor was it statistically significant in all periods. The relationship between our quantitative measure of sustainability to quintile portfolio volatility levels was extremely strong with statistically significant results across the entire panel of outcomes.

These results are generally consistent with a broad spectrum of the ESG literature where the relationship between sustainability, return and risk. The relationship to returns is more positive than negative, but not particularly strong. On the other hand, the hypothesis that firms viewed as sustainable realized lower risks to investors is very strongly supported. We emphasize the issue of realized risks, as the idea that investors would expect lower risk from "sustainable" firms is effectively a tautological outcome of our methodology.

In terms of making intuitive sense of these results our attention is first drawn to the Global Financial Crisis. Our tests were structured to include or exclude financial stocks because the sample period include the Global Financial Crisis of late 2007 through 2009. During this period, many large financial firms were saved from bankruptcy by government bailouts based on the "too big to fail" doctrine. In the USA, the federal government expenditure on assistance to companies was around $750 Billion or about 5% of annual GDP.

The resources for financial rescues concurrently announced in the United Kingdom by the Bank of England was truly staggering at 180 Billion GBP, or $900 Billion for a economy less about a quarter of the size of the USA. Smaller financial institutions were not afforded rescue and many did go bankrupt. The complex relationship between governments and banks is described in [Belev and diBartolomeo, 2019]. The post-GFC period involved increased regulation of financial institutions and significantly higher capital requirements to restore stability.

Our second consideration to put our results in context is the rise of ESG Investors. In 2006, the United Nations created their Principles on Responsible Investing (UNPRI) and encouraged financial firms to become signatories, pledging to pay lots of attention the societal and environmental impact of their investments. As of the end of our first sample period (2/2010), about 400 firms had signed up with UNPRI. During the second sample period (2/2010 to 12/2021) about 4500 more firms joined, bringing the total to nearly 5000 today.

Many organizations that are UNPRI signatories have been publicly criticized for not following their pledge in terms of investing practices. We speculate that many large investment entities are now abandoning investing in any firms that appear to be non-sustainable as "window dressing". This reduces the firm stock price becoming a negative feedback loop, in essence creating a self-fulfilling prophecy of poor performance for Q1 stocks [Ricketts, 2021].

13.6 CONCLUSIONS

This study suggests that the intuitive benefit of investing in firms that are "sustainable" is real. We provide three different empirical approaches to the question. Multiple concepts used to define what is (or is not) a sustainable firm have high correlation in practice creating consistent impacts on investment performance. Metrics of firm sustainability whether done from the perspective of ESG, fundamental analysis (e.g. moats), or purely quantitative models of firm survival all seem to be beneficial.

Of greatest interest is that our three different tests provided at first glance, conflicting evidence of different sources of benefit. In the "perfect foresight" test, the benefit of firm sustainability arose from superior long-term return performance, not from material declines in portfolio volatility. However, this result is quite narrow given the limitations of a structurally hypothetical analysis.

Our second test was to undertake a detailed comparison of two different sector classification schemes, in which one scheme was focused on the concept of sustainability and the other scheme is presumed to be indifferent. Due to a relatively short sample period available for empirical study our results are indicative rather than conclusive. Based on three different quantitative measures, we would conclude only "a null result." That is, there was no compelling evidence of an advantage to investors by using either scheme. As such, it may be argued that investors who emphasize sustainability in their investing practices are apt to do themselves no harm by utilizing the sector scheme focused on sustainability as opposed to a more traditional classification scheme (like GICS or the FTSE-Russell ICB scheme). This issue should be revisited as longer time samples become available.

The preponderance of our empirical analysis was based on the quantitative measure of sustainability introduced in [DiBartolomeo, 2010]. Over the sample period from January 1992 through December 2021, empirical data strongly supports the concept that investor expectations of company sustainability do have a material impact on stock returns and the risk of equity portfolios. Avoiding companies with low expectations of firm survival periods provides a material improvement in both returns and return/risk ratios (mostly through reductions in risk). The degree of improvement suggests that the expected life model captures a large portion of the return benefit that investors would enjoy with perfect foresight of whether firms did or did not survive into the long future. The empirical realizations of lower volatility for "sustainable" firms are very statistically significant and are highly economically material in terms of the related increase of geometric mean returns as compared to arithmetic mean returns as described in [Messmore, 1995].

An alternative explanation of some of our empirical results on sustainability would be simply that the equity risk model used as an input to the sustainability estimation is a good model and therefore our results are self-fulfilling. However, there appears to be a significant increase in investor emphasis on sustainability from the first part of the sample period to the second part which would not be explained merely by good initial estimates of equity risk as the underlying model was unchanged.

13.7 ACKNOWLEDGEMENTS

The authors acknowledge the research assistance of Alexander Pearce, Steven Dyer and Howard Hoffman. We thank Matarin Capital for a useful insight, and the Sustainable Accounting Standards Board for use of their sector data.

REFERENCES

Amiraslani, H., Lins, K., Servaes, H., and Tamayo, A. The Bond Market Benefits of Corporate Social Capital. *European Corporate Governance Institute Working Paper*.

Belev, E. and diBartolomeo, D. (2019). Finance meets macroeconomics: a structural model of sovereign credit risk. In *World Scientific Reference on Contingent Claims Analysis in Corporate Finance: Volume 4: Contingent Claims Approach for Banks and Sovereign Debt*, pages 433–461. World Scientific.

Black, F. and Cox, J. C. (1976). Valuing corporate securities: some effects of bond indenture provisions. *The Journal of Finance*, 31(2):351–367.

Bragdon, J. H. and Marlin, J. (1972). Is pollution profitable. *Risk Management*, 19(4):9–18.

DiBartolomeo, D. (2010). Equity risk, credit risk, default correlation, and corporate sustainability. *The Journal of Investing*, 19(4):128–133.

Dyer, S. and Zieff, W. (2017). Performance of ESG Sustainable Investing and Minimum Variance Strategies. https://www.northinfo.com/documents/771.pdf.

Friedman, M. (1970). A Friedman doctrine - The Social Responsibility of Business Is to Increase Its Profits. The New York Times. https://www.nytimes.com Accessed: 2023-05-27.

Henke, H.-M. (2016). The effect of social screening on bond mutual fund performance. *Journal of Banking & Finance*, 67:69–84.

Hotelling, H. (1931). The economics of exhaustible resources. *Journal of Political Economy*, 39(2):137–175.

Jobson, J. D. and Korkie, B. M. (1981). Performance hypothesis testing with the sharpe and treynor measures. *Journal of Finance*, 36(4):889–908.

Kantos, C. and DiBartolomeo, D. (2020). How the pandemic taught us to turn smart beta into real alpha. *Journal of Asset Management*, 21(7):581–590.

Kanuri, S. and McLeod, R. W. (2016). Sustainable competitive advantage and stock performance: the case for wide moat stocks. *Applied Economics*, 48(52):5117–5127.

Khan, M., Serafeim, G., and Yoon, A. (2016). Corporate sustainability: first evidence on materiality. *The Accounting Review*, 91(6):1697–1724.

Kurtz, L. (2012). Macro factors in corporate governance. Northfield Asia Research Seminar.

Kurtz, L. and DiBartolomeo, D. (1996). Socially screened portfolios: an attribution analysis of relative performance. *The Journal of Investing*, 5(3):35–41.

Kurtz, L. et al. (2011). The long-term performance of a social investment universe. *The Journal of Investing*, 20(3):95–102.

Ledoit, O. and Wolf, M. (2008). Robust performance hypothesis testing with the sharpe ratio. *Journal of Empirical Finance*, 15(5):850–859.

Leland, H. E. (1994). Corporate debt value, bond covenants, and optimal capital structure. *The Journal of Finance*, 49(4):1213–1252.

Leland, H. E. and Toft, K. B. (1996). Optimal capital structure, endogenous bankruptcy, and the term structure of credit spreads. *The Journal of Finance*, 51(3):987–1019.

Memmel, C. (2003). Performance hypothesis testing with the sharpe ratio. *Available at SSRN 412588*.

Merton, R. C. (1974). On the pricing of corporate debt: the risk structure of interest rates. *The Journal of finance*, 29(2):449–470.

Messmore, T. (1995). Variance drain. *Journal of Portfolio Management*, 21(4):104.

Plantinga, A. and de Groot, S. (2001). Risk Adjusted Performance Measures and Implied Risk Attitudes. *Capital Markets eJournal*.

Ricketts, D. (2021). Ex-BlackRock sustainable CIO on how ESG risks turning into a mis-selling scandal. Financial News. https://www.fnlondon.com/articles/ex-blackrock-sustainable-cio-on-how-esg-risks-turning-into-a-mis-selling-scandal-20210914 Accessed: 2023-05-27.

Yaksick, R. (1995). Expected optimal exercise time of a perpetual american option: a closed-form solution. In *Advances in Stochastic Modelling and Data Analysis*, pages 29–56. Springer.

CHAPTER 14

Hot Off the Press: Predicting Intraday Risk and Liquidity with News Analytics

Ryoko Ito

Goldman Sachs International, Global Markets

Giuliano De Rossi

Goldman Sachs International, Global Markets

Michael Steliaros

Goldman Sachs International, Global Markets

CONTENTS

DOI: 10.1201/9781003293644-14

14.1 INTRODUCTION

News analytics has emerged as one of the most successful applications of natural language processing methods in finance. In particular, signals based on news sentiment are widely used in trading strategies, typically over short investment horizons. We investigate whether it is possible to harvest the predictive power of news arrival intensity at intraday frequencies in order to forecast stock volatility and volume. Our goal is to obtain a news-driven predictive model for a large set of equities, exploiting potential spillover effects among single stocks, for use in optimizing single stock and portfolio execution.

We introduce a measure of news arrival intensity, i.e. the number of news stories about a stock divided by the total number of stories about stocks in the same sector. This proportion reflects media attention to a given stock in relative terms within its sector. Along the same lines, we allow for spillover effects within sectors by using a sector-level news intensity measure. It reflects media attention that focuses on the stock's sector relative to other sectors. These transformations of news analytics data are intuitive and efficient because proportions are bounded (between 0 and 1), which makes our model robust to outliers.

We find compelling evidence that intraday volatility and volume react to news arrival intensity. We also show that the impact of news spills over: Volatility and volume of a stock react to news about other companies in the same sector. The predictive power of our news variables is both statistically and economically significant out of sample. We show how this relationship can be used to produce forecasts in real time for a large portfolio using the Russell 1000, the STOXX 600 and the TOPIX 500 constituents.

We use the news analytics data provided by Refinitiv for stocks traded in the USA and Europe. The data is based on text in English. Our news-driven model successfully extract predictive power for stocks traded in these regions and show the robustness of the model output, as well as its economic and statistical significance.

We also consider whether news written in a language other than English can play an important role in driving intraday volume and volatility in different markets globally. For this purpose, we consider the News Dolphin news analytics provided by FTRI/Alexandria. This data is based on articles published by Nikkei, one of the world's largest business news agencies domiciled in Tokyo. We find strong evidence that our measure of news arrival intensity computed from this dataset improves forecasts for a striking majority of stocks in the TOPIX 500. Our results highlight the importance of news written in languages native to each market by locally domiciled news agencies when modelling risk and liquidity. Our model performance is shown to withstand the test of the historic turbulence in the first quarter of 2020.

We describe a number of important statistical features of the news analytics dataset. They are reflective of the fact that media attention is distributed unevenly across stocks and during the year. That is, most stocks typically receive little media coverage most of the time except for the earnings season during which stock-specific news tends to intensify for many stocks. Then it is tempting to think that news variables would matter only when big news strikes. However, we do not find clear evidence that the out-of-sample performance of our model depends on stock characteristics or time of the year. That is, the inclusion of news variables seems to be important for an econometric model even when there is no news for a given stock at a given point in time. This is because their inclusion allows us to capture extreme dynamics in the data and improves the precision of estimated coefficients on other variables.

In the existing literature, several papers have adopted univariate models of volatility such as the GARCH and HAR models in daily or monthly frequency with (normalized) news counts as an exogenous predictor. For example, [Engle et al., 2020] extend the GARCH family of models to capture the relationship between news arrival and volatility. They measure the rate of information arrival by counting the number of news items. They show that indicators of public information arrival explain on average 26% of changes in firm-specific return volatility. [Rahimikia and Poon, 2020] find that information about company news items and the depth of limit order book can be reflected in the HAR models to improve daily realized volatility forecasts. [Conrad and Engle, 2021] propose a multiplicative factor multi-frequency component GARCH model. Their model reflects that daily forecast errors of GARCH models tend to show counter-cyclical dynamics when the errors are averaged at a lower frequency. They find that the model's out-of-sample precision improves on the classic GARCH models and HAR-type models. Also see [Tetlock, 2010], [Allen et al., 2015], [Shi et al., 2016], and [Caporin and Poli, 2017].

It is also common to use news analytics data in high frequency to capture intraday equity dynamics. [Barunik et al., 2019] propose a sentiment-driven stochastic volatility model for forecasting volatility in high frequency using data from the NASDAQ news platform. [Cui and Gozluklu, 2021] explore after-hours trading (AHT) in the U.S. equity markets. The authors collect a large set of news releases during AHT and document their effect on AHT activity and market quality. They find that three news categories attract most AHT; they are earnings announcements, insider trades and index reconstitutions. [Song and Bhattacharyya, 2021] study the impact of major macroeconomic news in the USA on intraday asset price movements. They show that the impact of surprising macroeconomic news on asset prices is almost fully realized within the first few minutes of the news arrival. Also see [Gross-Klussmann and Hautsch, 2011], [Ho et al., 2013], [Uctum et al., 2017], [Riordan et al., 2013], [Clements et al., 2015], and [Kalev et al., 2004]. It is common in the literature to introduce a deterministic model of the intraday volatility pattern and use dummy variables to relate news arrival counts to volatility by identifying periods of heightened news intensity.

Moreover, news analytics data can be used to augment equity portfolio risk models. For instance, [Mitra et al., 2009] provide a theoretical framework in which equity portfolio risk estimates are allowed to be sensitive to changes in the market environment and investor sentiment. Their framework contrasts with traditional factor risk models that do not update quickly when new information becomes available. [Mitra et al., 2018] suggest a portfolio model in which long and short positions are influenced by market sentiment acquired from news wires and microblogs. They use the second-order stochastic dominance (SSD) as the choice criterion for determining positions and discuss computational techniques for solving the proposed model. [Creamer, 2015] extends the Black-Litterman model for portfolio optimization to incorporate news sentiment, which is assumed to be reflective of investors' expectations. More recently, [Engle et al., 2020] propose a procedure for constructing portfolios that are dynamically hedged against climate change risk. They use textual analysis of newspapers to extract innovations relating to news about climate change. Then they use a mimicking portfolio approach to build climate change hedge portfolios. They discipline the exercise by using third-party ESG scores of individual firms to model their climate risk exposures. In practical applications, the Northfield platform is a commercially available analytics for portfolio construction that incorporates risk scores based on textual news feeds. Also see Messina et al. (2020) for a survey of studies in this field.

The main empirical findings of the existing literature can be summarized as follows: News intensity is typically found to matter to predict volatility. However, it also highlights that it is important to filter news articles by relevance in order to mitigate the effect of noise in the data. Surprisingly, no clear evidence is found that negative news has a stronger effect compared to positive news. Our estimation results conform with these common findings in the literature.

The news-driven forecasting model we introduce contributes to the literature by finding strong evidence of the relation between news arrival intensity, volatility and volume in a global dataset at intraday frequencies. In particular, using our proposed measure of news arrival intensity, we find evidence of a spillover effect of news across stocks. Amongst other findings of import, the disentanglement of those effects around earnings periods is of particular interest. Our model is simple and can be applied to optimal execution both at the stock and the portfolio level.

The rest of this chapter proceeds as follows. Section 14.2 outlines important data features to motivate our approach and provide some intuition on the mechanism that links news to intraday market dynamics. Then Section

14.3 describes our model and necessary adjustments for different markets. We present the in-sample estimation results for stocks in the USA and Europe in Section 14.4. Based on the results for the USA and Europe, we extend the model to stocks in Japan in Section 14.5. Section 14.6 concludes.

14.2 DATA CHARACTERISTICS: UNDERSTANDING NEWS ANALYTICS

For stocks traded in the USA and Europe, we compute news arrival intensity using Refinitiv news analytics data. Our analysis for these regions is based on news written in English. Refinitiv applies a natural language processing algorithm to machine-readable news and assigns to each article topics, relevant stock IDs and numerical scores of relevance and sentiment amongst other information.

For stocks traded in Japan, we use the News Dolphin dataset based on Nikkei's news. Nikkei is one of the largest business news agencies in the world and highly influential in Japan, which makes this dataset particularly interesting when analysing stocks traded in Japan. News Dolphin produced by the Financial Technology Research Institute (FTRI) displays several interesting statistical features worth contrasting with Refinitiv's data. FTRI and Alexandria Technology collate and distribute News Dolphin in real time.

In order to select relevant news stories, we consider using the vendor-provided numerical scores (e.g. sentiment and relevance) and categories (e.g. company and topic). Moreover, since observations are irregularly spaced through time, we aggregate them in 30-minute buckets to reduce noise before estimation.

A few firms get the bulk of media attention

Figure 14.1 shows the distribution of news articles across stocks in the Russell 1000. Each entity in the Russell 1000 index attracts roughly 200 news articles on average per year. The distribution of the number of news articles depicted here indicates that most companies attract little media attention, whereas a few companies at the top of the ranking regularly attract wide media coverage (e.g. Amazon and Apple). The distribution of media attention appears to be heavy tailed. The Refinitiv dataset shows similar characteristics for the STOXX 600. The distribution of Nikkei's news articles across stocks in the TOPIX 500 also has similar characteristics. Entities typically attract about 400 news articles on average per year. Most stocks traded in Tokyo attract relatively little attention from Nikkei, whereas a few companies in the right tail regularly attract coverage (e.g. Toyota Motor Corporation and Mitsubishi UFJ Financial Group).

This heavy-tailed distribution in news articles represents an econometric challenge as sparse exogenous variables with few large data points can make

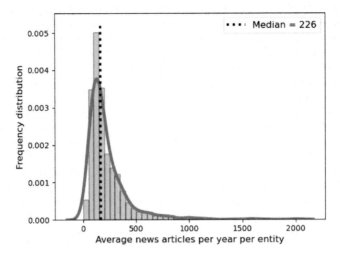

Figure 14.1 Distribution of media attention across firms. Average news articles per year per entity across stocks are shown along the x-axis. Refinitiv's data between 2014 and 2018 for stocks in the Russell 1000

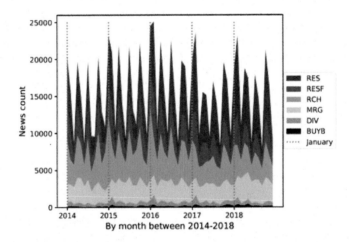

Figure 14.2 Quarterly seasonal patterns in news arrivals by category for the Russell 1000 constituents (Refinitiv's category codes in the legend are defined in the footnote)

models statistically unstable. In Section 14.3, we describe our proposed transformation of news analytics data into proportions, which is key to achieving robust results in this regard as proportions are bounded. This is also one motivation for keeping the aggregation interval relatively wide and to apply minimal filters.

Quarterly seasonal patterns due to earnings announcements

Figure 14.2 shows that news arrives throughout the year, but the rate of arrival peaks four times a year when companies tend to announce quarterly

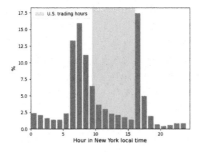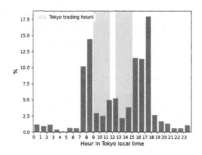

Figure 14.3 Intraday seasonal patterns in news arrivals. Percentage of news articles in the sample by hourly bin. The height of the bars sums to 100%. Refinitiv's data between 2014 and 2018 for stocks in the Russell 1000 (left). News Dolphin data between 2015 and 2020 for stocks in the TOPIX 500 (right).

earnings results. The timing depends on the region: they are in January, April, July and October for stocks in the Russell 1000, and February, May, August and November for stocks in the TOPIX 500. News relating to quarterly earnings, analyst forecasts, and company performance (topics at the top of Figure 14.2) appears more seasonal than other topics such as mergers and acquisitions (MRG). These quarterly seasonal patterns in new arrivals are effective in predicting spikes in volatility and volume that occur after earnings announcements. Our model also includes dummy variables for earnings announcement days as they help adjust the effect of news on those special days, which we describe further in the estimation section later on.

Intraday seasonal patterns and lunchtime news

Intraday patterns can also be detected in the data (Figure 14.3). Most news about stocks traded in the USA Europe and Japan arrives either just before or just after trading hours (shaded area). For the U.S. stocks, we detect a small peak when markets in Europe open and vice versa for the European stocks when the trading in New York opens. For Japanese stocks, there is also a small peak around noon in Tokyo local time as Tokyo Stock Exchange has a lunch break between 11.30 am and 12.30 pm.

During continuous trading, we aggregate data at a 30-minute frequency. We aggregate data outside trading hours into one bin to distinguish the overnight effect in volatility and volume. There is relatively little news relevant to stock trading over the weekend. Hence we aggregate news between Friday close and Monday open into one overnight bin, and we do not separately consider the weekend effect. For Japanese stocks, we also aggregate data during the lunch break into one lunch-hour bin in order to distinguish the effect of news in those periods.

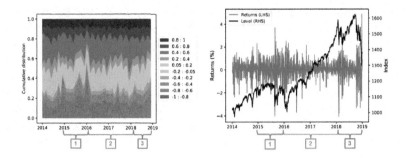

Figure 14.4 Distribution of news sentiment across news articles over time (left) and market dynamics in the USA (right). News articles with relatively neutral sentiment (between -0.05 and 0.05) are excluded for the purpose of this visualization, although they are included in the rest of the analysis. The daily historical data between 2014 and 2018 for stocks in the Russell 1000.

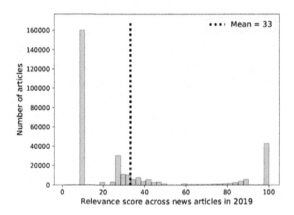

Figure 14.5 Histogram of relevance score across articles about stocks in the TOPIX 500 in 2019

Overall sentiment moves with the market

News stories classified as having positive sentiment tend to occur more often than negative ones (Figure 14.4). In addition, overall sentiment moves with the market. The proportion of negative news spiked when return volatility was high and the index fell in 2015–2016 (the area marked by [1]). Positive sentiment dominated and remained stable when the index rose in 2016 and 2017 (period [2]). Finally, negative sentiment grew again when the index plummeted in 2018 ([3]). We estimate a version of our model in which positive and negative news are treated separately so that we can test for potential asymmetric effects on volatility and volume.

Few news articles have high relevance score

Figure 14.5 shows the distribution of relevance scores across news about stocks in the TOPIX 500 in 2019. The score is computed by FTRI. A few articles are deemed highly relevant to a given stock according to this metric. There is also a large clustering at 10. This means that excluding news with a relevance score below 100 removes a large number of observation points, which may or may not be appropriate, depending on the use case.

14.3   MODEL FOR INTRADAY VOLATILITY AND VOLUME WITH NEWS

We adopt a predictive model given by the following linear regression:

$$y_{t,\tau,i} = \gamma_i + \gamma_i^N N_{t,\tau-1,i} + \gamma_i^{sec} N_{t,\tau-1,-i} + \gamma_i^D D_{t,\tau,i} + \varphi_i y_{t,\tau-1,i} + \sum_{j=1}^{p} \lambda_{i,j} y_{t-j,\tau,i} + \varepsilon_{t,\tau,i}.$$

(14.1)

Here $y_{t,\tau,i}$ is the logarithm of the variable to be forecast for stock i at intraday bin τ on trading day t. The length of each intraday bucket τ is 30 minutes. $N_{t,\tau,i}$ is a measure of news arrival intensity for stock i, and $N_{t,\tau,-i}$ is the same measure for all stocks in stock i's sector but excluding stock i. $D_{t,\tau,i}$ is a dummy variable that identifies special days (e.g. earnings announcement days) for stock i.

This linear regression model is fast and easy to estimate and implement, which is an important advantage when the panel is large compared with some state-of-the-art non-linear volatility models (e.g. GARCH and its variants). Despite the simplicity, in-sample and out-of-sample performance seems relatively robust and consistent across sectors.

Measuring relative media attention using news proportion

We quantify news arrival intensity for stock i in relation to other stocks in the same sector. Likewise, the sector-wide news arrival intensity is quantified in relation to other sectors in the entire index. Formally, the two metrics are given by

$$N_{t,\tau,i} = 100 \times \frac{C_{t,\tau,i}}{\sum_{j \in S(i)} C_{t,\tau,i}} \quad \text{and} \quad N_{t,\tau,-i} = 100 \times \frac{\sum_{j \in S(i), j \neq i} C_{t,\tau,i}}{\sum_j C_{t,\tau,i}},$$

(14.2)

where $C_{t,\tau,i}$ is the number of stories about stock i and $S(i)$ denotes the set of all stocks in stock i's sector. Thus $N_{t,\tau,i}$ captures media attention around stock i relative to other stocks in the same sector, and $N_{t,\tau,-i}$ captures media attention around stock i's sector relative to other sectors in the same index.

The intuition is as follows: Suppose that at a certain point in time there are ten news articles about Company A and 1,000 about Company B. Then the latter should be deemed to attract higher media attention than the former, and our news variables capture this relative weight. Since $N_{t,\tau,-i}$ excludes stories about stock i, it allows for spillover effects of news about other companies on stock i.

Seasonally adjusting intraday volume and volatility in one step

Intraday patterns in volatility and volume are estimated with other covariates in a one-step procedure using the lags of $y_{t,\tau,i}$ in 24-hour increments (i.e. $y_{t-j,\tau,i}$). The estimation results show that this approach captures seasonal patterns well. It contrasts with the more commonly adopted two-step procedures to seasonally adjust data before estimation, which are prone to biased results.

Sentiment effect, overnight effect and special day effect

We test four versions of each news variable ($N_{t,\tau,i}$ and $N_{t,\tau,-i}$) to distinguish the cases in which

- The news sentiment score is positive ($+$) or negative ($-$), and

- The bin is at the start of continuous trading (i.e. $\tau = 1$) or later in the day.

Each of the eight resulting variables can enter the model separately to allow for asymmetric effects. The distinction by sentiment score is motivated by the fact that overall sentiment appeared to move with the market as we saw in Section 14.2. However, in Section 14.4 we find that this distinction by sentiment may not be substantive as coefficients on positive and negative news seem to have the same signs and broadly similar magnitude. Hence we also estimate a simplified model that aggregates positive and negative news. The special day dummy ($D_{t,\tau,i}$) is also split into dummies for earnings announcement days, index rebalancing days, triple witching days (quarterly), option expiry days (monthly) and end-of-month dummies to assist in quantifying the specific impact of those events after accounting for the impact of news. The coefficients on the latter three dummy variables tend to be statistically significant but relatively small. Hence it may also suffice to include dummy variables only for earnings announcement and index rebalance days to simplify the model. We denote the variable for overnight news aggregated into one bin by "ON," and for the lunch hour by "NOON" for application to stocks in Japan. The data description section motivated these aggregations in more detail earlier.

Table 14.1 displays the summary statistics. It is clear from the table that news ratios are very sparse, i.e. a large number of observations are zero. Hence filtering excessively can introduce many zero-valued data points, which can make the model output unstable as we discussed in the data description section. Section 14.4 provides analysis on the robustness of the model relating to these points.

14.4 APPLICATION TO STOCKS IN THE USA AND EUROPE

First, we estimate the model for stocks in the USA and Europe to examine the significance of news analytics based on articles written in English. The sample

TABLE 14.1 Summary statistics of news variables. Stocks in the Financials sector of the Russell 1000

		Mean	Mean (x > 0)	Median	Median (x > 0)
$N_{t,\tau-1,i}$	newsRatio (+)	0.2	67.2	0	75
	newsRatio (-)	0.1	72.7	0	100
	newsRatio (+,ON)	0.04	5.2	0	3
	newsRatio (-,ON)	0.04	9.5	0	5
$N_{t,\tau-1,-i}$	newsRatio (+, sector)	12.3	38.1	0	30
	newsRatio (-, sector)	9.6	46.6	0	40
	newsRatio (+, ON, sector)	0.4	5.3	0	4
	newsRatio (-, ON, sector)	0.4	5.0	0	4

consists of 30-minute realized volatility and trade volume observed between January 2016 and December 2017. The model is estimated for the constituents of the Russell 1000 and the STOXX 600 using a panel data approach. In the following section, we report the results for the Financials sector of the Russell 1000 (Table 14.2). Results for other sectors are similar and lead to the same conclusions. They are reported in the appendix.

14.4.1 In-Sample Estimation Results

The effect of news spills over: The first thick box shows that intensified overnight news about other stocks reduces stock i's trade volume but increases stock i's return volatility. These findings are consistent across sectors in the Russell 1000 and the STOXX 600. The coefficients in the first box can be interpreted as follows: a 5 percentage point increase (which is the average according to Table 14.1) in overnight news proportion of other stocks ($N_{t,\tau-1,-i}$):

- Increases realized volatility in stock i's returns by 9.5%~23% (i.e. 1.9%~4.6% times 5), and

- Reduces stock i's trade volume by 8%~28.5% (i.e. 1.6%~5.7% times 5).

All coefficients are significant at the 5% confidence level except for the one in grey. Overnight variables (denoted by "ON") tend to be statistically significant across sectors. The coefficients on $N_{t,\tau,i}$ tend to be positive, implying that stock-specific news increases volatility and volume. The dummies for earnings announcement dates and index rebalance dates also tend to be statistically significant. In particular, the second thick box shows that one of the overnight dummies that allow for the impact of earnings announcement is negative in the model for realized volatility: This coefficient adjusts the impact of stock-specific overnight news so that the net overnight effect of earnings announcements is still positive.

TABLE 14.2 Representative regression results. Stocks in the Financials sector of the Russell 1000

	Sector	Financials			
	LHS	Log(Realized Volatility)		Log(Volume)	
	Statistic	Coef	p-values	Coef	p-values
$N_{t,\tau-1,i}$	newsRatio (+)	2.8E-04	0.004	3.6E-04	0.015
	newsRatio (-)	2.1E-04	0.088	0.001	0.000
	newsRatio (+,ON)	0.008	0.000	0.010	0.000
	newsRatio (-,ON)	0.003	0.000	0.008	0.000
$N_{t,\tau-1,-i}$	newsRatio (+, sector)	-1.8E-04	0.000	0.001	0.000
	newsRatio (-, sector)	-1.9E-04	0.000	2.2E-04	0.000
	newsRatio (+, ON, sector)	0.046	0.000	-0.057	0.000
	newsRatio (-, ON, sector)	0.019	0.000	-0.016	0.000
$D_{t,\tau,i}$	earningDates	0.284	0.000	0.363	0.000
	earningDates (ON)	-0.214	0.000	0.631	0.000
	lastDayOfMonth	-0.007	0.001	0.031	0.000
	tripleWitching	0.028	0.000	0.053	0.000
	optionExpiry	-0.051	0.000	-0.013	0.000
	indexRebalance	0.088	0.000	0.135	0.000
	indexRebalance (Close)	0.325	0.000	0.625	0.000
$y_{t,\tau-1,i}$	y_lag1	0.389	0.000	0.417	0.000
	y_season1	0.160	0.000	0.188	0.000
	y_season2	0.091	0.000	0.107	0.000
	y_season3	0.061	0.000	0.071	0.000
$y_{t-j,\tau,i}$	y_season4	0.058	0.000	0.062	0.000
	y_season5	0.050	0.000	0.055	0.000
	y_season6	0.029	0.000	0.046	0.000
	y_season7	0.018	0.000	0.050	0.000
	y_season8	0.053	0.000	0.067	0.000
	R-squared	0.62		0.56	
	AIC	0.427		1.403	
	BIC	0.431		1.408	

The overall headline results remain stable when the in-sample period is rolled forwards on a monthly basis over the subsequent 12-month period (Figure 14.6).

While the panel regression is useful when selecting the benchmark model, we use firm-level time series regressions to test the model's predictive ability out-of-sample. This is because forecast precision hinges on the autoregressive coefficients (i.e. the ones on lagged $y_{t,\tau,i}$), which tend to vary significantly across stocks as widely documented in the GARCH literature.

14.4.2 Out-of-Sample Performance

By how much can our measure of news arrival intensity improve forecasts? In order to address this question, we use our model to forecast intraday volatility and volume. The specification reported in Table 14.2 can be used for this purpose, but we need to contrast its performance against an alternative model without news arrival intensity to gauge its role. Thus, we drop news variables from this specification and re-estimate coefficients over the same in-sample

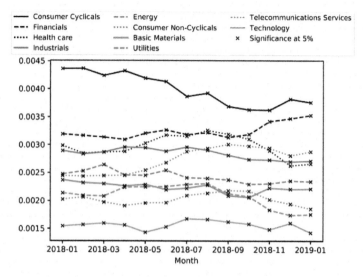

Figure 14.6 Estimated coefficient on $N_{t,\tau-1,i}$ $(-,\text{ON})$ when the in-sample period is rolled by month in 2018. $y_{t,\tau,i}$ is the logarithm of realized volatility. The x-axis represents the last month of the two-year in-sample period. Russell 1000 data.

period (i.e. 2016–2017). This gives us a model without news. We examine forecasts by the two models to see whether the one with news yields more accurate forecasts than the other without it. The out-of-sample period is 2018. Using mean absolute error (MAE) as an indicator of accuracy, we compute for each stock the percentage improvement in forecast precision due to our news variables.

Risk and liquidity in the USA and Europe react to English language news

Figure 14.7 shows the results across stocks in the Russell 1000. We find that our news variables typically improve forecast precision of intraday volume by about 12% in the first half-hour of the trading period. During the last half-hour of trading, the improvement in precision is about 3%. It is worth emphasizing that the distributions in Figure 14.7 lie mostly on the positive side, meaning that news variables would improve forecasts for most stocks, in varying degrees. To be more specific, our news variables improve forecast precision for 91% and 89% of the stocks during the first half-hour and the last half-hour, respectively. The distribution in grey for the last 30-minute bin is more concentrated around the median than during the first 30-minute bin, reflective of the high volatility at the open.

Figure 14.7 also shows the results for volatility prediction. The precision improves typically by about 3% and 7% in the first and the last hour of trading, respectively. These numbers are relatively moderate compared to volume

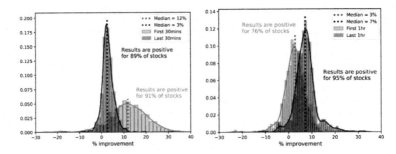

Figure 14.7 Volume forecasting (left) and volatility forecasting (right) over 2018 for stocks in the Russell 1000. Distribution across firms of the percentage improvement in precision during the first 30 minutes and the last 30 minutes of trading period when news variables are included

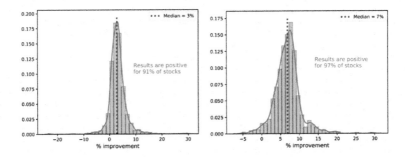

Figure 14.8 Volume forecasting (left) and volatility forecasting (right) over 2018 for stocks in the Russell 1000. Distribution across firms of the percentage improvement in precision per day when news variables are included

prediction, likely because intraday realized volatility is more volatile than intraday volume. But what is striking is that news variables again improve the precision for the majority of stocks: 76%~95% of the distributions lie on the positive side.

News variables improve precision during the day too: The overall daily forecast precision improves for 91%~97% of the Russell 1000 constituents (Figure 14.8). The magnitude of the improvement seems typically larger for volatility than for volume.

Figure 14.9 illustrates how news arrival intensity can explain volatility by way of example taken from our out-of-sample results. The figure shows the realized 30-minute volatility of Apple (solid black line) on November 27th, 2018, when the company attracted media attention after the U.S. President Donald Trump was quoted by the Wall Street Journal as saying that iPhones imported from China could be hit by tariffs. The solid blue line depicts the out-of-sample forecast generated by our preferred news-driven model. The dashed blue line is obtained by ignoring the contribution of news variables to the

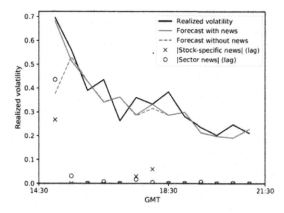

Figure 14.9 An out-of-sample outcome for a selected stock showing the intraday interval when news explained volatility. The realized 30-minute volatility of Apple Inc. on November 27th, 2018

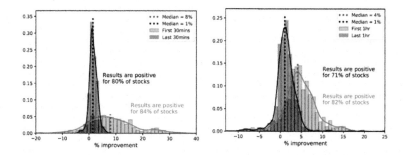

Figure 14.10 Volume forecasting (left) and volatility forecasting (right) over 2018 for stocks in the STOXX 600. Distribution across firms of the percentage improvement in precision during the first 30 minutes and the last 30 minutes of trading period when news variables are included

model's predictions. The figure shows that by incorporating our news arrival intensity metric we can move closer to the actual realization as the increased arrival intensity results in higher predicted volatility.

Figures 14.10–14.11 show the out-of-sample results for the constituents of the STOXX 600. The results are similar to the ones we discussed for the Russell 1000 (albeit somewhat weaker): News variables improve forecast precision for the majority of stocks in all cases. The magnitude of the improvement tends to be higher during the early hours of trading than towards the end of the session. The difference in forecasting accuracy between Europe and the USA is likely due to the greater heterogeneity that characterizes European equity markets and the fact that our dataset is only based on English-language news.

We used the Diebold-Mariano (DM) statistic to formally test the statistical significance of the differences in the MAE values between the two competing

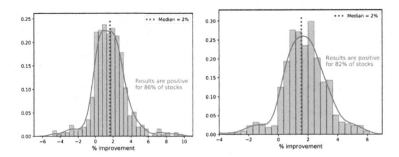

Figure 14.11 Volume forecasting (left) and volatility forecasting (right) over 2018 for stocks in the STOXX 600. Distribution across firms of the percentage improvement in precision per day when news variables are included

specifications (i.e. the model with news and the alternative without). The level of significance is at 5% in the following discussions. In terms of volatility prediction for stocks in the Russell 1000, the DM test implied that forecast precision statistically significantly improves for 28%~95% of firms depending on the time of the day when news variables are included. As for volume prediction, the precision improves for 42%~80% of firms in the Russell 1000, according to the test. These results are displayed in Table 14.3. The picture for the STOXX 600 is similar (Table 14.4).

The DM statistic tends to be conservative when the underlying data is volatile as it takes into account the degree of uncertainty around the results. For instance, the DM results appear to be weaker in the first one-hour trading window than in the last one-hour window. Arguably this result reflects the fact that the market fluctuates significantly at the open as the impact of overnight news propagates and the initial price discovery phase unfolds. The DM test translates our out-of-sample results with caution given the varying magnitude of the improvement in forecast precision across firms.

14.5 APPLICATION TO STOCKS IN JAPAN

We showed so far that intraday prediction of volatility and volume improves for a majority of stocks in the Russell 1000 and STOXX 600 when variables reflective of news arrival are amongst the explanatory variables in our predictive model. We found that the results for STOXX 600 are similar to Russell 1000 albeit somewhat weaker, which can potentially be attributed to the greater heterogeneity of European markets and the fact that the news analytics dataset we applied was based only on English text.

In order to test the power of news written in native languages, we apply the model to TOPIX 500 stocks and the News Dolphin news analytics based on news published by Nikkei. We follow the same estimation approach and

TABLE 14.3 Percentage of stocks for which news variables improve forecast precision according to the Diebold-Mariano test. Stocks in the Russell 1000

Volatility prediction (the Russell 1000 constituents)			
MAE computed every day for entities in the following:	Bins in first 1hr window	Intraday bins in-between	Bins in last 1hr window
Basic Materials	15%	96%	94%
Consumer Cyclicals	24%	96%	93%
Consumer Non-Cyclicals	35%	96%	91%
Energy	36%	98%	95%
Financials	22%	97%	93%
Health care	55%	92%	86%
Industrials	31%	90%	83%
Technology	38%	98%	87%
Telecommunications Services	13%	88%	88%
Utilities	8%	97%	97%
Average	28%	95%	91%

Volume prediction (the Russell 1000 constituents)			
MAE computed every day for entities in the following:	Bins in first 1hr window	Intraday bins in-between	Bins in last 1hr window
Basic Materials	45%	57%	89%
Consumer Cyclicals	66%	30%	80%
Consumer Non-Cyclicals	35%	22%	71%
Energy	45%	39%	82%
Financials	74%	50%	74%
Health care	56%	39%	74%
Industrials	56%	60%	87%
Technology	62%	51%	82%
Telecommunications Services	25%	25%	75%
Utilities	47%	42%	89%
Average	51%	42%	80%

backtesting methodology as the other markets. We use MAE as an indicator of accuracy and compute the percentage improvement in forecast precision due to the inclusion of our news variables for each stock.

14.5.1 In-Sample Estimation Results

Table 14.5 shows the panel regression results for stocks in the Financials sector when the in-sample window is 2018–2019. As before, the results for other sectors are similar and lead to the same conclusions. They are provided in the appendix. The magnitude and the sign of coefficients in this table are similar to the results for the USA and Europe in Table 14.2. For instance, coefficients in the bold box show the spillover effect of news: Intensified overnight news about other financial stocks reduces stock i's traded dollar volume but increases stock i's return volatility, which is inline with our previous findings. In particular, a 10 percentage point increase in overnight news proportion of other stocks $(N_{t,\tau-1,-i})$ seems to increase realized volatility in stock i's returns by about 19% and reduce stock i's traded dollar volume by about 7%. As for the direct effect of stock-specific news, the positive coefficients on $N_{t,\tau-1,i}$

TABLE 14.4 Percentage of stocks for which news variables improve forecast precision according to the Diebold-Mariano test. Stocks in the STOXX 600

Volatility prediction (The STOXX 600 constituents)			
MAE computed every day for entities in the following:	Bins in first 1hr window	Intraday bins in-between	Bins in last 1hr window
Basic Materials	36%	62%	81%
Consumer Cyclicals	39%	68%	74%
Consumer Non-Cyclicals	8%	87%	70%
Energy	0%	86%	68%
Financials	34%	86%	76%
Health care	44%	68%	69%
Industrials	27%	74%	72%
Technology	29%	54%	50%
Telecommunications Services	23%	68%	61%
Utilities	11%	83%	94%
Average	**25%**	**74%**	**71%**

Volume prediction (The STOXX 600 constituents)			
MAE computed every day for entities in the following:	Bins in first 1hr window	Intraday bins in-between	Bins in last 1hr window
Basic Materials	77%	53%	62%
Consumer Cyclicals	79%	54%	79%
Consumer Non-Cyclicals	69%	46%	71%
Energy	50%	23%	50%
Financials	78%	43%	77%
Health care	65%	53%	59%
Industrials	70%	54%	82%
Technology	48%	56%	45%
Telecommunications Services	45%	41%	61%
Utilities	50%	25%	69%
Average	**63%**	**45%**	**65%**

suggest that news about stock i increases its dollar volume and return volatility, as expected. All coefficients are significant at the 5% confidence level.

The dummy variables for earnings announcement and index rebalance days are statistically significant, and their coefficients are also markedly similar to the ones in Table 14.2 for the USA and Europe. The coefficients on dummy variables for earnings announcement adjust the impact of stock-specific news $(N_{t,\tau-1,i})$, which tends to spike on those special days.

The goodness-of-fit statistics at the bottom of Table 14.5 are also similar to the ones we reported earlier for the other markets.

Figure 14.12 shows that coefficients are fairly stable when the in-sample period is rolled on a monthly basis 12 times from 2017–2018 until 2018–2019, similar again to our previous findings.

14.5.2 Out-Of-Sample Performance

Model performance during the 2020Q1 sell-off

Figure 14.13 shows the out-of-sample results for 2020Q1 when the in-sample period is 2018–2019. It shows the average percentage improvement in MAE

TABLE 14.5 Representative regression results based on the 20180–2019 in-sample window. Stocks in the Financials sector of the TOPIX 500

Sector		Financials			
LHS		Log(Realized Volatility)		Log($ Volume)	
Statistic		Coef	p-values	Coef	p-values
$N_{t,\tau-1,i}$ newsRatio		7.2E-04	0.000	1.2E-03	0.000
newsRatio (ON)		0.005	0.000	0.004	0.000
newsRatio (NOON)		0.001	0.001	0.001	0.000
$N_{t,\tau-1,-i}$ newsRatio (sector)		-7.6E-05	0.040	0.000	0.000
newsRatio (ON, sector)		0.019	0.000	-0.007	0.000
newsRatio (NOON, sector)		0.004	0.000	0.005	0.000
$D_{t,\tau,i}$ earningDates		0.212	0.000	0.337	0.000
indexRebalance		0.083	0.000	0.110	0.000
$y_{t,\tau-1,i}$ y_lag1		0.269	0.000	0.347	0.000
y_season1		0.177	0.000	0.203	0.000
y_season2		0.113	0.000	0.109	0.000
y_season3		0.082	0.000	0.077	0.000
$y_{t-j,\tau,i}$ y_season4		0.057	0.000	0.061	0.000
y_season5		0.057	0.000	0.057	0.000
y_season6		0.045	0.000	0.038	0.000
y_season7		0.033	0.000	0.044	0.000
y_season8		0.043	0.000	0.064	0.000
R-squared		0.53		0.56	
AIC		0.519		1.410	
BIC		0.522		1.413	

of realized volatility forecasts per 30-minute bin across stocks by including news variables in the model. The distribution in Figure 14.13 lies mostly on the positive side, meaning that news variables would have improved forecasts for a large majority (92%) of stocks in the TOPIX 500 during 2020Q1. The magnitude of improvement varies widely across stocks. The mode of the distribution is about 5%, meaning that forecasts typically improve by 5% or more for about half of the stocks in the TOPIX 500. Figure 14.13 also dissects the results into the morning session and the afternoon session to gauge whether

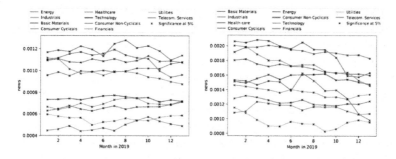

Figure 14.12 Rolling estimated coefficients month by month in 2019 when the LHS is log(realized volatility) (left) and log(dollar volume) (right). Stocks in the TOPIX 500

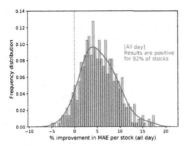 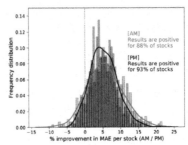

Figure 14.13 Forecasting realized volatility in 2020Q1 for stocks in the TOPIX 500. Distribution across stocks of the percentage improvement in precision per 30-minute bin when news variables are included. Results are for all intraday bins (left), which is dissected into the morning (AM) session and the afternoon (PM) session (right).

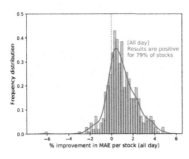 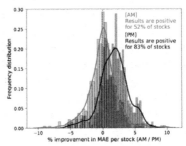

Figure 14.14 Forecasting dollar volume in 2020Q1 for stocks in the TOPIX 500. Distribution across stocks of the percentage improvement in precision per 30-minute bin when news variables are included. Results are for all intraday bins (left), which is dissected into the morning (AM) session and the afternoon (PM) session (right).

the predictive power lies at specific times of the day. In both sessions, the inclusion of news variables improves forecast precision for a large majority (about 90%) of the stocks, indicating that the model steadily performs during the day.

As for intraday prediction of dollar volume, Figure 14.14 shows that news variables again improve the precision for the majority (79%) of the stocks. This predictive power of news appears to lie mostly in the afternoon session in this case (although our analysis presented later reveals that this is not always the case). These results for dollar volume prediction are overall strong and in favour of the inclusion of news variables, although they are somewhat weaker than the results for volatility prediction.

The predictive performance depicted in Figures 14.13–14.14 is striking considering that markets were particularly turbulent in 2020Q1. It is interesting

TABLE 14.6 Summary of forecast improvement for alternative estimation and prediction windows. Stocks in the TOPIX 500

	Bins	Out-of-sample window			
		2020Q1	2019	2018	2017
Realized volatility	All day	92.2	99.8	99.0	95.3
	AM	88.4	99.6	99.0	95.1
	First 1hr	75.5	95.0	96.9	85.8
	Last 1hr	94.4	99.8	99.6	97.4
	PM	92.6	99.6	98.5	92.5
	First 1hr	95.3	99.0	96.2	74.9
	Last 1hr	83.9	99.4	98.3	95.1
$ volume	All day	79.2	73.4	82.4	85.6
	AM	51.7	70.2	71.3	82.1
	First 1hr	47.2	59.8	70.3	80.9
	Last 1hr	73.7	89.1	69.5	69.9
	PM	83.3	59.8	82.0	72.9
	First 1hr	74.9	78.9	77.1	71.1
	Last 1hr	81.0	44.1	75.3	66.5

to note that, during this quarter, the number of news articles per stock published by Nikkei was lower than the same quarter in the previous years. The reduction in the number of news is thought to be due to limited resources and journalists working from home during the pandemic as well as more macro-driven than stock-specific-driven newsflow.

Stability of the performance across time

The above results are based on the 2018-2019 in-sample window and the 2020Q1 out-of-sample window. In order to test robustness, we rolled these windows year by year starting from 2015–2016 as the in-sample and 2017 as the out-of-sample. Table 14.6 summarizes the outcomes.

The 2020Q1 column includes the results shown in Figures 14.13–14.14. In addition, the two separate trading sessions are further dissected into the first hour and the last hour, respectively. The motivation here is to identify any regularities in predictive performance during the day.

The table reveals that the performance of the model is stable when the in-sample and out-of-sample windows are shifted across time. The model is also robust to extreme market movements as evidenced in the 2020Q1 out-of-sample results. The model performance is also reasonably stable across intraday bins, particularly for volatility. News seems to show more power in predicting volatility than dollar volume in this exercise. Before the pandemic, news variables improved volatility forecasts for almost all stocks.

14.5.3 Additional Findings

In Section 14.2, we explained that media attention is highly uneven across stocks and during the year because there are strong seasonal patterns in news

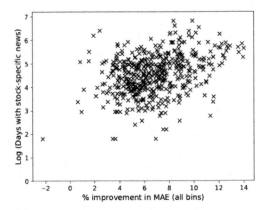

Figure 14.15 Extent of media coverage against predictive power of news. The y-axis is the number of days (in log-scale) with stock-specific news. The x-axis is the percentage improvement in MAE in realized volatility forecasts. The out-of-sample results for 2019 for stocks in the TOPIX 500. News Dolphin data with the relevance score at least 90 (out of 100) are used.

arrival intensity at daily, weekly and quarterly frequencies. It might be tempting to think that news variables matter only when big news strikes. Then it is worthwhile investigating whether our model tends to extract stronger predictive power of news for stocks with frequent coverage than stocks without, or during earning announcement seasons. Figure 14.15 presents the frequency of news coverage (given by the number of days with stock-specific news in log-scale) for each stock in the TOPIX 500 against the percentage improvement in MAE by including news variables in the model. We take the realized volatility prediction for the 2019 out-of-sample period as an example. There does not seem to be a strong relationship between the predictive power of news and the extent of media coverage. We make a similar observation if we look at forecast outcomes for dollar volume, other out-of-sample windows, or the other regions. We also found no evidence that our out-of-sample results depend on stock characteristics (e.g. market cap and the total number of shares) or special days (e.g. earning announcement days). This suggests that the performance of our model would remain reasonably stable and consistent for a given basket of stocks over time.

In fact, news variables are important for a model when there is no news too because their inclusion improves the precision of estimated coefficients on other variables, the most notable ones being the autoregressive coefficients on the lags of $y_{t,\tau,i}$. As our in-sample results showed, our news variables are relevant explanatory covariates whose coefficients are economically non-negligible. Hence the inclusion of news variables ensures that the overall dynamics of the model are more unbiased throughout the out-of-sample period.

Filtering the data: Noise reduction versus robustness

Filtering news by the relevance score can help remove noise from the data, but it reduces sample size, as discussed in the data description section. Hence it is worthwhile assessing whether the in-sample and out-of-sample results are affected by the use of the relevance filter.

For this purpose, we re-estimated the model and reproduced forecasts using news of all relevance scores for the application in Japan. We find that in-sample coefficients become markedly more stable than in Figure 14.12 if we do not filter the data by relevance. We also find that the predictive power of news we presented in the previous section is not affected if we do not filter the data.

These findings point to the fact that filters can be applied more coarsely depending on the application and, in this instance, perhaps because News Dolphin is relatively homogeneous in the sense that it is based solely on Nikkei's news. Refinitiv's dataset considered for markets in the USA and Europe was based on news published by a variety of news agencies in which case noise reduction becomes more important.

14.6 CONCLUDING REMARKS

In this chapter, we introduced a news-driven model for forecasting intraday volatility and volume. We presented the evidence of the robustness as well as the statistical and economic significance of our news-signal metrics to reinforce their usefulness in optimal execution globally. Using the Refinitiv and News Dolphin datasets, we showed that our news-driven model can improve intraday volatility and volume forecasts for a large majority of the stocks in the Russell 1000, the STOXX 600 and the TOPIX 500 indices. The model is shown to withstand the market turbulence of 2020Q1.

We described at the beginning of this chapter that it is common to use numeric scores of relevance and sentiment in financial applications of this kind as those scores can help reduce noise and detect signal. See, for instance, [Boudoukh et al., 2019], [Caporin and Poli, 2017], [Shi et al., 2016] and references therein for some recent examples. However, we found that our news-driven model without the use of those scores also seems to be able to extract notably strong predictive power in the application for stocks in Japan while maintaining robustness to outliers over time. This is presumably due at least in part to the fact that Nikkei (the most influential business news agency in Japan) is the sole source of news in the News Dolphin dataset.

We think that the inclusion of news variables has an important indirect effect on our risk and liquidity forecasts through parameter estimation. Ignoring news in the predictive model could potentially lead to distorted estimates of the other parameters and hence affect forecast accuracy.

We captured news arrival intensity using proportion metrics. Our simple and intuitive transformation of seemingly heavy-tailed data was the key to the robustness of our results across time and markets. Although we have focussed on modelling risk and liquidity, it may be possible to use our measures of news arrival intensity to extract directional signals in price movements and signed volume if sentiment scores are also used to distinguish news with positive and negative sentiment. This is an interesting application for further research.

DISCLAIMER

APPENDIX

Appendix 14.A   IN-SAMPLE RESULTS FOR OTHER SECTORS IN RUSSELL 1000

TABLE 14.A.1   In-sample estimation results for other sectors in the Russell 1000. The LHS of the model is log(realized volatility).

Sector	Basic Materials		Consumer Cyclicals		Consumer Non-Cyclicals		Energy		Health care		Industrials		Technology		Telecommunications Services		Utilities	
Statistic	Coef	p-values	Coef	p-values	Coef	p-values	Coef	p-values	Coef	p-values	Coef	p-values	Coef	p-values	Coef	p-values	Coef	p-values
newsRatio (+)	5.8E-04	0.000	7.0E-04	0.000	0.001	0.000	2.9E-04	0.021	7.4E-04	0.000	3.8E-04	0.000	7.4E-04	0.000	7.0E-04	0.004	3.7E-04	0.021
newsRatio (−)	0.001	0.000	0.001	0.000	0.001	0.000	0.000	0.006	0.001	0.000	0.001	0.000	8.6E-04	0.000	7.7E-04	0.003	7.9E-04	0.012
newsRatio (+, ON)	0.004	0.000	0.005	0.000	0.004	0.000	0.004	0.000	0.004	0.000	0.006	0.000	0.005	0.000	0.003	0.000	0.003	0.000
newsRatio (−, ON)	0.002	0.000	0.004	0.000	0.002	0.000	0.002	0.000	0.003	0.000	0.003	0.000	0.002	0.000	0.002	0.000	0.002	0.000
newsRatio (+, sector)	-5.0E-05	0.285	-1.1E-04	0.000	-2.1E-04	0.000	-1.1E-04	0.001	-7.5E-05	0.005	1.1E-05	0.612	-2.2E-04	0.000	2.5E-04	0.217	-2.5E-04	0.000
newsRatio (−, sector)	-1.5E-04	0.003	-6.5E-05	0.000	2.3E-04	0.000	9.7E-05	0.003	-1.6E-04	0.000	-1.6E-05	0.455	-5.3E-05	0.015	4.7E-05	0.798	-4.4E-04	0.000
newsRatio (+, ON, sector)	0.080	0.000	0.045	0.000	0.044	0.000	0.043	0.000	0.055	0.000	0.065	0.000	0.053	0.000	0.145	0.000	0.078	0.000
newsRatio (−, ON, sector)	0.039	0.000	0.016	0.000	0.048	0.000	0.036	0.000	0.023	0.000	0.012	0.000	0.028	0.000	0.079	0.000	0.065	0.000
earningDates	0.368	0.000	0.357	0.000	0.361	0.000	0.277	0.000	0.321	0.000	0.378	0.000	0.333	0.000	0.354	0.000	0.222	0.000
earningDates (ON)	-0.323	0.000	-0.355	0.000	-0.466	0.000	-0.205	0.000	-0.281	0.000	-0.381	0.000	-0.287	0.000	-0.521	0.000	-0.244	0.000
lastDayOfMonth	-0.011	0.005	-0.020	0.000	0.015	0.000	-0.025	0.000	-0.025	0.000	-0.024	0.000	-0.016	0.000	-0.027	0.004	-0.003	0.480
tripleWitching	-0.001	0.862	0.013	0.001	0.035	0.000	-0.008	0.104	-0.001	0.827	0.045	0.000	0.006	0.127	0.003	0.826	0.041	0.000
optionExpiry	-0.039	0.000	-0.029	0.000	-0.021	0.000	-0.054	0.000	-0.013	0.000	-0.051	0.000	-0.036	0.000	0.011	0.266	-0.048	0.000
IndexRebalance	0.079	0.000	0.055	0.000	0.141	0.000	0.026	0.000	0.067	0.000	0.102	0.000	0.083	0.000	0.061	0.022	0.081	0.000
IndexRebalance (Close)	0.281	0.000	0.314	0.000	0.242	0.000	0.298	0.000	0.378	0.000	0.254	0.000	0.345	0.000	0.285	0.000	0.397	0.000
y_lag1	0.378	0.000	0.408	0.000	0.386	0.000	0.347	0.000	0.428	0.000	0.404	0.000	0.430	0.000	0.368	0.000	0.331	0.000
y_season1	0.166	0.000	0.165	0.000	0.186	0.000	0.186	0.000	0.156	0.000	0.159	0.000	0.169	0.000	0.172	0.000	0.204	0.000
y_season2	0.087	0.000	0.082	0.000	0.089	0.000	0.100	0.000	0.077	0.000	0.077	0.000	0.083	0.000	0.083	0.000	0.105	0.000
(...omitted)	...		...		...		...		...		...		...		...		...	
R-squared	0.64		0.65		0.65		0.73		0.65		0.60		0.66		0.60		0.65	
AIC	0.494		0.431		0.430		0.181		0.472		0.559		0.447		0.523		0.297	
BIC	0.499		0.436		0.434		0.185		0.477		0.563		0.451		0.531		0.301	

TABLE 14.A.2 In-sample estimation results for other sectors in the Russell 1000. The LHS of the model is log(volume).

Sector	Basic Materials		Consumer Cyclicals		Consumer Non-Cyclicals		Log(Realized Volatility) Energy		Health care		Industrials		Technology		Telecommunications Services		Utilities	
Statistic	Coef	p-values	Coef	p-values	Coef	p-values	Coef	p-values	Coef	p-values	Coef	p-values	Coef	p-values	Coef	p-values	Coef	p-values
newsRatio (+)	0.001	0.000	0.001	0.000	0.002	0.000	4.3E-04	0.033	0.001	0.000	0.001	0.000	0.001	0.000	0.002	0.000	0.001	0.003
newsRatio (−)	0.002	0.000	0.002	0.000	0.002	0.000	0.001	0.006	0.003	0.000	0.001	0.000	0.002	0.000	0.001	0.013	0.001	0.006
newsRatio (+, ON)	0.005	0.000	0.013	0.000	0.004	0.000	0.002	0.008	0.008	0.000	0.011	0.000	0.014	0.000	−0.001	0.239	−0.002	0.017
newsRatio (−, ON)	0.004	0.000	0.015	0.000	0.004	0.000	0.004	0.000	0.012	0.000	0.010	0.000	0.011	0.000	0.001	0.130	0.002	0.044
newsRatio (+, sector)	2.4E-06	0.974	8.6E-05	0.001	4.8E-04	0.000	8.1E-05	0.153	4.0E-04	0.000	2.1E-04	0.000	4.0E-04	0.000	0.001	0.000	3.4E-04	0.000
newsRatio (−, sector)	1.7E-04	0.031	−1.3E-04	0.000	0.001	0.000	2.4E-04	0.000	−5.5E-05	0.076	3.7E-04	0.000	2.9E-04	0.000	0.001	0.002	4.9E-04	0.000
newsRatio (+, ON, sector)	−0.089	0.000	−0.041	0.000	−0.055	0.000	−0.040	0.000	−0.057	0.000	−0.081	0.000	−0.058	0.000	−0.147	0.000	−0.072	0.000
newsRatio (−, ON, sector)	−0.054	0.000	−0.018	0.000	−0.042	0.000	−0.033	0.000	−0.030	0.000	−0.009	0.000	−0.021	0.000	−0.012	0.595	−0.104	0.000
earningDates	0.467	0.000	0.474	0.000	0.424	0.000	0.346	0.000	0.404	0.000	0.484	0.000	0.477	0.000	0.518	0.000	0.256	0.000
earningDates (ON)	0.741	0.000	0.737	0.000	0.833	0.000	0.597	0.000	0.844	0.000	0.923	0.000	0.686	0.000	0.877	0.000	0.323	0.000
lastDayOfMonth	0.041	0.000	−0.004	0.268	0.046	0.000	−0.009	0.071	0.002	0.604	0.012	0.004	0.032	0.000	0.047	0.002	0.094	0.000
tripleWitching	0.067	0.000	0.031	0.000	0.049	0.000	0.037	0.000	0.055	0.000	0.036	0.000	0.039	0.000	0.019	0.459	0.124	0.000
optionExpiry	−0.017	0.014	−0.001	0.780	0.033	0.000	−0.027	0.000	−0.027	0.000	−0.002	0.637	0.001	0.740	0.033	0.055	−0.046	0.000
indexRebalance	0.148	0.000	0.132	0.000	0.135	0.000	0.010	0.446	0.080	0.000	0.129	0.000	0.111	0.000	0.056	0.178	0.057	0.002
indexRebalance (Close)	0.635	0.000	0.608	0.000	0.595	0.000	0.626	0.000	0.670	0.000	0.611	0.000	0.734	0.000	0.753	0.000	0.550	0.000
y_lag1	0.429	0.000	0.477	0.000	0.447	0.000	0.431	0.000	0.489	0.000	0.443	0.000	0.472	0.000	0.451	0.000	0.370	0.000
y_season1	0.189	0.000	0.188	0.000	0.204	0.000	0.186	0.000	0.180	0.000	0.190	0.000	0.189	0.000	0.189	0.000	0.200	0.000
y_season2	0.092	0.000	0.092	0.000	0.098	0.000	0.102	0.000	0.091	0.000	0.096	0.000	0.092	0.000	0.086	0.000	0.111	0.000
(...omitted)	...		...		...		...		...		...		...		...		...	
R-squared	0.57		0.59		0.59		0.61		0.58		0.56		0.59		0.62		0.56	
AIC	1.435		1.353		1.286		1.103		1.367		1.464		1.371		1.458		1.341	
BIC	1.440		1.357		1.290		1.108		1.371		1.469		1.375		1.467		1.345	

Appendix 14.B IN-SAMPLE RESULTS FOR STOXX 600

TABLE 14.B.1 In-sample estimation results for sectors in the STOXX 600. The LHS of the model is log(realized volatility).

Sector	Basic Materials		Consumer Cyclicals		Consumer Non-Cyclicals		Energy		Financials		Health care		Industrials		Technology		Telecommunications Services		Utilities	
Statistic	Coef	p-values	Coef	p-values	Coef	p-values	Coef	p-values	Coef	p-values	Coef	p-values	Coef	p-values	Coef	p-values	Coef	p-values	Coef	p-values
newsRatio (+)	0.001	0.000	1.1E-03	0.000	5.4E-04	0.000	3.1E-04	0.001	5.1E-04	0.000	8.7E-04	0.000	0.001	0.000	6.3E-04	0.000	5.0E-04	0.001	5.7E-04	0.000
newsRatio (−)	0.001	0.000	0.002	0.000	1.2E-03	0.000	3.7E-04	0.000	6.3E-04	0.000	1.1E-03	0.000	0.002	0.000	1.5E-03	0.000	1.1E-03	0.000	0.001	0.000
newsRatio (+, ON)	0.003	0.000	0.004	0.000	0.003	0.000	0.002	0.000	0.004	0.000	0.003	0.000	0.006	0.000	0.003	0.000	0.003	0.000	0.003	0.000
newsRatio (−, ON)	0.002	0.000	0.005	0.000	0.003	0.000	0.002	0.000	0.003	0.000	0.003	0.000	0.004	0.000	0.003	0.000	0.003	0.000	0.002	0.000
newsRatio (+, sector)	5.0E-05	0.243	1.5E-04	0.000	−1.2E-04	0.018	−1.5E-04	0.004	1.9E-04	0.000	3.6E-05	0.490	−2.6E-04	0.000	−2.1E-04	0.017	3.2E-04	0.001	−2.1E-04	0.004
newsRatio (−, sector)	4.3E-05	0.302	−6.2E-05	0.007	−2.3E-04	0.000	−9.4E-05	0.075	8.5E-05	0.000	−4.0E-05	0.436	−1.1E-05	0.602	−3.4E-06	0.972	−1.7E-04	0.048	1.6E-04	0.022
newsRatio (+, ON, sector)	0.007	0.000	0.006	0.000	0.025	0.000	0.007	0.000	0.006	0.000	0.007	0.000	0.007	0.000	0.016	0.000	0.008	0.000	0.008	0.000
newsRatio (−, ON, sector)	0.004	0.000	0.006	0.000	0.035	0.000	0.010	0.000	0.004	0.000	0.007	0.000	0.004	0.000	0.008	0.000	0.015	0.000	0.012	0.000
earningDates	0.308	0.000	0.339	0.000	0.380	0.000	0.207	0.000	0.234	0.000	0.366	0.000	0.353	0.000	0.399	0.000	0.321	0.000	0.183	0.000
earningDates (ON)	−0.152	0.000	−0.281	0.000	−0.483	0.000	−0.150	0.000	−0.204	0.000	−0.236	0.000	−0.245	0.000	−0.303	0.000	−0.252	0.000	−0.115	0.001
lastDayOfMonth	0.026	0.000	0.024	0.000	0.031	0.000	0.040	0.000	−0.034	0.000	0.034	0.000	0.050	0.000	0.033	0.000	2.2E-02	0.000	−0.021	0.014
optionExpiry	−0.007	0.181	0.011	0.001	9.5E-03	0.038	−0.009	0.081	0.005	0.054	0.001	0.920	0.023	0.000	0.002	0.691	0.020	0.001	−0.007	0.218
indexRebalance	0.022	0.004	0.000	0.998	0.022	0.003	−0.001	0.866	0.003	0.589	0.032	0.001	−0.014	0.005	−0.010	0.341	0.005	0.620	0.005	0.600
indexRebalance (Close)	−0.021	0.221	0.060	0.000	0.083	0.000	−0.019	0.464	0.021	0.075	0.015	0.514	0.048	0.000	0.102	0.000	0.017	0.535	0.047	0.095
y_lag1	0.329	0.000	0.359	0.000	0.349	0.000	0.353	0.000	0.330	0.000	0.316	0.000	0.335	0.000	0.354	0.000	0.351	0.000	0.341	0.000
y_season1	0.136	0.000	0.134	0.000	0.132	0.000	0.151	0.000	0.129	0.000	0.134	0.000	0.130	0.000	0.144	0.000	0.132	0.000	0.140	0.000
y_season2	0.085	0.000	0.083	0.000	0.085	0.000	0.088	0.000	0.082	0.000	0.081	0.000	0.086	0.000	0.085	0.000	0.086	0.000	0.091	0.000
(...omitted)	...		...		...		...		...		...		...		...		...		...	
R-squared	0.56		0.57		0.55		0.63		0.53		0.53		0.54		0.55		0.56		0.56	
AIC	1.387		1.382		1.336		1.134		1.484		1.432		1.441		1.548		1.305		1.211	
BIC	1.390		1.384		1.339		1.137		1.486		1.435		1.444		1.551		1.308		1.214	

TABLE 14.B.2 In-sample estimation results for sectors in the STOXX 600. The LHS of the model is log(volume).

Log(Realized Volatility)

Sector	Basic Materials		Consumer Cyclicals		Consumer Non-Cyclicals		Energy		Financials		Health care		Industrials		Technology		Telecommunications Services		Utilities	
Statistic	Coef	p-values	Coef	p-values	Coef	p-values	Coef	p-values	Coef	p-values	Coef	p-values	Coef	p-values	Coef	p-values	Coef	p-values	Coef	p-values
newsRatio (+)	0.001	0.000	0.002	0.000	0.001	0.000	4.9E-04	0.000	0.001	0.000	0.001	0.000	0.001	0.000	0.002	0.000	0.001	0.000	0.001	0.000
newsRatio (−)	0.001	0.000	0.003	0.000	0.002	0.000	5.0E-04	0.000	0.001	0.000	0.002	0.000	0.002	0.000	0.003	0.000	0.002	0.000	0.001	0.000
newsRatio (+, ON)	0.004	0.000	0.009	0.000	0.003	0.000	4.0E-04	0.412	0.009	0.000	0.003	0.000	0.011	0.000	0.002	0.000	0.002	0.001	0.001	0.027
newsRatio (−, ON)	0.003	0.000	0.010	0.000	0.005	0.000	4.7E-04	0.353	0.010	0.000	0.003	0.000	0.011	0.000	0.004	0.000	0.002	0.001	0.002	0.013
newsRatio (+, sector)	0.000	0.001	0.000	0.000	0.000	0.000	4.4E-04	0.000	2.0E-04	0.000	5.6E-5	0.344	3.0E-05	0.240	-8.8E-05	0.421	0.001	0.000	7.8E-05	0.408
newsRatio (−, sector)	-1.0E-04	0.040	1.3E-05	0.638	9.7E-05	0.107	1.7E-04	0.006	6.5E-05	0.001	2.7E-04	0.000	1.1E-04	0.000	-1.2E-04	0.321	3.0E-05	0.770	4.1E-04	0.000
newsRatio (+, ON, sector)	-0.008	0.000	-0.008	0.000	-0.026	0.000	-0.012	0.000	-0.009	0.000	-0.009	0.000	-0.011	0.000	-0.022	0.000	-0.010	0.000	-0.013	0.000
newsRatio (−, ON, sector)	-0.009	0.000	-0.008	0.000	-0.056	0.000	-0.013	0.000	-0.002	0.000	-0.009	0.000	-0.007	0.000	-0.013	0.000	-0.025	0.000	-0.013	0.000
earningDates	0.360	0.000	0.345	0.000	0.437	0.000	0.260	0.000	0.257	0.000	0.370	0.000	0.370	0.000	0.473	0.000	0.360	0.000	0.191	0.000
earningDates (ON)	0.393	0.000	0.239	0.000	0.433	0.000	0.385	0.000	0.210	0.000	0.218	0.013	0.200	0.000	0.473	0.000	0.322	0.000	-0.025	0.824
lastDayOfMonth	0.044	0.000	0.044	0.000	0.060	0.000	0.046	0.000	0.022	0.000	0.040	0.000	0.059	0.000	0.028	0.000	0.067	0.000	0.001	0.884
optionExpiry	-0.019	0.001	-0.002	0.641	0.004	0.466	-0.005	0.478	0.000	0.967	0.004	0.556	0.004	0.283	0.023	0.002	-0.013	0.105	-0.026	0.001
indexRebalance	0.051	0.000	0.014	0.046	0.050	0.000	0.011	0.397	0.029	0.000	0.093	0.000	0.038	0.000	-0.007	0.600	0.064	0.000	0.069	0.000
indexRebalance (Close)	0.255	0.000	0.296	0.000	0.346	0.000	0.173	0.000	0.247	0.000	0.242	0.000	0.286	0.000	0.388	0.000	0.290	0.000	0.186	0.000
y-lag1	0.432	0.000	0.481	0.000	0.456	0.000	0.461	0.000	0.452	0.000	0.424	0.000	0.454	0.000	0.459	0.000	0.459	0.000	0.454	0.000
y-season1	0.154	0.000	0.153	0.000	0.157	0.000	0.150	0.000	0.144	0.000	0.156	0.000	0.153	0.000	0.153	0.000	0.154	0.000	0.148	0.000
y-season2	0.083	0.000	0.077	0.000	0.077	0.000	0.074	0.000	0.076	0.000	0.088	0.000	0.081	0.000	0.083	0.000	0.080	0.000	0.078	0.000
(...omitted)	...	...	...	...	...	...	...	...	...	...	...	...	...	...	...	...	...	...	...	...
R-squared	0.57		0.59		0.59		0.61		0.56		0.58		0.56		0.59		0.62		0.56	
AIC	1.435		1.353		1.286		1.103		1.403		1.367		1.464		1.371		1.458		1.341	
BIC	1.440		1.357		1.290		1.108		1.408		1.371		1.469		1.375		1.467		1.345	

Appendix 14.C   IN-SAMPLE RESULTS FOR TOPIX 500

TABLE 14.C.1   In-sample estimation results for other sectors in the TOPIX 500. The LHS of the model is log(realized volatility).

Sector	Basic Materials		Consumer Cyclicals		Consumer Non-Cyclicals		Energy		Health care		Industrials		Technology		Telecommunications Services		Utilities	
							Log(Realized Volatility)											
Statistic	params	pVals	params	pVals	params	pVals	params	pVals	params	pVals	params	pVals	params	pVals	params	pVals	params	pVals
newsRatio	9.6E-04	0.000	1.1E-03	0.000	0.001	0.000	1.1E-03	0.000	1.1E-03	0.000	8.8E-04	0.000	7.2E-04	0.000	4.9E-04	0.000	5.9E-04	0.000
newsRatio (ON)	3.3E-03	0.000	5.9E-03	0.000	0.005	0.000	1.4E-03	0.000	3.3E-03	0.000	4.9E-03	0.000	4.7E-03	0.000	3.0E-03	0.000	2.0E-03	0.000
newsRatio (NOON)	0.001	0.000	0.001	0.000	0.000	0.273	0.000	0.604	0.001	0.000	0.001	0.000	0.001	0.018	0.001	0.037	0.001	0.072
newsRatio (sector)	3.3E-05	0.489	1.3E-04	0.000	2.3E-04	0.000	5.5E-04	0.096	-4.6E-05	0.521	1.2E-04	0.000	-1.3E-04	0.000	-1.3E-04	0.524	-2.0E-04	0.083
newsRatio (ON, sector)	0.038	0.000	0.019	0.000	0.020	0.000	0.064	0.000	0.049	0.000	0.016	0.000	0.030	0.000	0.042	0.000	0.034	0.000
newsRatio (NOON, sector)	0.004	0.000	0.004	0.000	0.004	0.000	-0.003	0.349	0.006	0.000	0.004	0.000	0.004	0.000	0.007	0.000	0.006	0.000
earningDates	0.384	0.000	0.350	0.000	0.357	0.000	0.245	0.000	0.342	0.000	0.303	0.000	0.350	0.000	0.172	0.000	0.252	0.000
earningDates (ON)	-0.139	0.000	-0.085	0.000	-0.061	0.000	-0.196	0.000	-0.101	0.000	-0.082	0.000	-0.162	0.000	-0.219	0.000	-0.079	0.001
indexRebalance	0.114	0.000	0.160	0.000	0.123	0.000	0.053	0.018	0.125	0.000	0.109	0.000	0.094	0.000	0.140	0.000	0.146	0.000
y_lag1	0.258	0.000	0.300	0.000	0.268	0.000	0.229	0.000	0.250	0.000	0.302	0.000	0.291	0.000	0.325	0.000	0.220	0.000
y_season1	0.194	0.000	0.178	0.000	0.185	0.000	0.218	0.000	0.207	0.000	0.178	0.000	0.183	0.000	0.212	0.000	0.195	0.000
y_season2	0.110	0.000	0.098	0.000	0.110	0.000	0.126	0.000	0.110	0.000	0.101	0.000	0.105	0.000	0.108	0.000	0.120	0.000
y_season3	0.087	0.000	0.070	0.000	0.073	0.000	0.090	0.000	0.083	0.000	0.073	0.000	0.080	0.000	0.088	0.000	0.094	0.000
(...omitted)	...	...	...	...	...	...	...	...	...	...	...	...	...	...	...	...	...	...
R-squared	0.55		0.55		0.54		0.59		0.54		0.58		0.58		0.63		0.55	
AIC	0.513		0.498		0.584		0.436		0.571		0.443		0.515		0.362		0.351	
BIC	0.516		0.501		0.587		0.444		0.575		0.446		0.518		0.371		0.355	

TABLE 14.C.2 In-sample estimation results for other sectors in the TOPIX 500. The LHS of the model is log(dollar volume).

Sector	Basic Materials		Consumer Cyclicals		Consumer Non-Cyclicals		Energy		Health care		Industrials		Technology		Telecommunications Services		Utilities	
																	Log(Dollar volume)	
Statistic	params	pVals	params	pVals	params	pVals	params	pVals	params	pVals	params	pVals	params	pVals	params	pVals	params	pVals
newsRatio	0.002	0.000	0.002	0.000	0.001	0.000	9.8E-04	0.001	0.002	0.000	0.001	0.000	0.001	0.000	0.001	0.000	0.001	0.000
newsRatio (ON)	0.003	0.000	0.008	0.000	0.003	0.000	0.000	0.676	0.003	0.000	0.007	0.000	0.003	0.000	0.000	0.717	0.000	0.170
newsRatio (NOON)	0.001	0.000	0.001	0.000	0.001	0.011	0.002	0.076	0.001	0.001	0.001	0.004	0.001	0.006	0.001	0.077	0.001	0.108
newsRatio (sector)	3.9E-04	0.000	4.4E-04	0.000	1.2E-03	0.000	1.0E-03	0.039	−1.5E-04	0.119	4.3E-04	0.000	3.4E-05	0.479	−0.001	0.005	−3.7E-04	0.025
newsRatio (ON, sector)	−0.017	0.000	−0.007	0.000	−0.007	0.000	−0.015	0.003	−0.018	0.000	−0.005	0.000	−0.007	0.000	0.000	0.997	−0.014	0.000
newsRatio (NOON, sector)	0.006	0.000	0.004	0.000	0.004	0.000	−0.005	0.329	0.008	0.000	0.004	0.000	0.005	0.000	0.008	0.000	0.011	0.000
earningDates	0.480	0.000	0.526	0.000	0.557	0.000	0.333	0.000	0.452	0.000	0.451	0.000	0.549	0.000	0.357	0.000	0.425	0.000
earningDates (ON)	−0.051	0.038	0.049	0.006	0.069	0.002	0.004	0.957	−0.027	0.392	0.046	0.002	0.120	0.000	−0.004	0.948	0.047	0.279
indexRebalance	0.143	0.000	0.168	0.000	0.203	0.000	0.004	0.920	0.153	0.000	0.135	0.000	0.138	0.000	0.163	0.000	0.164	0.000
y_lag1	0.378	0.000	0.389	0.000	0.368	0.000	0.368	0.000	0.373	0.000	0.381	0.000	0.406	0.000	0.429	0.000	0.350	0.000
y_season1	0.215	0.000	0.221	0.000	0.216	0.000	0.194	0.000	0.220	0.000	0.213	0.000	0.214	0.000	0.233	0.000	0.216	0.000
y_season2	0.098	0.000	0.103	0.000	0.102	0.000	0.115	0.000	0.099	0.000	0.098	0.000	0.093	0.000	0.097	0.000	0.111	0.000
y_season3	0.070	0.000	0.066	0.000	0.070	0.000	0.069	0.000	0.077	0.000	0.074	0.000	0.068	0.000	0.057	0.000	0.077	0.000
(...omitted)	...	...	...	...	...	...	...	...	...	...	...	...	...	...	...	...	...	...
R-squared	0.57		0.58		0.56		0.58		0.57		0.57		0.59		0.65		0.59	
AIC	1.253		1.170		1.312		1.164		1.218		1.186		1.171		0.691		1.089	
BIC	1.256		1.173		1.316		1.171		1.221		1.189		1.174		0.700		1.093	

REFERENCES

Allen, D. E., McAleer, M. J., and Singh, A. K. (2015). Machine News and Volatility: The Dow Jones Industrial Average and the TRNA Real-Time High-Frequency Sentiment Series. *Academic Press San Diego*, pages 327–344.

Barunik, J., Chen, C. Y.-H., and Vecer, J. (2019). Sentiment-driven stochastic volatility model: a high-frequency textual tool for economists. *arXiv preprint arXiv:1906.00059*.

Boudoukh, J., Feldman, R., Kogan, S., and Richardson, M. (2019). Information, trading, and volatility: evidence from firm-specific news. *The Review of Financial Studies*, 32(3):992–1033.

Caporin, M. and Poli, F. (2017). Building news measures from textual data and an application to volatility forecasting. *Econometrics*, 5(3):35.

Clements, A., Fuller, J., Papalexiou, V., et al. (2015). Public news flow in intraday component models for trading activity and volatility. Technical report, National Centre for Econometric Research.

Conrad, C. and Engle, R. F. (2021). Modelling volatility cycles: the MF2-GARCH model. *Working Paper series 21-05, Rimini Centre for Economic Analysis*.

Creamer, G. G. (2015). Can a corporate network and news sentiment improve portfolio optimization using the Black–Litterman model? *Quantitative Finance*, 15(8):1405–1416.

Cui, B. and Gozluklu, A. E. (2021). News and trading after hours. *Available at SSRN 3796812*.

Engle, R. F., Giglio, S., Kelly, B., Lee, H., and Stroebel, J. (2020). Hedging climate change news. *The Review of Financial Studies*, 33(3):1184–1216.

Gross-Klussmann, A. and Hautsch, N. (2011). When machines read the news: using automated text analytics to quantify high frequency news-implied market reactions. *Journal of Empirical Finance*, 18(2):321–340.

Ho, K.-Y., Shi, Y., and Zhang, Z. (2013). How does news sentiment impact asset volatility? Evidence from long memory and regime-switching approaches. *The North American Journal of Economics and Finance*, 26:436–456.

Kalev, P. S., Liu, W.-M., Pham, P. K., and Jarnecic, E. (2004). Public information arrival and volatility of intraday stock returns. *Journal of Banking & Finance*, 28(6):1441–1467.

Mitra, G., Erlwein-Sayer, C., Valle, C. A., and Yu, X. (2018). Using market sentiment to enhance second-order stochastic dominance trading models. In *High-Performance Computing in Finance*, pages 25–48. Chapman and Hall/CRC.

Mitra, L., Mitra, G., and Dibartolomeo, D. (2009). Equity portfolio risk estimation using market information and sentiment. *Quantitative Finance*, 9(8):887–895.

Rahimikia, E. and Poon, S.-H. (2020). Big data approach to realised volatility forecasting using har model augmented with limit order book and news. *Available at SSRN 3684040*.

Riordan, R., Storkenmaier, A., Wagener, M., and Zhang, S. S. (2013). Public information arrival: price discovery and liquidity in electronic limit order markets. *Journal of Banking & Finance*, 37(4):1148–1159.

Shi, Y., Ho, K.-Y., and Liu, W.-M. (2016). Public information arrival and stock return volatility: evidence from news sentiment and markov regime-switching approach. *International Review of Economics & Finance*, 42:291–312.

Song, S. and Bhattacharyya, R. (2021). The intraday impact of macroeconomic news on price moves of financial instruments. *Available at SSRN 3798844*.

Tetlock, P. C. (2010). Does public financial news resolve asymmetric information? *The Review of Financial Studies*, 23(9):3520–3557.

Uctum, R., Renou-Maissant, P., Prat, G., and Lecarpentier-Moyal, S. (2017). Persistence of announcement effects on the intraday volatility of stock returns: evidence from individual data. *Review of Financial Economics*, 35:43–56.

CHAPTER 15

Exogenous Risks, Alternative Data Implications for Strategic Asset Allocation – Multi-Subordination Levy Processes Approach

Boryana Racheva-Iotova

FactSet

CONTENTS

15.1 INTRODUCTION

The beginning of the 21st century had triggered the first transformational force toward exiting the framework of Modern Portfolio Theory (MPT) in investment management via the introduction of tail-risk measures and advanced extreme dependence modeling, which for first went outside of the territory of academic research and discussions following the Global Financial Crisis. While early systematic academic attempts in that space started

DOI: 10.1201/9781003293644-15

as soon as 2000 [Rachev and Mittnik, 2000] and have since then been rigorously developed within both risk management application expanding beyond market risk [Schoutens and Cariboni, 2010] and derivatives pricing domains [Schoutens, 2003], [Rachev et al., 2011] its wide-spread adoption across the industry has been facilitated by the FactSet's release of Fat-Tailed Risk Model [Racheva-Iotova et al., 2018] for the buy-side community and is somewhat propelled on the sell-side via the Expected Tail Loss (ETL)-based Trading Book risk assessment requirements [Basel Committee et al., 2014]. Abandoning the Gaussian normal bell-shaped paradigm for financial returns modeling via the introduction of fat tails for risk management has immensely important implications, including the ability to assess portfolio potential losses during upcoming market-stress events by means of incorporating higher than Gaussian (normal) probability of extreme events, as well as capturing of the dynamic nature of correlations [Racheva-Iotova, 2020a]. In a similar fashion, it consistently improves the risk-adjusted performance of quantitatively constructed strategies [Radev et al., n.d.]. What is critical in achieving those benefits is for the approach to be consistent with the previously available theory, that is, to offer an extension, not a replacement. That matters for at least three reasons: (a) to be able to explain and be consistent; (b) to be able to extend the entire rational-finance framework – market and credit risk assessment, derivatives pricing and portfolio optimization, instead of having inconsistent partial and siloed solutions for each of the areas; (c) to stay as close as possible to the theoretical foundations that have been proven in providing a robust framework, that is, keep the benefits of the previously existing paradigm.

The approach to achieve this was found to be via the introduction of time-subordinated models represented by Lévy processes, where the core process, typically being Brownian motion, is time subordinated by a second fully skewed to the right stochastic process representative of the "market clock" resembling the intensity and speed of the news hitting the market as compared to the physical clock [Racheva-Iotova et al., 2018] [Schoutens, 2003]. The most well-known return distributions arising from Lévy processes are the stable Paretian distributions [Rachev and Mittnik, 2000], and the class of tempered stable distributions [Schoutens, 2003], [Rachev et al., 2011] achieved by "tempering" the characteristic function of the stable law at the origin via a suitably selected function so that the tails still have power decay while the second moment becomes finite.

The second decade of the 21st century was marked by the advancements of behavioral finance and had led to the rise of goal-based investing in the wealth space and the emergence of the total portfolio approach for the institutional segments of the buy-side [Watson, 2019]. Similarly, we look at the necessity to model the behavioral phenomena and complex investors, preferences not as a replacement of the available paradigm but as an extension to it. Such bridge between behavioral finance and rational finance approaches is not trivial, but is achievable [Racheva-Iotova, 2020b] [Rachev et al., 2017]. One rigorous approach is based on "reshaping" the returns distribution via a function representing the behavioral views (that is a weighting function introduced in such a way that the resultant distribution stays in the same class) and selecting the objective function for the portfolio construction to be based on a risk measure that is capable of capturing the value function of the investor via a combination of multiple expected tail-risk and expected tail return components [Rachev et al., 2017]. Reshaping the returns distribution in the above-mentioned way is possible in and only in the domain of the Lévy processes and is analytically tractable for the stable and tempered stable processes in particular [Racheva-Iotova, 2020b].

The first two years of the 2020s presented yet another novelty layer to the investment portfolio management domain – the need to incorporate exogenous factors, which rose to the forefront with the global concerns around climate change and the global Covid pandemic.

In a broad sense, exogenous risks arise from environmental-, governance-, healthcare-, political-, policy- and technology-related and other similar potential disruptions. Those risks are also referred to as novel, and this is not without a reason – (1) they are novel both in terms of their existence and impact; (2) the lack of unified and standardized quantification and measurement remains a fact; and furthermore (3) the markets most probably have not yet priced those risk properly. The recent academic literature suggests that there are parallels among environmental, social and governance (ESG) risk, climate change risk, cybersecurity risk and geopolitical risk in terms of measurement challenges, and insufficient and non-comparable disclosures [Karagozoglu, 2021]. Thus, it is fundamental for any novel risks model to account for the data uncertainty and to possess solid theoretical and economic foundations instead of relying purely on data mining approaches.

(1), (2), and (3) above are typical situations where traditional approaches toward incorporating such risks via regression techniques would yield relatively low explanatory power and level of robustness.

Related to the most critical type of novel risk – the climate-change risk, although there are some studies showing that in developed markets the carbon-intensive companies underperformed during the last seven years [Kazdin et al., 2021] have seen a relative downward trend in their Price to Book valuation [Giese et al., 2021], and that linear factors could be constructed to incorporate climate risk and used within the standard arbitrage pricing theory and MPT-based framework for portfolio construction [Roncalli et al., 2021]; we think more time needs to pass for those to become reliable based on standardization of the reporting and ability of market practitioners to price those risks so that to offer robust regression-based analysis.

With regard to the impact of climate risk on asset allocation, recent attempts have been focused on modeling the impact on the expected returns of the asset classes and regions [Tokat-Acikel et al., 2021], but no advancements have been published related to climate risk assessment within a holistic asset-allocation framework.

We learned that the world is not driven by the bell-shaped distribution, that our objectives expand beyond risk-return profiles and that the factors that affect the path toward reaching goals expand far outside the financial markets' domain. We will aim at proposing a modeling framework for this third layer stepping upon what [Racheva-Iotova, 2020b] has regarding the extreme and behavioral phenomena modeling, with the objective to build an integrated and cohesive approach toward the modeling of fundamental phenomena related to quantitative modeling of financial asset management. We will do that via the introduction of multi-subordinated Lévy processes. With that, we will extend the paradigm of extreme events modeling to the incorporation of such novel factors in a top-down framework. We will illustrate the approach through an example of integrating market and climate transition risk in an asset-allocation setting.

The chapter is organized as follows: Section 15.2 provides a general set-up for multi-subordinated processes. In Section 15.3, we apply the framework for the special case of the stable Paretian family to model the return distribution of market variables exposed to both systematic and idiosyncratic market and novel risks. We then illustrate the framework via a stylized asset-allocation example, choosing carbon intensity as a measure for climate transition as one particular source of novel risk. Section 15.4 discusses other datasets that can be used within the same approach for incorporation of ESG, political and cyber risks. The final section concludes and provides an outline for future work.

15.2   THE RISK MODELING FRAMEWORK – GENERAL SET-UP

Let's recall that time-subordinated Levy processes are defined as follows: Let S be a risky asset with price process $\mathbf{S} = (S_t, t \geq 0, S_0 > 0)$ and log-price process $\mathbf{X} = (X_t = lnS_t, t \geq 0)$. We define the price and log price of $\mathbf{S}$ as

$$S_t = S_0 e^{X_t}, t \geq 0, S_0 > 0 \tag{15.1}$$

$$X_t = X_0 + \mu t + \sigma B_{T(t)}, t \geq 0, \mu \in R, \sigma \in R^+ \tag{15.2}$$

Where $\mathbf{B} = (B_t, t \geq 0)$ is a standard Brownian motion and $\mathbf{T} = (T_t, t \geq 0, T_0 = 0)$ is a Levy subordinator. We view X_t as the return of a single subordinated log-price process. $\mathbf{B}$ and $\mathbf{T}$ are independent processes. The trajectories of $\mathbf{T}$ are assumed to be right-continuous with left limits. Since $\mathbf{B}$ is a Levy process and $\mathbf{T}$ is a Levy subordinator, the resulting process $\mathbf{X}$ is also a Lévy process [Ken-Iti, 1999].

The time-subordinating process $T(t)$ represents the "news" intensity or the market clock as a random process of the physical time t: As philosophically stated by Clark in 1973, "On the days when no new information is available, trading is slow and the price process evolves slowly. On days when new information violates old expectations, trading is brisk and the price process evolves much faster."

We can now start thinking of exogenous factors in the same way – as a new type of risk emerges, the more "intense" its impact is, the faster the market clock will have to run in order to absorb it. To do so, we will use the notion of multiple-subordinated models defined as follows:

Let S be a risky asset with price process $\mathbf{S} = (S_t, t \geq 0, S_0 > 0)$ and log-price process $\mathbf{X} = (X_t = lnS_t, t \geq 0)$ which now follows

$$X_t = X_0 + \mu t + \sigma B_{U(T(t))}, t \geq 0, \mu \in R, \sigma \in R^+ \qquad (15.3)$$

where the triplet $(B_s, T(s), U(s), s \geq 0)$ are independent processes with $\{B_s, s \geq 0\}$ being a standard Brownian motion. $\{T(s), s \geq 0\}$ and $\{U(s), s \geq 0\}$ are Levy subordinators. Then the log-process $X = (X_t = lnS_t, t \geq 0)$ is a double-subordinator process.

Similarly, we can define n-subordinated process as

$$X_t = X_0 + \mu t + \sigma B_{U_{n-1}(...U_1(T(t)))}, t \geq 0, \mu \in R, \sigma \in R^+ \qquad (15.4)$$

where the set $(B_s, T(s), U_1(s), ..., U_{n-1}(s) s \geq 0)$ are independent processes with $B_s, s \geq 0$ being a standard Brownian motion. $T(s), s \geq 0$ and $U_i(s), s \geq 0$, for all $i = 1, ..., n - 1$, are Levy subordinators.

Depending on the type of the novel risk factor under consideration, we need to assign proper observable variables to model the new "market-clock speed" process, that is, the subordinator. For transitional risk impact, this can be carbon intensity – the distribution of the carbon intensity in a given sector relative to another can teach us how much "carbon-development news" it needs to absorb per unit of time as compared to the same observation for other sectors, and thus the relative impact on the price-process characteristics. ESG-type market impact measures can be extracted based on natural language processing (NLP) approaches related to the news articles tagged to ESG categories, see, for example, [FactSet, n.d.]. Cyber risk can be incorporated based on security-relevant events observed [Bilarev et al., n.d.], aggregated in a form of scores, etc. What associates these various types of alternative data is that (1) we cannot directly measure their monetary impact as they are not priced by the market; (2) they are all representative of "news" events the intensity of which across populations of businesses can speak toward the vulnerability of those business to such events; (3) they are all positive random variables when looked on cross-sectional or through-time basis. In general, the framework allows for the incorporation of any such alternative datasets. Note that for most of the applications related to portfolio optimization, in case of incorporating one novel risk type, the scale of the transition of the observed variable to market returns volatility impact doesn't matter as long as it is linear. The same would apply for risk-budgeting applications. For applications requiring absolute levels of risk estimation, the approach may rely on experts' opinions to define the worst- and best-case scenarios and then fit the distribution around those levels by means of considering them as two extreme suitably selected quantiles.

Most of the novel factors being exogenous in nature have a systematic component and thus shock the market as a whole and then have sector-specific implications that can be of different nature and severity. One exception would be the domain of operational risks, cyber being one such example, yet even in such domains, depending on the case, there could be a market-systemic impact. We use sector here without loss of generality; the approach can be applied to any other segregation criteria such as region, industry, credit rating or a combination of these.

With that in mind, let's consider a sector log-return process being driven by

$$R_{i,t} = \beta_i R_{m,t} + C_{i,t} \tag{15.5}$$

where $R_{i,t}$ is the log-return process of the $i-th$ sector, $R_{m,t}$ is the log-return of the market, $C_{i,t}$ is the sector-specific log-return process and $R_{m,t}$ follows a process of type (4), $C_{i,t}$ follows a driftless process of type (4), $R_{m,t}$ and $C_{i,t}$ are independent by construction and $\beta_{(i)} \in R$ is a constant measuring the exposure of the $i-th$ sector to the market returns. Let's note that the first subordinator $T(t)$ reflects extreme *market* risks, which are already priced by the market, while the next set of subordinators $U_1(s), ...U_{n-1}(s)$ represent the subordinators accounting for novel risk factors and thus can't be deducted from the observed market prices. Rather, they need to be modeled based on suitably selected exogenous variables as discussed above.

Now, assuming for simplicity and without loss of generality only one novel risk factor, if the return process of the $i-th$ sector specific component and the market, respectively, are:

$$C_{i,t} = \sigma_i B_{i,U_i(T_i(t))} \tag{15.6}$$

and

$$R_{m,t} = R_0 + \mu_m t + \sigma_m B_{m,U_m(T_m(t))} \tag{15.7}$$

then the novel-factor-adjusted log-return process would be

$$R_{i,t} = \beta(R_0 + \mu_m t + \sigma_m B_{m,U_m(T_m(t))}) + \sigma_i B_{i,U_i(T_i(t))} \tag{15.8}$$

where $\sigma_m \in R^+$ and $\sigma_i \in R^+$ are the scale parameters of the log-return distributions of the market and the sector-specific return, respectively; $\mu_m\ in R$ is the drift of the market log-return process.

15.3 METHODOLOGY FOR THE STABLE PARETIAN CASE

We will herein further present the possible implementation of the framework in the stable Paretian case; however, generalization to any Lévy process is viable as per [Stein et al., 2009]. For introduction to stable Paretian distributions, see [Samorodnitsky and Taqqu, 2017]. We will denote the class of stable Paretian random variables with tail-index parameter α, skewness parameter β, scale parameter σ and a shift parameter μ as $S_\alpha(\sigma, \beta, \mu)$, where $0 < \alpha \leq 2, -1 \leq \beta \leq 1, \sigma > 0$ and $\mu \in R$. Recall that the lower the tail-index α is the fatter are the tails of the distribution and the normal distribution emerges as a special case of $\alpha = 2, \beta = 0$ or more precisely $N(\mu, 2\sigma^2) \sim S_2(\sigma, 0, \mu)$.

Proposition 1 Let $X \sim S_{\alpha'}(\sigma, 0, 0)$ with $0 < \alpha' \leq 2$ and let $0 < \alpha' \leq 2$ and $0 < \alpha < \alpha'$.

Let A be α/α' – stable random variable totally skewed to the right with scale parameter $\sigma_A = cos(\frac{\pi\alpha}{2\alpha'})^{\alpha'/\alpha}$, i.e., $A \sim S_{\alpha/\alpha'}(cos(\frac{\pi\alpha}{2\alpha'})^{\alpha'/\alpha}, 1, 0)$ and assume X and A are independent. Then

$$Z = A^{1/\alpha'}X \sim S_\alpha(\sigma, 0, 0) \tag{15.9}$$

See [Samorodnitsky and Taqqu, 2017], Proposition 1.3.1 for the proof.

Proposition 2 Let's define a random variable r_j representing the log-return increments of either the market or sector/business segment/j specific returns per unit of time as

$$r_j = U_j^{1/\alpha_j} Z_j = U_j^{1/\alpha_j} T_j^{1/2} B_j \tag{15.10}$$

where $Z_j \sim S_{\alpha'_j}(\sigma'_j, 0, 0)$, $U_j \sim S_{\alpha'_j}(\sigma'_j, 1, 0)$ are independent. Furthermore, T_j and B_j are independent, and $B_j \sim N(0, 2\sigma_j^2), T_j \sim S_{\alpha'_j/2}(cos(\frac{\pi\alpha'}{2\alpha'})^{2/\alpha'}), 1, 0)$.

Then $r_j \sim S_{\alpha'_j \alpha_j}((\frac{\sigma}{cos(\frac{\pi\alpha_j}{2})})^{\frac{1}{\alpha_j}})^{1/\alpha'_j}$ and can be looked as increments of a double-subordinated Lévy process.

Proof 1 Follows directly from applying twice Proposition 1 and definition (3).

Now we can define double-subordinated increments of the i-th sector log-returns process as

$$r_i = \beta_i U_m^{1/\alpha_m} Z_m + U_i^{1/\alpha_i} Z_i = \beta_i U_m^{1/\alpha_m} T_m^{1/2} B_m + U_i^{1/\alpha_i} T_i^{1/2} B_i \tag{15.11}$$

where the constant $\beta_i \in R$ is the i-the sector exposure to the market, Z_m is a stable Paretian distribution random variable which is fit on observed market returns, B_m is the implied normal driver, T_m is the implied market-risk return's subordinator, and U_m is the novel-risk factor subordinator to be fit on a suitably selected exogenous random variable. Similarly, Z_i is a stable Paretian distribution random variable which is fit on observed sector-specific returns, B_i is the implied normal sector-specific driver, T_i is the implied sector specific-returns subordinator and U_i is the novel-risk factor subordinator to be fit on a suitably selected exogenous random variable representative of the sector specifics.

Equation (6) can be generalized to multiple-subordination paradigm following (4). Alternatively, we can break down the effect of one type of novel factor to it's subcategories via a generalization of the form

$$r_i = \beta_i U_m{}^{1/\alpha_m} Z_m + U_i{}^{1/\alpha_i} Z_i = \beta_i U_m{}^{1/\alpha_m} T_m{}^{1/2} B_m + \Sigma_k U_{i,k}{}^{1/\alpha_{i,k}} T_i{}^{1/2} B_i \quad (15.12)$$

where index k iterates across all categories of the novel factor.

15.4 RESEARCH RESULTS – APPLICATION TO TRANSITIONAL CLIMATE-RISK-ADJUSTED ASSET ALLOCATION

Transition risks are business-related risks that follow societal and economic shifts toward a low-carbon and more climate-friendly future. These risks can include policy and regulatory risks, technological risks, market risks reputational risks and legal risks. These risks are interconnected and often top of mind for investors as they attempt to navigate an increasingly aggressive low-carbon agenda that can create capital and operational consequences to their assets. There is at present no standardized way of modeling transitional risk as this is a relatively new and complex phenomenon. While several approaches exist [Bingler and Colesanti Senni, 2020], carbon emissions is a key input to any such approach. In what follows, we illustrate a top-down approach to use carbon emissions data to construct a transitional-risk aware asset allocation.

We now provide an illustration via a hypothetical asset allocation exercise across ten global industries represented by the MSCI US Total Return Sectors Indices using minimum risk optimization as of September 30, 2021. Real estate is being excluded due to bimodal nature of the carbon intensity data, which requires further investigation.

We first start by fitting Equation (5) on 180 monthly returns and extracting the residuals C_i time-series for each sector index $i = 1, .., 10$ against MSCI US Index.

Regression coefficients are reported in Table 15.1.

Symmetric stable parameters of the sectors' residuals time-series and the market represented by the MSCI US Index are then calculated based on

TABLE 15.1 Sector indices' regression coefficients against MSCI US, 180 monthly returns, September 30, 2021

Sector	Beta
Communication Services	0.736467
Consumer Discretionary	1.141101
Consumer Staples	0.599757
Energy	1.184209
Financials	1.298694
Health Care	0.742408
Industrials	1.178396
Information Technology	1.09318
Materials	1.208156
Utilities	0.4762

TABLE 15.2 Symmetric stable parameters of the sectors' residuals and the market,represented by the MSCI US indices, 180-monthly observations as of September 30, 2021

	Tail Parameter (Alpha)	Scale Parameter (Sigma)
Utilities	1.86	0.0239
Materials	1.91	0.0187
Energy	1.64	0.0287
Consumer Staple	1.98	0.0153
Industrials	1.89	0.0130
Consumer Discretionary	1.75	0.0122
Health Care	1.95	0.0163
Information Technology	1.94	0.0155
Communication Services	1.98	0.0242
Financials	1.56	0.0172
MSCI USA (market)	1.59	0.0247

180 monthly observations as of September 30, 2021 via fitting maximum likelihood method [Rachev and Mittnik, 2000]. Shift Parameter is ignored as it is irrelevant for minimum-risk optimization. The parameters are reported in Table 15.2.

The transitional climate risk subordinator random variable is modeled via the cross-sectional carbon intensity distribution at the selected point of time within the entire market. The universe is based on the ISS carbon intensity data as available within FactSet Workstation. All calculations are done as of September 30, 2021. The carbon-based subordinators for each sector are

TABLE 15.3 Stable parameters of carbon intensity subordinators as of September 30, 2021, cross-sectional estimation

	Tail Parameter (Alpha)	Scale Parameter (Sigma)
Utilities	0.5	16.0735
Materials	0.9	3.76942
Energy	0.99	3.24022
Consumer Staple	0.9	2.07895
Industrials	0.99	0.775805
Consumer Discretionary	0.99	0.42445
Health Care	0.99	0.117783
Information Technology	0.6	0.285669
Communication Services	0.99	0.0851
Financials	0.99	0.0288
MSCI USA (market)	0.4	6.8

modeled cross-sectionally within the companies belonging to the given sector via fitting maximum likelihood method [Rachev and Mittnik, 2000]. The subordinator for the entire market is cross-sectionally estimated using the full universe of the MSCI US stocks.

The parameters of the subordinators modeled directly over the time-series of the market intensity metrics for each sector are reported in Table 15.3.

The parameters of the double-subordinated sectors' residuals and the market following carbon-intensity second subordination are estimated based on Proposition 2 and are provided in Table 4. Such estimated parameters for U_i and U_m for each of the sectors and the market in (6), combined with the parameters estimate for Z_i and Z_m as provided in Table 2, allow us to apply a double subordinated model of type (6).

The final parameters of the double subordinated model of (6) for each r_i, representing the i-th sector log-returns, are provided based on Proposal 1 in Table 15.4.

We then contrast the optimal asset allocation policies with and without the climate-risk adjustment. We do so based on minimum 99% expected tail risk, one-month horizon. The optimization process follows [Rachev et al., 2010] and uses FactSet's optimizer engine [Bilarev et al., n.d.]. The scenarios generation used in the optimization process is based on [Rachev et al., 2010], Section 2.3 applied for the symmetric stable case. In the case of no carbon adjustment, we simulate according to Equation (5), as per regression betas as reported in Table 1 and stable parameters as presented in Table 2. Note that in this case the subordinator of each sector residual initial normal distribution is implicitly modeled, see [Rachev et al., 2010], Section 2.3. In the case of carbon

TABLE 15.4 Double-subordinated sectors' index parameters as of September 30, 2021. For practical purposes, min Alpha is set to 1.4.

	Tail Parameter (Alpha)	Scale Parameter (Sigma)
Utilities	1.4	6.483407
Materials	1.721261	5.880124
Energy	1.628394	26.19311
Consumer Staple	1.786493	4.08414
Industrials	1.875437	8.01046
Consumer Discretionary	1.73638	6.709481
Health Care	1.932886	2.86739
Information Technology	1.4	0.827474
Communication Services	1.965149	2.392578
Financials	1.540045	1.516934
MSCI USA (market)	1.4	4.76

TABLE 15.5 Optimal weights based on 99% ETL minimization

	(I) No Carbon Adjustment Optimal Weights	(II) Carbon Adjustment Optimal Weights
Communication Services	10.36	12.95
Consumer Discretionary	3.50	3.50
Consumer Staples	35.00	35.00
Energy	3.50	3.50
Financials	3.50	3.50
Health Care	19.13	27.55
Industrials	3.50	3.50
Information Technology	3.50	3.50
Materials	3.50	3.50
Utilities	14.51	3.50

adjustment, we simulate according to Equation (6), as per regression betas as reported in Table 1 and stable parameters as presented in Table 5, which represent double-subordinated parameters staring from normal distribution with the first subordinator being time-series market subordinator and the second – the cross-sectional carbon-intensity subordinator. We impose min 3.5% and max 35% weight constraint. Table 5 shows the optimal weights with and without carbon Adjustment.

The optimal allocations following carbon adjustment exit the carbon-heavy utility sector putting it to the minimal threshold of 3.5% and spread its weight to healthcare and communication services.

We must note that using carbon intensity only as an input to the second subordinator restricts the climate impact measurement to only its negative implications as it captures the potential policy impact via carbon taxation only. Other factors related to legal, reputational and technology can be added, the later potentially being able to add positive effects to carbon-intensive businesses in case of substantial disruptive innovation in the space. Such impacts will be better analyzed on security-by-security level, and we provide alternative datasets applicable for such deeper analysis in the following section.

15.5 SUGGESTED FURTHER WORK AND ALTERNATIVE DATASETS

As discussed, the framework can be extended to multiple novel risk factors via (4) or (7), as long as we have alternative datasets that can represent intensity of novel risk drivers. We note that the paper does not aim to discuss the datasets standardization approaches that would be required to incorporate multiple subordinators. The data can be extracted in numerous ways, including based on unstructured text or news data, observing operational flaws, etc., with the only hard requirement to be totally skewed to the right random variables.

What follows is a discussion of such notable alternative datasets.

Truvalue Labs ESG scores for assessment of ESG-type novel risks

This dataset can be used for incorporation of ESG-vulnerability novel risk driver, as well as reputational and technology aspects of climate risk.

Truvalue Labs is among the first companies to apply AI to uncover timely, objective ESG data that identifies both risks and opportunities at the speed of current events. Truvalue Labs focuses on company ESG behavior from external sources and includes both positive and negative events that go beyond traditional sources of ESG risk data. Using machine learning, Truvalue Labs aggregates, extracts, analyzes and generates scores on millions of documents each month, as events happen. They quantify ESG information from unstructured text to deliver a third-party perspective on company ESG performance. Starting from more than 100,000 data sources, such as local, national and international news, non-governmental organizations (NGOs), watchdog groups, trade blogs, industry publications and social media based on NLP, the technology produces on average 300,000 signals on a monthly basis, which are then aggregated to four scores, Insight, Pulse, Momentum and Volume, each one having 26 Sustainability Accounting Standards Board (SASB™) categories and 16 Sustainable Development Goal (SDG) categories.

Pulse measures near-term performance changes that respond to news as it happens. It highlights both opportunities and controversies, enabling real-time monitoring of companies. Insight is an exponentially weighted moving average of the pulse score and measures a company's longer term ESG track record, equivalent to an ESG rating. It is less sensitive to daily events and reflect the enduring performance record of a company over time. Both pulse and insight have values between zero and one hundred. Momentum measures the trend of a company's insight score and is used to identify companies with improving or deteriorating ESG performance. Volume measures the information flow or number of articles about a company over the past 12 months. For more information on TVL methodology, see [FactSet, n.d.].

Applications as discussed in the current chapter can be based on pulse, insight or volume score, being used as input data for calibration of the second novel risk factor subordinator. Equation (7) can be used to break down the effects based on each of the SASB^{TM} or SDG Categories in this way modeling out the effects of the different sustainability factors. The data can be used for asset-allocation purposes on a sector, or other group-based level, following cross-sectional analysis as illustrated in the previous section, or on asset-by-asset level for portfolio construction using the time-series of scores per company.

GeoQuant for assessment on political risk

GeoQuant provides high-frequency political risk assessment scores based on a political science-infused machine learning algorithm that combines low-frequency, structural country level data with high-frequency data derived from news media. GeoQuant's approach is based on a fundamental model of political risk based on 22 factors, which are then combined into bottom-up, modular political risk models designed for adaptation and customization by asset-type, sector and/or geography. The first layer of the factor hierarchy has three categories – Governance, Social and Security. Governance is further broken into Government, Institutional and Policy; Social – into Social Polarization and Human Development and Security – into Internal and External factors. Each measure is modeled via a structural and a high-frequency component, the prior based on quantitative measures and the latter on unstructured text from traditional and social media sources and is analyzed near real-time via NLP and ML. Each score is available on a country level. For more information on Geo-Quant Methodology, see [GEOQUANT, 2020].

This dataset can be used to define political risk subordinators following Equation (7), where subordinators are fit on the time-series of the scores and used for region-based asset-allocation and portfolio construction.

BitSight for assessment of cyber security risk

BitSight provides cyber security ratings for corporations. Their methodology is based on complex technology and infrastructure collecting cyber-risk-related data from Internet. They use extensive network of sensors deployed at key locations across the Internet with which they can monitor communications from compromised systems, DNS queries and responses, malicious traffic, attempts at brute force attacks, etc., which signals are then mapped to assets belonging to organizations. All observations are further mapped to twenty-three risk vectors each of which measures a particular area of security performance. Within the risk vector, the finding is then assigned a grade or a severity weight. That information is then aggregated following specific normalization procedures to arrive at the final cyber security score. For more information on BitSight methodology, see [Bilarev et al., n.d.].

As in the previous examples, the normalized and size-weighted factor scores can be used for modeling cyber-risk-associated subordinators per organization.

15.6 SUMMARY

We have proposed a general framework for modeling exogenous novel risk factors in an integrated framework via the notion of multi-subordinated Lévy processes. The approach introduces a unified framework for consistent integration of traditional and novel types of risk and can serve for both risk budgeting and asset-allocation applications.

We provided an analytical expression of double-subordinated parameters, distribution for the stable Paretian case, which offers convenient, easy to apply and robust approach. We applied it to the case of climate risk adjustment for asset-allocation based on carbon intensity data and illustrated the differences in the optimal allocations with and without the inclusion of this novel factor.

Finally, we have provided three other examples of datasets that could be used within similar frameworks for assessment of ESG risks, political and cyber risks. Future work should include deeper consideration and calibration approaches for combining multiple sources of novel risk and application to other representatives from the Lévy-processes family.

The chapter offers a new and unique approach to integrate traditional and novel exogenous risk factors into one cohesive modeling approach for asset allocation and portfolio construction. The Covid pandemic's increasing political turbulence and climate challenges demonstrated that the financial system is open and highly susceptible to exogenous influences that need to find their way to quantitatively enter investment decision-making. The chapter offers a substantial step in that direction.

REFERENCES

Basel Committee et al. (2014). Fundamental review of the trading book: A revised market risk framework. https://www.bis.org/publ/bcbs265.pdf.

Bilarev, T., Mitov, G., Racheva-Iotova, B., and Stefanov, I. (n.d.). Multi-asset class portfolio construction. Technical report, FACTSET.

Bingler, J. A. and Colesanti Senni, C. (2020). Taming the green swan: how to improve climate-related financial risk assessments. *Available at SSRN 3795360*.

FactSet. (n.d.). ESG Data and Analytics from Truvalue Labs. Technical report, FactSet.

GEOQUANT (2020). The year ahead 2020 data-driven predictions for global investors. Technical report, GEOQUANT.

Giese, G., Nagy, Z., and Rauis, B. (2021). Foundations of climate investing: how equity markets have priced climate-transition risks. *The Journal of Portfolio Management*, 47(9):35–53.

Karagozoglu, A. K. (2021). Novel risks: a research and policy overview. *The Journal of Portfolio Management*, 47(9):11–34.

Kazdin, J., Schwaiger, K., Wendt, V.-S., and Ang, A. (2021). Climate alpha with predictors also improving company efficiency. *The Journal of Impact and ESG Investing*, 2(2):35–56.

Ken-Iti, S. (1999). *Levy Processes and Infinitely Divisible Distributions*. Cambridge University Press.

Rachev, S., Fabozzi, F. J., Racheva-Iotova, B., and Shirvani, A. (2017). Option pricing with greed and fear factor: the rational finance approach. *arXiv preprint arXiv:1709.08134*.

Rachev, S. T., Kim, Y. S., Bianchi, M. L., and Fabozzi, F. J. (2011). *Financial Models with Lévy Processes and Volatility Clustering*. John Wiley & Sons.

Rachev, S. T. and Mittnik, S. (2000). *Stable Paretian models in finance*, volume 7. Wiley, ISBN: 978-0-471-95314-2.

Rachev, S. T., Racheva-Iotova, B., Stoyanov, S. V., and Fabozzi, F. J. (2010). Risk management and portfolio optimization for volatile markets. In *Handbook of Portfolio Construction*, pages 493–508. Springer.

Racheva-Iotova, B. (2020a). Back-testing studies on the factset short-term risk model during the covid-19 outbreak. *Factset Insight Article*.

Racheva-Iotova, B. (2020b). Wealth Management Next Frontiers-The Inevitable Need to Meet Behavioral and Quantitative Approaches. In *Handbook of Applied Investment Research*, pages 479–509. World Scientific.

Racheva-Iotova, B., Mitov, I., Mossessian, D., and Bodurov, V. (2018). Factset fat-tail multi-asset class model - introducing turbulence adjusted risk. Technical report, FACTSET.

Radev, N., Mitov, G., and Racheva-Iotova, B. (n.d.). Minimum tail risk portfolio construction. Technical report, FACTSET.

Roncalli, T., Le Guenedal, T., Lepetit, F., Roncalli, T., and Sekine, T. (2021). The market measure of carbon risk and its impact on the minimum variance portfolio. *The Journal of Portfolio Management*, 47(9):54–68.

Samorodnitsky, G. and Taqqu, M. S. (2017). *Stable Non-Gaussian Random Processes: Stochastic Models with Infinite Variance: Stochastic Modeling*. Routledge.

Schoutens, W. (2003). *Lévy Processes in Finance: Pricing Financial Derivatives*. Wiley Online Library.

Schoutens, W. and Cariboni, J. (2010). *Lévy Processes in Credit Risk*. John Wiley & Sons.

Stein, M., Rachev, S. T., and Stoyanov, S. (2009). R ratio optimization with heterogeneous assets using genetic algorithm. *Investment Management and Financial Innovations*, 6(3-1):117–125.

Tokat-Acikel, Y., Aiolfi, M., Johnson, L., Hall, J., and Jin, J. Y. (2021). Top-down portfolio implications of climate change. *The Journal of Portfolio Management*, 47(9):69–91.

Watson, W. T. (2019). Total portfolio approach (tpa) - a global asset owner study into current and future asset allocation practices. Technical report, ThinkingAhead Institute.

(IV.C)

Case Studies on ESG

CHAPTER 16

ESG Controversies and Stock Returns

Tiago Quevedo Teodoro

MarketPsych Data

Joshua Clark-Bell

MarketPsych Data

Richard L. Peterson

MarketPsych Data

CONTENTS

16.1 INTRODUCTION

In a 1970 New York Times opinion piece, economist Milton Friedman criticized the concept of "social responsibility," partly due to its "analytical looseness and lack of rigor" [Friedman, 1970]. Friedman, who won the Nobel prize for economics in 1976 and was one of the most influential economists of the 20th century, suggested that businessmen who invested in such efforts would be "spending someone else's [stockholders and clients] money." The Friedman doctrine [Friedman, 1962] remained quantitatively unchallenged in the following decades. With the concept of social responsibility being ill-defined, sparse

DOI: 10.1201/9781003293644-16

availability of datasets and studies difficult to replicate, the debate remained largely qualitative.

In the mid-2000s the media increasingly turned its focus to both the environmental impacts of human activities (most prominently global warming) and the societal impacts of corporate practices. Notably, in a 2004 United Nations (UN) report, the acronym ESG was coined to raise awareness of sustainable environmental, social and corporate governance (ESG) issues facing stakeholders [Compact, 2004]. The interest of balancing shareholder returns with stakeholder outcomes re-emerged under the rubric of "stakeholder capitalism" [Schwab, 2021]. The growth of investment funds managed using ESG guidelines has expanded rapidly since then. According to a 2018 report by Morningstar Research, assets under management (AUM) in portfolios guided by sustainable investment directives were at $23 trillion, an increase of more than 600% over the previous decade [Hale, 2018]. Bloomberg Intelligence projects $53 trillion in global assets will be invested according to ESG principles by 2025 [Bloomberg Intelligence, 2021]. This in turn has created a large demand for new, higher quality ESG-based metrics. More rigorous metrics now allow researchers and practitioners to more robustly test Friedman's hypothesis.

16.1.1 Do ESG Ratings Predict Stock Returns?

Currently, the bulk of academic and industry research indicates that corporate social responsibility is aligned with the creation of value for shareholders. An extensive 2015 academic review of more than 2000 papers reported that "roughly 90% of studies find a non-negative ESG–CFP (Corporate Financial performance) relationship" [Friede et al., 2015]. A joint academic-industry review of 200 academic studies by industry-affiliated academics at the University of Oxford found that "80% of the reviewed studies demonstrate that prudent sustainability practices have a positive influence on investment performance" [Clark et al., 2015]. Practitioners have also reported positive relationships between ESG scores and stock price returns [Belsom and Lake, 2021]. However, the lack of standardization in ESG reporting and the well-known low correlation between standard ESG ratings is a still a concern [Brandon et al., 2021; Mackintosh, 2018; Subramanian et al., 2019; Lester and McKnett, 2018].

16.1.2 Weaknesses in ESG Datasets

Market-leading ESG datasets are distributed by large financial data providers including MSCI, Refinitiv and S&P Global. These traditional ESG ratings are largely based on the inside-out perspective, *i.e.*, company-reported ESG information. However, despite having overlapping information sources, the methodologies differ, and the traditional ESG ratings can still diverge considerably across companies [Dorfleitner et al., 2015]. For example, industry researchers found that "[i]n the United States, a portfolio selecting the top 50% of stocks with the highest social score based on Provider 1's data produces excess returns that have a correlation of only 0.51 with the portfolio selecting the top 50% of stocks based on Provider 2's social score" [Li and Polychronopoulos, 2020]. The divergences are largely attributable to differences in scope, weight and measurement methodologies employed during construction of the ratings [Friede et al., 2015]. Furthermore, ratings based on company reports can be updated as infrequently as every 12 months. As a result, the ratings can be considerably backward looking and represent ESG issues and virtues that have long been taken into account by market participants. Hence, as the ESG space evolves, alternative datasets will become ever more important.

16.1.3   ESG Alt-Data

Online social media allows concerned citizens, employees and activists to speak out publicly about illicit corporate activities. For example, concerned citizens discuss toxic consumer products in specific Reddit threads. Activists reveal corporate misdeeds on blogs. Employees whistleblow illicit management activities on job review sites. Sometimes, such concerns attract the attention of mainstream journalists through whom they reach wider news audiences. As a result, news and social media information flow can provide a valuable source of ESG data. This is specially the case for ESG controversies, which are more likely to be discreetly buried in company-reported ESG data. For example, in Figure 16.1 we show the number of references in the media about accounting controversies surrounding Wirecard AG. The Financial Times (FT) was the first media outlet to report suspicions of accounting fraud on January 30, 2019. The FT article suspected "falsification of accounts" [McCrum, 2019]. Wrongdoings were denied by the company until June 18, 2020, when the CEO revealed a €1.9 billion hole in the books.

Additionally, while investors have been found to react to both positive and negative contents that pass through their social networks, negative stories (including ESG content) are shown to have the largest impact on share prices [Uhl, 2014; Li, 2006; Subramanian et al., 2019].

In this paper, a new point-in-time ESG dataset based on the media information flow, the Refinitiv MarketPsych ESG Analytics (henceforth RM-ESG), is utilized to examine the effect of ESG controversies on individual equity returns.

16.1.4   Refinitiv MarketPsych ESG

The Refinitiv MarketPsych Analytics utilizes natural language processing (NLP) to convert large volumes of textual data posted in news and social

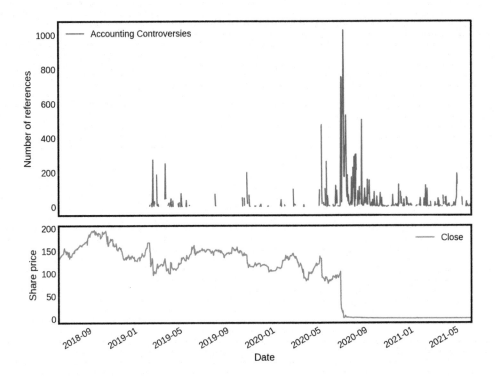

Figure 16.1 The number of media references to accounting controversies surrounding Wirecard AG from 2018 to 2020 in the news and social media (upper plot) and the share price (lower plot)

media into structured ESG ratings. Each day, approximately one million relevant ESG-related articles and comments in 13 languages are filtered out from the most influential media sources. The data history extends back to 1998 with coverage of over 110,000+ companies. The Refinitiv MarketPsych Analytics is used in the portfolio selections of sustainability ETFs and quantitative investors [Wong, 2021], to drive global fixed income allocations [Utkarsh and van Broekhuizen, 2021], and in consulting and research applications.

The RM-ESG scores are divided into two groups, the so-called Advanced and Core scores. The **Advanced** variant covers 80 ESG topics, with 40 being related to controversies.[1] In aggregate, the scores quantify the hopes, concerns and criticisms of millions of online voices about specific aspects of each company. The RM-ESG scores are updated every minute according to the content found and analyzed in the previous 60 seconds. It is thus a near real-time dataset that quickly reacts to breaking controversies reported in the media.

The **Core** dataset is composed of 17 high level ESG ratings, of which one – the RM-ESG Controversies score – is an aggregation of all controversies in the

---

[1]The ESG themes scored in the data are displayed in the Appendix. More information is available via the data brochure and whitepaper [Refinitiv, 2021].

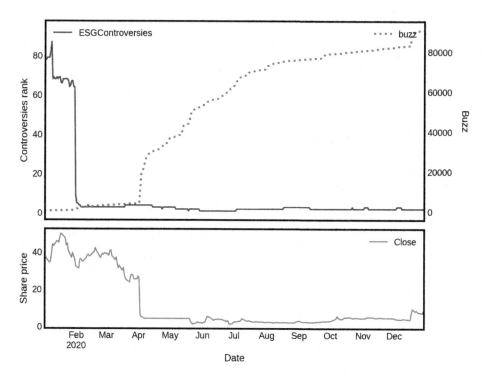

Figure 16.2 RM ESG Controversies score (blue, right axis), Buzz (magenta, dashed, left axis) and Luckin Coffee share price in 2020 (lower subplot)

Advanced feed. Core scores consist of intra-industry rankings from 1 (worst) to 100 (best) based on the 365-day exponentially averaged values (with a 90-day half-life) in the Advanced set. Another important score for the present study is the so-called Buzz. This value (in the Core dataset) refers to the amount of media references to a given company related to ESG topics in the previous 365 days. It is thus a proxy for the media's ESG-related references to a company. Figure 16.2 shows the reaction of the RM-ESG Controversies score and Buzz value for Luckin Coffee in 2020, when the company was found to be involved in accounting gimmicks. Similarly to what was observed in the Wirecard case, the first reports of wrongdoing anticipated the disclosed internal investigation. In this case, the accounting fraud story was first mentioned by the short-seller research firm Muddy Waters, whose report on Luckin Coffee was released to clients on January 31, 2020. On the same day, the RM-ESG Controversies scores plummeted due to the increasing chatter in the news and social media about the Mudddy Waters report. The results of the investigation were only announced by the company in April of the same year, at which time the story became a trend in global news, as depicted by the Buzz value. This was followed by a drop in share price of 80% and the halting of trading for several days.

In this study, we show that the Wirecard and Luckin Coffee cases are not isolated. By analyzing the cross-section of a large sample of companies (the Russell 3000 components), we found a positive relationship between the RM-ESG Controversies score and forward stock price performance. The methodology applied in the study is detailed in the next section. We then present the results and discuss what they mean in terms of portfolio allocation. Conclusions are outlined in the final section.

16.2 METHODOLOGY

In this study, we analyze the impact of ESG-related controversies on stock returns. The regressor is the RM-ESG Controversies score sampled on the last trading day of each month. The evaluated period is from January 2006 to September 2020. The universe of stocks comprises the components of the Russell 3000 (point-in-time). Returns are computed as the price change, after adjusting for dividends and splits, from close to close on the last trading day of each month.

To account for the varying levels of media chatter about companies of different sizes, MarketPsych applies a Buzz-weighting method in the computation of the RM-ESG Controversies scores. Companies with a large volume of media controversies are punished more than those with only a few media reports. This is done by adjusting the raw scores of companies with low Buzz [Refinitiv, 2021]. Hence, there is small negative correlation between the RM-ESG Controversies scores and the companies' market capitalization (−8% in this study's universe). To reduce the company size effect in the conclusions to be drawn in this study, we opted for further filtering the base universe (the Russell 3000 components) by only selecting companies with the largest Buzz. Figure 16.3 displays the correlation between the Controversies score and the market capitalization of only the top Nth percentile of companies (by Buzz). As the selection becomes stricter, the correlation approaches zero, as expected. It then turns positive for the 90th percentile and above which may be coincidental given the small number of companies that would be left (less than 300). We thus choose a threshold of 80% for the trade-off between reduced correlation and still having a large number of companies in the universe. Hence, each month, the filtered universe corresponds to a maximum of about 600 companies, which are to the top 20% of companies in the Russell 3000 that had ESG-related media content in the previous 365 days.

The cross-sectional analysis uses rotating portfolios where the universe of stocks are distributed into quantiles (with equal weight of the components) based on their end of the trading month RM-ESG Controversies score. Further variations are discussed in the following sections.

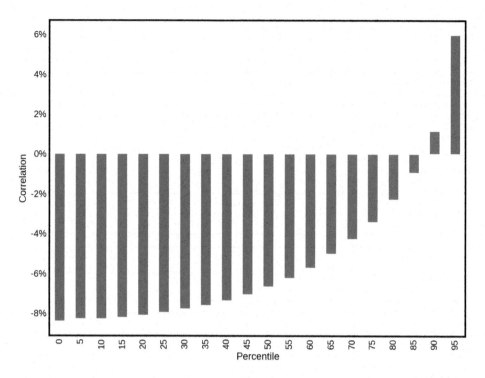

Figure 16.3 Correlation between the end-of-the-month RM-ESG Controversies score and the market capitalization of Russell 3000 companies when selecting (point-in-time) only the top percentile (x-axis) of companies according to the Buzz value

16.3 RESULTS

Table 16.1 displays the statistics of the theoretical portfolios constructed as described in the previous section. The results indicate that the least controversial companies compose the best performing portfolio in both absolute returns and with the lowest variance. On the other hand, companies ranked among the most controversial form the worst performing portfolio (in absolute terms as well as on a risk-adjusted basis). The statistics of the long-short portfolio, P5–P1, henceforth referred to as CNT, is also shown.

Figure 16.4 displays the equity curve of portfolios 1, 5 and MKT. At a first look, it may seem like the outperformance of the least controversial companies only appears after the early 2010s, which would be in line with research by Bank of America Merrill Lynch [Subramanian et al., 2019]. On the other hand, Figure 16.5, which shows the per-year performance of the CNT portfolio, tells a more nuanced story. It is true that in the first five years of this analysis, 2006 to 2010, the P1 portfolio had outperformed P5 by only 5%, while the outperformance in the following quinquennial periods was 52% (2011–2015) and 15% (2016–2020), respectively. However, the initial underperformance from

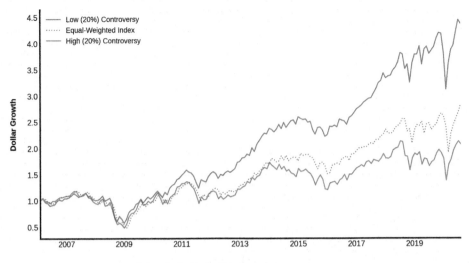

Figure 16.4 Monthly rotation model depicting the equity growth of theoretical portfolios composed Russell 3000 companies with the highest 20% Buzz values. Companies were ranked into quintiles based on their past-month RM-ESG Controversies score. The forward average return of the most controversial 20% of companies (P1, red) is compared against the bottom 20% (P5, green) and an equally weighted portfolio (MKT, grey dotted line) from January 2006 to September 2020.

2006 to 2010 is primarily caused by the 2009 market rally following the global financial crisis.

Interestingly, the worst and third worst yearly performance of the CNT portfolio occurred in 2009 and 2020, respectively, both of which experienced US stock bear market bottoms [Chiappini et al., 2021]. To better understand this behavior, we analyzed the average daily volatility of the stocks composing P1 and P5. As Figure 16.6 shows, the quintile P5 is less volatile than P1, and the individual components are also less volatile. Considering this finding and taking into account the leverage effect [Bouchaud et al., 2001], the 2009 and 2020 strategy underperformance could indicate that the ESG scores are somewhat reactive to past performance. This would be expected, as the market will have at least partially incorporated ESG controversies that have recently appeared in the media. To test this relation, we perform a linear regression with the performance of the CNT portfolio as a dependent variable and two other long-short portfolios, SIZ and MOM, as regressors:

$$CNT_t = \alpha + \gamma SIZ_t + \delta MOM_t + \epsilon_t \qquad (16.1)$$

TABLE 16.1 This table displays the average one-month forward performance (%) of hypothetical portfolios from January 2006 to September 2020 constructed by first selecting companies in the Russell 3000 with the highest values of Buzz (top 20%). The equally weighted performance of the stocks in the resulting universe is in the MKT row. The indices CNT, MOM, and SIZ refer to portfolios that were long-short in the highest and lowest quintiles (respectively) according to the RM-ESG Controversies score, prior-month performance, or market capitalization, respectively. The performance of all quintiles when dividing by the RM-ESG Controversies scores is also shown as P1 (most controversial) up to P5 (least controversial). Zero trading costs are assumed.

	Mean	Std	Min	50%	Max
P1	0.66	7.0	−23.0	1.4	25.2
P2	0.85	7.2	−22.3	1.5	40.6
P3	0.67	6.3	−20.3	1.3	22.8
P4	0.75	5.7	−21.6	1.2	19.5
P5	1.01	5.7	−25.6	1.7	19.7
CNT	0.35	3.0	−11.9	0.4	12.2
MOM	−0.13	5.0	−25.1	0.3	13.2
SIZ	0.69	6.2	−34.0	0.7	21.3
MKT	0.78	6.2	−21.6	1.4	26.1

SIZ represents the market neutral portfolio formed from the top and bottom quintiles by market capitalization, while MOM uses a similar pair of quintiles but formed on basis of the prior month's stock price performance.

Results displayed in Table 16.2 indicate that changes in the momentum (MOM) factor indeed partly relate, although weakly, to the CNT returns. The results also indicate that despite reducing the effect of size by selecting only companies with high buzz, SIZ is still a significant explanatory variable. Nevertheless, the coefficient of the regression is still about 55% of the average return of the CNT portfolio.

16.3.1 Exclusion by Controversies

The results shown so far indicate a positive relation between the RM-ESG Controversies and future returns.[2] To demonstrate how this affects portfolio construction, we take the perspective of an equity ESG fund that "greens" its portfolio by excluding highly controversial companies. Figure 16.7 shows

---

[2]As a reminder, a high RM-ESG Controversies rating indicates that the company has been *less* involved in ESG controversies than its peers. This is for consistency with other core ratings produced by MarketPsych.

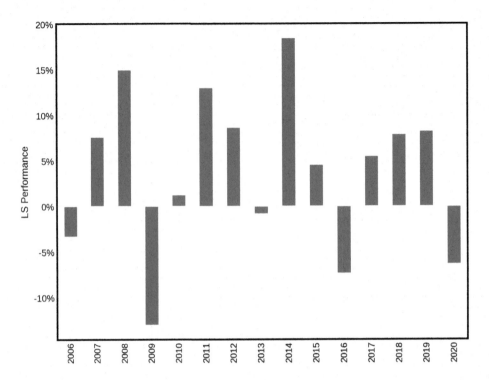

Figure 16.5 Annual difference in performance between portfolios P5 and P1 as depicted in Figure 16.4, which is also the long-short RM-ESG Controversies portfolio, CNT

TABLE 16.2 Result of the linear regression in Eqn. 16.1. The dependent variable is the CNT portfolio as described in Table 16.1. SIZ and MOM represent two long-short portfolios constructed by the highest vs. the lowest quintile according to the market capitalization and the prior month's return, respectively. $R^2 = 0.34$; F-statistic = 44.37; N. observations = 174.

	Coeff.	Std	t
c	0.0019	0.0019	0.9973
SIZ	0.2589	0.0344	7.5251
MOM	0.0583	0.0422	1.3815

the results of such an experiment. The leftmost point in the plot refers to the average performance of a portfolio that selects the top 20% of companies by Buzz in the Russell 3000 (point-in-time). As one moves to the right in the plot, the results refer to portfolios that remove the worst offenders (according to the level of RM-ESG Controversies) in the quantile shown in the x-axis. For example, a vertical line on the x-axis representing the 80th quantile intersects with the performance of the P1 portfolio in Figure 16.4. As expected

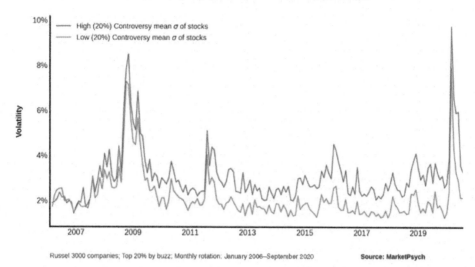

Figure 16.6 Monthly rotation model depicting the average volatility of theoretical portfolios composed of Russell 3000 companies with the highest 20% Buzz values. Companies were ranked by their past-month RM-ESG Controversies score. The forward average standard deviation of the most controversial 20% of companies (P1, red) is compared against the bottom 20% (P5, green) over 2006 to 2020.

given the prior results, excluding the worst offenders boosts the average return while reducing the volatility of the portfolio. This is the case with up to the 80th quantile. After that point, while returns still rise until the 90th quantile, volatility slightly increases. The peak risk-adjusted return would occur at the 90th quantile. However, at this point, the portfolio would be composed of less than 60 companies. On the other hand, the construction of the RM-ESG Controversies scores ensures a high level of sector diversification. Thus, despite the low number of stocks in the 90th quantile portfolio, volatility is still lower than when all $\sim$ 600 stocks were included. The risk-adjusted return of the 90th quantile portfolio is 46% higher than that of the MKT portfolio. Finally, after the 90th quantile, as fewer stocks are included, the performance quickly degrades as volatility rises steeply and the average return falls.

16.4 DISCUSSION

The argument over the share price impact of corporate social responsibility has remained unresolved since the 1970s. While recent research seems to indicate a positive relationship, methodological differences, data quality issues,

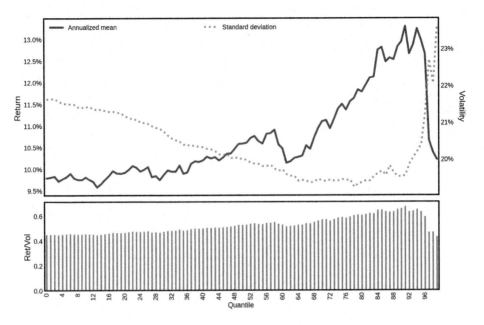

Figure 16.7 Average annualized return (left, blue), annualized standard deviation (right, dashed red) and sharpe ratio (lower subplot, assuming 0% interest rate during the entire period) of a long only portfolio which uses the top 20% Buzz companies in the Russell 3000 as a universe while excluding the respective quantile (x-axis) according to the level of the RM-ESG Controversies. All portfolios assume zero trading costs.

publication bias and industry bias all confounded reliable data construction and research.

In this paper, we studied the Refinitiv MarketPsych ESG Analytics, a dataset which uses NLP to score ESG-themed content in digital media articles. The data is generated with a consistent methodology applied to all ESG media articles from 1998 to the present. Furthermore, we focused on ESG controversies, as negative news has been shown to be more impactful on share prices. Additionally, rigorous research about the impacts of ESG controversies is still lacking.

In studying this dataset, results show that among US stocks, there is a relation between the percentage of ESG controversies reported in the media coverage of a company and its future share price performance. A portfolio composed of less controversial companies outperforms a variant portfolio composed of more controversial ones. Underperformance occurred in only five of the 15 years analyzed in this study: 2006, 2009, 2013, 2016 and 2020, which suggests that the effect is not concentrated only in recent years, as some ESG industry research indicates. It is important to note that the RM-ESG

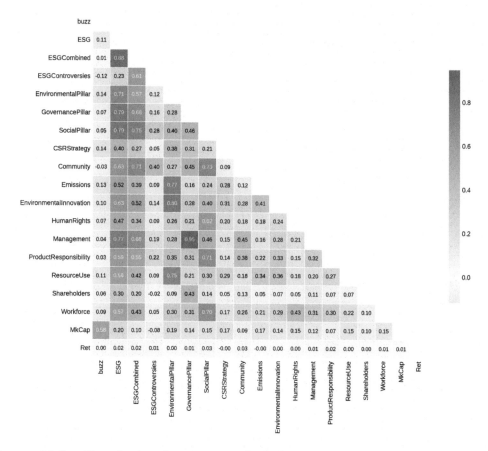

Figure 16.8 Correlation between end-of-the-month RM-ESG Core scores, contemporaneous monthly return, and market capitalization of Russell 3000 components point-in-time from January 2006 to September 2020.

Controversies score applied in this study uses intra-industry rankings, which avoids industry biases that are commonly found in other ESG studies.

Finally, we are aware that there is an obvious and inescapable conflict of interest in this paper. All authors were involved in the creation of the RM-ESG dataset and financially benefit from its sale. Nonetheless, the hope is that the information provided here is useful for informing and guiding future, more objective, third-party research. For example, while we focused only on the RM-ESG Controversies scores, MarketPsych offers a plethora of data on high-level ESG ratings and granular ESG themes. In Figure 16.8, we show the correlation between the core scores, market capitalization and contemporaneous returns of the companies in the Russell 3000. The RM-ESG Controversies scores generally display low correlation with the individual category scores. The lowest correlation is -2% (with Shareholders) and the highest is 40% (with Community). The correlation with the so-called Pillar scores and the ESG score itself is also relatively low (from 12% to 28%). As a teaser, we show

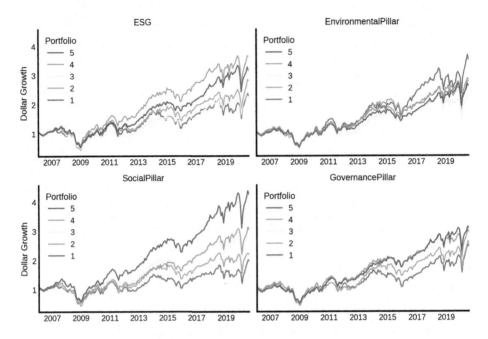

Figure 16.9 Monthly rotation models depicting the equity growth of theoretical portfolios composed of Russell 3000 stocks with the highest 20% Buzz values in each month. Companies were ranked into quintiles according to the ESG (top left), Environmental Pillar (top right), Social Pillar (bottom left) and Governance Pillar (bottom right) scores. The quintiles constituents are equally weighted, and each study assumes zero transaction costs. The period analyzed is from January 2006 to September 2020.

in Figure 16.9 results for the three Pillar and overall ESG scores using the same methodology applied to derive the results in Table 16.1. The spread between the top and bottom quintiles with the Social pillar scores is interesting. Social controversies are particularly difficult to quantify with more traditional datasets, while being particularly fit for data structured from news and social media. Also, as seen in Figure 16.8, Social scores have the highest correlation (among the Pillar scores) with the Controversies scores, which gives an indication of a starting point in a further study about controversies.

16.5 CONCLUSION

Based on this research, there is evidence of consistent share price underperformance for US companies entangled in ESG controversies reported in news and social media.

The effect of excluding highly controversial companies on the expected portfolio performance is nearly monotonic. We found that a long-only portfolio aimed at reducing the risk from such controversies by excluding highly controversial companies could have achieved an increase in risk-adjusted returns of up to 46% in the period from 2006 to 2020. These findings indicate that funds and companies with a focus on social responsibility (or at least avoiding controversial behavior), regardless of their industry, achieve higher share price performance, adding to the volume of evidence negating Friedman's hypothesis.

REFERENCES

Belsom, T. and Lake, L. (2021). ESG factors and equity returns – a review of recent industry research. *PRI Blog*.

Bloomberg Intelligence (2021). ESG assets may hit $53 trillion by 2025, a third of global AUM. *Bloomberg*.

Bouchaud, J.-P., Matacz, A., and Potters, M. (2001). Leverage effect in financial markets: the retarded volatility model. *Physical Review Letters*, 87(22):228701.

Brandon R. G., Krueger, P., and Schmidt, P. S. (2021). ESG rating disagreement and stock returns. *Financial Analysts Journal*, 77(4):104–127.

Chiappini, H., Vento, G., and De Palma, L. (2021). The impact of COVID-19 lockdowns on sustainable indexes. *Sustainability*, 13(4).

Clark, G., Feiner, A., and Viehs, M. (2015). From the stockholder to the stakeholder: how sustainability can drive financial outperformance. *Arabesque*.

Compact, U. (2004). Who cares wins: connecting financial markets to a changing world. *International Finance Corporation*.

Dorfleitner, G., Halbritter, G., and Nguyen, M. (2015). Measuring the level and risk of corporate responsibility–an empirical comparison of different esg rating approaches. *Journal of Asset Management*, 16(7):450–466.

Friede, G., Busch, T., and Bassen, A. (2015). ESG and financial performance: aggregated evidence from more than 2000 empirical studies. *Journal of Sustainable Finance & Investment*, 5(4):210–233.

Friedman, M. (1962). *Capitalism and freedom*. University of Chicago Press.

Friedman, M. (1970). A Friedman doctrine - the social responsibility of business is to increase its profits. *The New York Times*.

Hale, J. (2018). Sustainable funds, U.S. landscape report. *Morningstar Research*.

Lester, A. and McKnett, C. (2018). Harnessing ESG as an alpha source in active quantitative equities. *State Street Global Advisors.*

Li, F. (2006). Do stock market investors understand the risk sentiment of corporate annual reports? *Available at SSRN 898181.*

Li, F. and Polychronopoulos, A. (2020). What a difference an ESG ratings provider makes. *Research Affiliates.*

Mackintosh, J. (2018). Is Tesla or Exxon more sustainable? It depends whom you ask. *Wall Street Journal.*

McCrum, D. (2019). Executive at Wirecard suspected of using forged contracts. *Financial Times.*

Refinitiv (2021). Refinitiv Marketpsych ESG Analytics: Quantifying sustainability in global news and social media.

Schwab, K. (2021). Stakeholder Capitalism: A Global Economy that Works for Progress, People and Planet. *John Wiley & Sons.*

Subramanian, S., Kabra, M., Chopra, S., Yeo, J., Strzelinska, P., and Huang, L. (2019). ESG matters ESG from A to Z: a global primer. *Bank of America Merrill Lynch Research.*

Uhl, M. W. (2014). Reuters sentiment and stock returns. *Journal of Behavioral Finance,* 15(4):287–298.

Utkarsh, A. and van Broekhuizen, R. (2021). Responsible investing and asset allocation. *NN Investment Partners.*

Wong, G. (2021). Quantitative investing gets sentimental. *AGF Investments.*

CHAPTER 17

Oil and Gas Drilling Waste – A Material Externality

J. Blake Scott

President, Waste Analytics LLC

CONTENTS

DOI: 10.1201/9781003293644-17

17.1   INTRODUCTION

In 2015, the United Nations adopted the 2030 Agenda for Sustainable Development which has 17 Sustainable Development Goals (SDGs) at its center. These 17 SDGs address issues for all people and the planet in order to reduce poverty and inequality, improve education and health and sustainably increase economic growth. Restated, economic sustainability translates into making sound environmental decisions by companies. In particular, SDG 12 – "Ensure Sustainable Consumption and Production," provides guidance by addressing waste. The importance of the topic of waste is critical because almost all pollution is the result of waste. Even though SDG 12 provides excellent targets, we must further determine which wastes cause the most harm. The answer lies in determining externalities and materiality.

In economics, externalities are consequences by companies to third parties that are not in the costs of those goods that services. Even though externalities can be good or bad, the consequences of most recent concern in environmental, social, or governance (ESG) discussions are negative because they relate to environmental issues. However, not all negative externalities are the same. As an example, disposing of one gallon of used motor oil in a backyard is very different from an oil change service provider disposing of 20,000 gallons of used motor oil in a pit outside its facility. We need to be able to determine which externalities are of concern – meaning which ones are material.

A general definition of materiality usually means that something is relevant or significant. Given a specific topic, what I might find significant, may differ from what someone else may find significant. The answer to the question of relevancy or significance depends upon the topic and who is asking the question. From a securities perspective, the United States Supreme Court (SCOTUS) says that something is material when there is "a substantial likelihood that the disclosure of the omitted fact would have been viewed by the reasonable investor as having significantly altered the 'total mix' of information made available." However, from a sustainability aspect, materiality refers to issues that have repercussions to the company regardless of how the investor may view it. Issues of materiality can arise in many situations, but an emerging area concerns corporations and their environmental impact on stakeholders. As an example, communities in which a corporation operates may think odor from a concentrated animal feeding operation is material, but the company and its investors might not. If the community with the concern creates enough issues, the corporation may reverse its view of materiality of the odor.

Put simply, to address the targets in SDG 12, it is imperative to identify wastes that are material externalities regardless of the type of materiality discussed. One of the wastes that needs discussion is oil and gas drilling waste even though it is not often discussed in the media. To understand this better, it is necessary to discuss drilling waste in more detail.

17.1.1 Drilling Waste

Upstream oil and gas creates three classes of waste streams: produced water, drilling waste and associated waste. Of these, produced water is the largest total volume, drilling waste is the largest solid waste stream, and associated waste is the smallest volume and includes tank bottoms, filter socks, etc.

Drilling waste consists of rock cuttings and drilling fluids generated from drilling an oil or gas well. Think of drilling waste as being analogous to the wood shavings left behind when drilling a hole in a piece of wood, but rather than wood; it's comprised of soil and rock particles and coated in drilling fluids. The drilling fluids remove cuttings from the borehole, cool the drill bit, prevent sluffing of the hole, prevent formation swelling, etc.

The United States Environmental Protection Agency (USEPA) currently exempts drilling wastes from hazardous waste regulations (known as the E&P exemption), but the European Union (EU) and the United Kingdom (UK) consider certain drilling wastes to be hazardous. Almost all governmental bodies consider drilling waste to be a waste stream requiring some form of management because of the volumes generated and its contaminants.

17.1.2 Types of Drilling Waste

Drilling waste can typically be broken down into two main categories, water based and oil based, and they depend on the type of drilling fluid used to drill a particular section of the wellbore. A well can generate both types of drilling waste, and it can be broken down further into liquid and solid portions.

Drilling fluids (muds) contain many additives and are usually specific to the types of geological formations that are drilled through. The additives serve many purposes, but the overall goal is to provide a means for efficient drilling to the target formation with an intact wellbore. To be clear, the main purpose of the drilling fluid is for drilling a hole. Any contaminants from the drilling fluid and drilling formations that must be disposed are a secondary thought because millions of dollars are spent getting to the target formation for oil and gas production.

17.1.3   Volumes of Drilling Waste

The volume of drilling waste is directly related to the number of wells drilled each year, and the number of wells drilled is directly related to the price of oil and gas. Therefore, the volume of drilling waste is a function of the price of oil and gas.

Since horizontal drilling of shale plays and tight sands began around 2000, more wells are drilled to maintain the same level of production because of the high decline rates in these unconventional wells. Conventional vertical wells have much lower decline rates, resulting in fewer wells drilled over time. The increased decline rate in unconventional wells requires oil and gas companies to continuously drill if they intend to maintain production or even increase production. The result is more drilling waste.

17.1.4   Contaminants in Drilling Waste

Drilling waste contaminants come from both the drilling fluids used and the geologic formations drilled through, and these wastes often contain salts, heavy metals and hydrocarbons. Also, because there are no limitations on what chemicals are used in drilling fluids, drilling waste often contains various emerging contaminants.

These four broad categories of contaminants can be found in both types of drilling waste – water based and oil based. However, the percentages of the broad categories and the specific types of contaminants change, based on the type of drilling fluids and the formations drilled through.

17.1.5   Drilling Waste Management Methods

Oil and gas wells are drilled either offshore or on land with the vast majority being drilled on land. Regardless, most drilling waste is managed onshore. The onshore management methods are onsite burial, land application, road application, landfilling, injection, incineration and recycling. All of these methods, except for injection, are surface management methods because the solid content is too high for injection for most subsurface formations.

Each governmental body that regulates drilling waste disposal and recycling usually has some form of closure criteria. In most cases, these closure criteria include physical and/or chemical parameters. More stringent management criteria result in higher costs, but they are usually more protective of the environment. In other words, proper management costs more today, but is much cheaper than managing these wastes and their pollution in the future.

17.1.6   Reporting of Drilling Waste

In many parts of the world, drilling waste is not reported by companies to regulatory agencies or to stakeholders. This means that the volumes of drilling waste generated and their contaminants are unknown, resulting in voids of knowledge about this waste stream and its impacts. Further, if externalities occur from poor waste management, the true costs of the management of these wastes are unknown to all stakeholders.

17.1.7 Research Questions

- Is drilling waste an externality?

- Is drilling waste a material issue?

- Is drilling waste addressed in corporate sustainability reporting?

- What are the potential impacts if drilling waste is not addressed?

17.2 LITERATURE REVIEW

17.2.1 Sustainable Development Goal 12

One of the greatest environmental challenges is achieving economic growth sustainably [Programme, n.d.]. SDG 12, Sustainable Consumption and Production, specifically addresses this and provides specific targets to meet the goal. All of the targets are important, but targets 12.4, 12.5, 12.6, 12.8 and 12.C provide guidance on identifying industrial wastes, detailing how these wastes are managed, stating who should report on these wastes and how to ensure that inefficient fossil fuel subsidies are addressed to reflect their environmental impact [Nations, n.d.]. In other words, SDG 12 wants impactful waste streams reported by those companies that generate the waste and, specifically, to eliminate any fossil-fuel subsidies that contribute to these large impactful wastes. The UN considers this foundational for a sustainable world.

17.2.2 Externalities

According to the International Monetary Fund (IMF), consumption, production and investment decisions often affect people not directly involved in the transactions [Helbling, 2020]. This usually results in externalities. Examples of negative externalities are usually environmental issues in the form of pollution. If these externalities were included in the costs of the goods, the prices of the goods would change. When considered for commodities, inclusion of externalities would change the profits of goods because corporations cannot change the price of commodity goods. Of course, if taken into account, large externalities would create higher priced products, or lower profits in the case of commodities, because they would include the costs to society of the production of these goods. Economists consider externalities a form of "market failure"

[Helbling, 2020]. This is a moral hazard because the decision-makers increase their benefits through damaging others that don't have full information [Helbling, 2020]. A well-known example would be greenhouse gas emissions (GHG) because they affect everyone through climate change, but the emitters, in most countries, bear no cost [Helbling, 2020] (International Monetary Fund, Finance and Development, no date). Another example would be environmental liabilities associated with different surface-waste disposal methods for drilling waste that offer very different pollution controls [U.S. Department of the Interior, Environmental Protection Agency, Office of Solid Waste and Emergency Response, 1987].

17.2.3 Materiality

If an externality exists, the real question is whether it is small or large. In other words, is the externality material? From a legal perspective, SCOTUS has provided a definition which basically says that something is material if the fact would give an investor pause [Bagnoli et al., 2021]. The Financial Accounting Standards Board (FASB) revised its definition of materiality to align with the SCOTUS definition in 2018 [Bagnoli et al., 2021]. From an application perspective for the United States Securities and Exchange Commission (USSEC), materiality often refers to instances of misstatements in financial reporting. However, a $10M liability to a company with $100B in revenue is not the same as a $10M liability to a firm with $20M in revenue. Therefore, auditors often look at some percentage of revenue, income, or liability to determine what is material to a particular firm. From the Norwegian Research Council, single-item materiality rules are 5% of pre-tax income, 0.5% of total assets, 1% of shareholder's equity, or 1% of total revenue [Institute, n.d.].

Financial materiality does not always include issues that might be of concern to all stakeholders because it only relies on currently measured items for the corporation. Citizens of a community might have concerns that are not captured in the financial statements. To address this, stakeholder concerns should be taken into consideration when addressing material issues. However, there is no universal standard for determining these material topics. Therefore, stakeholder input and feedback are necessary to address materiality issues [Bartels, 2014]. This information can then be used for risk assessment and inclusion in corporate sustainability reports. The result is that all stakeholders are better informed about the issues that are relevant [Bartels, 2014]. Also, guidance for these types of issues can be found in international frameworks for environmental, social and governance issues and from the UN's SDGs.

17.2.4 Upstream Oil and Gas Waste

According to the USEPA in 1987, upstream oil and gas waste is categorized as produced water, drilling waste, or associated waste [U.S. Department of the Interior, Environmental Protection Agency, Office of Solid Waste and Emergency Response, 1987]. These same categories of upstream oil and gas wastes are stated by the Society of Petroleum Engineers (SPE) [Reis, 1993]. Another example of these broad waste categories being recognized is from Malaysia [Lodungi et al., 2016]. In each of these citations, produced water is generally defined as the salty water brought to the surface with oil and gas, drilling waste includes the rock cuttings and fluids that are produced from drilling a new well, and associated wastes are small volume waste streams that cover all other types of wastes not found in produced water or drilling waste [American Petroleum Institute, 2000]. From a 1995 survey by the American Petroleum Institute (API) in the United States, it was estimated that 17,910 M barrels of produced water was disposed, 148 M barrels of drilling waste was disposed, and 20.6 M barrels of associated waste was disposed [American Petroleum Institute, 2000]. A further study of these numbers by API shows that the vast majority of upstream oil and gas waste is injected, but the solid waste, mostly drilling waste by a large percentage, is managed by other means on the surface [American Petroleum Institute, 2000].

17.2.5 Regulatory Framework of Upstream Oil and Gas Waste in the United States

In 1976, USEPA was authorized to create a national framework under the Resource Conservation and Recovery Act (RCRA) [U.S. Department of the Interior and Recovery, 2019]. This federal act governs hazardous wastes under Subtitle C and non-hazardous under Subtitle D. In 1978, USEPA proposed regulation of certain "special wastes" be deferred until further investigation could be done [U.S. Department of the Interior and Recovery, 2019]. This included certain upstream oil and gas wastes in which drilling waste was included.

In 1980, the United States Congress amended RCRA temporarily exempting certain upstream oil and gas waste. This is known as the Bentsen Amendment. The temporary exemption required USEPA to determine if these exempted oil and gas wastes should be regulated under Subtitle C [U.S. Department of the Interior and Recovery, 2019]. USEPA provided its report to Congress in 1987, and the report concluded that regulation under Subtitle C was not warranted. However, UESPA stated that this does not mean that these wastes can't cause harm to human health and the environment [U.S. Department of the Interior and Recovery, 2019]. The result is that these exempted oil and gas wastes are regulated under RCRA Subtitle D which means

TABLE 17.1 Summary of Estimated Drilling Wastes by Type of Waste From 1995 E&P Operations Survey [American Petroleum Institute, 2000]

Estimated Total Liquid Wastes	109,443,000	73.6%^{a}
Estimated Total Solid Wastes	39,257,000	26.4%
Estimated Total Drilling Waste Volume	148,700,000	100%
a – Estimated relative percentages of solid and liquid drilling waste are derived from Appendix E, Tables 8 and 9, "Estimated Volumnes of Liquid and Solid Drilling Waste for States with Reported Data."		

that the wastes' management is mostly governed by state laws because they have primacy over non-hazardous wastes.

In 2019, USEPA reported that it reviewed upstream oil and gas wastes' exemption to Subtitle C and came to the same conclusion as in 1987 which included, in both cases, a review of state regulations. This means that a patchwork of regulation exists for this "special waste" that differs from state to state. However, USEPA has stated that the differences in scope, specificity and language used by each state have created uncertainty about the regulatory coverage of oil and gas waste by some states [U.S. Department of the Interior and Recovery, 2019].

17.2.6 Types of Drilling Waste

Drilling waste is categorized as liquid or solid and by the base fluid (mud) type used to drill a particular section of the hole. Both approaches are important in trying to understand how the waste is managed and how it potentially impacts the environment. According to API's 1995 survey, approximately 74% of drilling waste was liquid and 26% was solid – mainly drill cuttings as shown in Table 17.1. This same paper references that their 1985 survey showed that almost 90% of drilling waste was liquid.

In their oil and gas waste review in 2019, USEPA used the same relationship of waste generated per foot from API's 1985 survey and API's drilling waste liquid to solid ratio from 1995, to determine the volumes of drilling waste for 2016 [U.S. Department of the Interior and Recovery, 2019]. In this same report, USEPA states that waste volumes are not routinely collected [U.S. Department of the Interior and Recovery, 2019].

In 1987, USEPA stated that there are two broad categories of drilling muds: water-based muds and oil-based muds [U.S. Department of the Interior, Environmental Protection Agency, Office of Solid Waste and Emergency Response, 1987]. Water-based muds are generally either freshwater muds or saltwater muds, and oil-based muds are usually either an oil or a synthetic oil. Typically, water-based liquids are disposed after each use and oil-based

TABLE 17.2 Estimated percentage of drilling wastes by base drilling fluid [American Petroleum Institute, 2000]

State	Estimated Volume Drilling Wastes Reported by Respondents (barrels)	Percentage of Wastes By Base Fluid				No. Of Responses
		Freshwater	Saltwater	Oil	Synthetic	
AK	111,000	70%			30%	1
CA	506,000	98%		1.5%	0.5%	8
CO	63,000	100%				2
IL	28,000	100%				2
KS	922,000	99%	1%			6
LA	342,000	93%		7%		4
MI	13,000	100%				1
MT	89,000	14%	86%			2
NM	21,000	82%	16%		2%	3
OK	95,000	63%		37%		2
TX	2,812,000	93%	7%			8
UT	131,000	100%				3
WY	30,000	100%				1
Estimated Survey Total	5,163,000	92.5%	5.5%	1.3%	¡ 0.6%	43

Estimated volume of drilling waste for rotary drilling w/ open mud system. Air/pneumatic drilling is excluded. Estimate is based on reported footage drilled multiplied by appropriate factors for drilling waste volume per foot by depth category. This is not and should not be interpreted as a total volume of drilling waste (see Table 3.9). This is based on the respondents only and was used solely to calculate national average distribution of drilling wastes by base fluid.

liquids are recycled due to their expense. The result is that drilling waste is usually made up of some form of water-based liquids, water-based solids and oil-based solids. This is seen in results from the 1995 API survey in the Table 17.2.

USEPA's 2019 report did not address any changes in the percentages of the types of drilling muds used since 1995. However, the 2019 report did acknowledge that horizontal drilling was becoming more prevalent. In order to drill the curve and horizontal sections of unconventional wells, drilling muds that reduce friction and provide lubricity must be used. The muds that most exhibit these properties are saltwater muds and oil-based muds.

The major component in either type of mud system, water based or oil based, is the base fluid followed by barite and then other additives such as brine, clays, polymers, gellants, etc. [IPEICA, 2009]. The base fluid choice is extremely important depending upon the hole section, but barite is important in the vast majority of the hole because it is a weighting agent used to prevent blowouts.

Usually, the initial base fluid for drilling mud used when drilling a well is freshwater because most regulatory agencies require freshwater to drill through the groundwater zone to reduce groundwater contamination. However, water-based drilling fluid has several chemical additions after the "surface section" in order to increase rate of penetration (ROP) of the wellbore. An example is in the Permian Basin where salt zones must be drilled through [Hovorka, 1997]. In this region, saltwater mud must be used to drill the salt sections to equalize osmotic pressure. Oil-based muds are often used for these same Permian wells after the salt section in order to increase ROP, reduce friction and support other well construction needs. Most wells drilled today create different drilling wastes from different sections of the wellbore. In fact, many wells have freshwater drilling waste, saltwater drilling waste and oil-based drilling waste. Also, water-based muds can contain emulsified oils and oil-based muds often contain emulsified brines [Jha et al., 2014].

17.2.7 Drilling Waste Volumes (Tons) Generated

Drilling waste is considered a large-volume waste, particularly by organizations such as the International Petroleum Industry Environmental Conservation Association (IPIECA) [IPIECA, n.d.-c]. One report estimated that more than 392,000,000 barrels of drilling waste were generated onshore in the United States in 2014 alone [Hale, 2017]. Of course, the amount of drilling waste generated each year depends on the number of wells drilled. Therefore, the more wells drilled, the more waste is generated. When looking at well density worldwide, it is clear that North America has far greater well density than any other geographic area [IPIECA, n.d.-d]. Further, USEPA estimated in 2018 that 3.6 M wells had been drilled in the United States to that date [U.S. Department of the Interior, 2018]. This drilling activity in North America will most likely continue due to the proliferation of horizontal drilling [U.S. Department of the Interior and Recovery, 2019].

To understand the volumes of drilling waste generated per well, USEPA estimated in 1987 from the API survey in 1985, the last estimate made, that 1.21 barrels of drilling waste are generated for every foot drilled [U.S. Department of the Interior, Environmental Protection Agency, Office of Solid Waste and Emergency Response, 1987].

For comparison, the United States produced approximately 79 million tons of coal ash in 2019 [American Coal Ash Association, n.d.]. If we assume that each cubic foot of drilling waste weighs 90 pounds (a conservative estimate between the USEPA [U.S. Department of the Interior and Recovery, 2019] and Tuncan [Tuncan et al., 1996], then each well drilled produces 5,524 tons of

TABLE 17.3 Inorganic elements in drilling fluid (mg/l) [U.S. Department of the Interior and Recovery, 2019]

Constituent	Vertical Wells			Horizontal Wells		
	n	50th	90th	n	50th	90th
Arsenic	5 / 8	0.01	0.02	10 / 12	0.03	0.18
Barium	8 / 8	1.3	4.9	32 / 32	23.8	1,810
Boron	8 / 8	0.85	6.1	32 / 32	2.5	15.1
Chloride	8 / 8	2,000	33,000	35 / 35	17,000	89,000
Chromium	4 / 8	0.05	0.16	13 / 21	0.25	1.3
Copper	4 / 8	0.01	0.03	12 / 20	0.17	0.53
Lead	2 / 8	0.07	1.0	12 / 13	0.05	0.30
Manganese	8 / 8	0.19	5.6	32 / 32	2.9	13
Molybdenum	6 / 8	0.13	0.20	11 / 13	0.11	0.41
Nickel	2 / 8	0.05	0.15	13 / 19	0.20	0.39
Sodium	8 / 8	2,100	16,000	33 / 33	11,400	33,900
Strontium	8 / 8	4.1	223	35 / 35	63	1,558
Zinc	5 / 8	0.07	0.20	18 / 25	0.09	1.7

n = Number of Samples Detected / Total

waste. Using 30,000 wells drilled per year in the United States, approximately 165 million tons of drilling waste are generated yearly. Based on these numbers, drilling waste is a large volume waste stream.

17.2.8 Contaminants in Drilling Waste

Drilling waste contains salts, metals, hydrocarbons and other contaminants [Ramirez Jr, 2009]. Some of these contaminants come from the drilling fluid (mud) and some from the formations. The literature sources detailed below regarding specific contaminants are supported by the broad components of drilling fluids previously mentioned.

In UESPA's 2019 report, they provided information on both drilling fluid and drilling solid contaminants. Table 17.3 below from the report shows differences in drilling fluids from vertical and horizontal wells for inorganics.

Another example of inorganic contaminants in drilling waste is in Table 17.4. This referenced paper does not distinguish between drilling fluids and solids because it discusses analytical results of drilling waste as a whole. The paper also details chlorides and hydrocarbons present in drilling waste in the written section. Specifically, the paper states that there are between 4,000 and 170,000 pounds of chlorides per well and between a trace amount and 168,000 pounds of total petroleum hydrocarbons, in the form of diesel, per well.

There are many other reports on the contaminants in drilling waste. One example is a presentation from the University of Colorado School of Law. In

TABLE 17.4   RCRA metal characteristics in solid drilling waste [Keller, 2017]

RCRA Metal	National Primary Drinking Water Standard[10] (mg/L)[a]	Concentrations Observed in Solid Drilling Waste (mg/kg)	Weight per well generated (lbs)
Arsenic	0.01	Trace to 211	Trace to 530
Barium	2.0	Trace to 400,000	5,000 to >500,000
Cadmium	0.005	Trace to 16	Trace to 35
Chromium	0.1	Trace to 160	15 to 350
Lead	0[b]	Trace to 270	Trace to 600
Mercury	0.002	Trace to 1.9	Trace to 4
Selenium	0.05	Trace to 27	Trace to 57
Silver	0.1	Trace to 7.2	Trace to 15

a: USEPA National Primary Drinking Water Standards https://www.epa.gov/ground-water-and-drinking-water/table-regulated-drinking-water-contaminants
b: Lead is regulated by a treatment technique that requires systems to control the corrosiveness of their water. If more than 10% of tap water samples exceed the action level, water systems must take additional steps. For lead, the action level is 0.015 mg/L[10].

TABLE 17.5   Characteristics of different mud systems from various fields* [Scott, 2010]

Characteristic	FWMC**	FWMC***	SWMC	OBC
pH (S.U.)	8.9	10	7.2	10.5
EC (mmhos/cm)	4.26	18	120,000	8.23
ESP (%)	1.3	61	Not Analyzed	2.23
TPH (mg/kg)	1570	114	61,000	156,000
Arsenic (mg/kg)	13.1	92.8	31	72.8
Barium (mg/kg)	5970	148	143	215
Cadmium (mg/kg)	0.343	0.511	0.342	1.22
Chromium (mg/kg)	30.9	72.6	27.6	15.5
Lead (mg/kg)	70.2	390	120	285
Mercury (mg/kg)	0.140	0.970	0.566	1.56
Selenium (mg/kg)	0.552	0.876	0.419	2.13

*   This data is not intended to be considered an average of the specified analytes from the mud types.
**   This FWMC was used on the top section of the hole through the fresh-water zone.
***   This FWMC was used during the entire hole depth.

this presentation, hydrocarbons and salts are shown with heavy metals from different mud types from various oil and gas fields in Table 17.5.

There are also potential contaminants that are emerging that are found in drilling fluids, e.g., PFAS [Norwegian Environment Agency, n.d.]. These "forever" chemicals are being used as de-foaming agents in water-based and oil-based fluids.

USEPA stated in 2019 that the constituents in multiple types of oil and gas waste, including drilling waste, have the potential to introduce contaminants in the environment. They then state that there appears to be substantial information not in the public domain, and because they had limited data, it would not likely change the characterization of the concentration and activity of contaminants in the waste [U.S. Department of the Interior and Recovery, 2019]. However, several of the sources cited here were not included in USEPA's review even though several citations were available at the time of USEPA's report. It is clear from the cited reports that drilling waste, both liquid and solid, has salts, metals and hydrocarbons in substantial concentrations.

17.2.9 Drilling Waste Management Methods

In USEPA's 1987 report, they stated that drilling waste was mainly managed onsite by reserve pits (burial), landspreading, annular disposal, solidification of reserve pits, treatment and disposal of liquid wastes to surface water and closed treatment systems [U.S. Department of the Interior, Environmental Protection Agency, Office of Solid Waste and Emergency Response, 1987]. In this same report, USEPA states that offsite drilling waste management consists of centralized treatment facilities, commercial landfarming, and reconditioning and reuse of drilling media [U.S. Department of the Interior, Environmental Protection Agency, Office of Solid Waste and Emergency Response, 1987].

From the API survey in 1995, volumes of waste and percentages of waste management methods used for liquid and solid drilling waste were provided as shown in Table 17.6 and Table 17.7.

The results of this 1995 API survey show that almost half of the liquid drilling waste is injected and the largest reported management used for drilling solids is burial. However, both tables report that the results include only states with reported data [by the operators]. Other sources also state that burial is the most used method for drill cuttings management.

In 2019, USEPA reported that a comprehensive review of state permits for different waste management units was not feasible at the time because the number of permits and associated documentation was enormous [U.S. Department of the Interior and Recovery, 2019]. Therefore, the best estimate for the management methods used onshore for drilling waste in the public record for the United States is from the API survey in 1995.

17.2.10 Environmental Damage from Drilling Waste

The main environmental effects from drilling waste are to soil, groundwater, or surface water due to the management methods used for these liquid and solid wastes. Therefore, most literature on this subject discusses these three receptors individually or collectively.

Part of USEPA's review in 1987 focused on potential environmental (resource) harm from drilling waste. Their risk model for drilling waste was leaching to groundwater from onsite reserve pits. Their conclusion was that

TABLE 17.6 Estimated volume of liquid drilling wastes disposed by method; estimated for states with reported data [American Petroleum Institute, 2000]

State	Est. Total Volume Drilling Wastes for State	Injection	Injected Down Annulus	Treat & Discharge	Road Spread	Evaporate On or Offsite	Land Spread Onsite	Land Spread Offsite	Reuse for Drilling	Other
AK	1,605,000	803,000								
AR	2,643,000	1,850,000								
CA	1,826,000					347,000	804,000	1,000	22,000	256,000
CO	3,582,000	1,290,000				1,863,000				
IL	382,000	317,000								
KS	4,810,000	48,000				914,000			1,780,000	
LA	22,477,000	1,573,000	10,115,000			225,000	2,245,000		112,000	
MI	1,731,000	697,000					1,385,000			
MT	1,366,000	297,000								137,000
NM	7,421,000					5,788,000	223,000			
OK	13,162,000	654,000				132,000			6,449,000	
TX	65,367,000				65,000	53,601,000		588,000	1,307,000	
UT	1,561,000					609,000			78,000	
WY	4,374,000					2,187,000				
Total Volume (excludes Appalachian States):	132,307,000	7,529,000	10,115,000	0	65,000	65,666,000	4,657,000	589,000	9,885,000	256,000
Appalachian States:										
KY	2,095,000			1,341,000			126,000			
NY	138,000									
OH	1,721,000	1,274,000								
PA	2,426,000	170,000								777,000
VA	57,000						32,000	2,000		
WV	856,000			188,000			111,000			
Total Volume (Appalachian States Only):	7,295,000	1,444,000	0	1,529,000	0	0	269,000	2,000	0	777,000
National Est. Total Volume[a]:	139,602,000	8,973,000	10,115,000	1,529,000	65,000	65,666,000	4,926,000	591,000	9,885,000	1,033,000
National Total, % by Disposal Method	Liquid Wastes = 73.6%	6%	7%	1%	0.05%	47%	3.50%	0.4%	7%	0.7%

[a] Table only includes states with reported data. Therefore, the estimated total volume of drilling waste is less than the 148.7 million barrels discussed in the text. The estimated total volume reported in the text, 148.7 million barrels, includes drilling waste volumes estimated for states for which no responses to drilling related questions were received on the 1995 survey.

chlorides and other highly mobile ions had the potential to alter groundwater but not significantly on a national basis [U.S. Department of the Interior, Environmental Protection Agency, Office of Solid Waste and Emergency Response, 1987]. However, earlier in the 1987 report, USEPA stated their risk analysis was only based on water-based drilling waste for a 200-year timeframe [U.S. Department of the Interior, Environmental Protection Agency, Office of Solid Waste and Emergency Response, 1987]. Further, the report stated that there was virtually no difference in leaching from reserve pits that were lined versus ones that were unlined at 200 years in the model, but more mobile contaminants would leach earlier in a 50-year range [U.S. Department of the Interior, Environmental Protection Agency, Office of Solid Waste and Emergency Response, 1987]. A striking part of the 1987 report states that USEPA

TABLE 17.7 Estimated volume of solid drilling wastes disposed by method; estimated for states with reported data [American Petroleum Institute, 2000]

State	Est. Total Volume Drilling Wastes for State[a]	Buried Onsite	Land Spread Onsite	Land Spread Offsite	Commercial Disposal Facility	Industrial or Municipal Landfill	Reuse or Recycle	Other
AK	1,605,000							803,000
AR	2,643,000	793,000						
CA	1,826,000	383,000	5,000	2,000	4,000	2,000		
CO	3,582,000		107,000	322,000				
IL	382,000	4,000	61,000					
KS	4,810,000	2,020,000						
LA	22,477,000	4,495,000	899,000		2,922,000			
MI	1,731,000	346,000						
MT	1,366,000	533,000						
NM	7,421,000	965,000	223,000					
OK	13,162,000	6,581,000						
TX	65,367,000	8,533,000	197,000	65,000			394,000	65,000
UT	1,561,000	874,000						
WY	4,374,000	2,187,000						
Total Volume (excludes Appalachian States):	132,307,000	26,060,000	1,492,000	389,000	2,926,000	2,000	394,000	868,000
Appalachian States:								
KY	2,095,000	587,000	42,000					
NY	138,000		138,000					
OH	1,721,000	448,000						
PA	2,426,000	49,000	1,432,000					
VA	57,000	23,000						
WV	856,000	565,000						
Total Volume (Appalachian States Only):	7,295,000	1,672,000	1,612,000					
National Est. Total Volume[a]:	139,602,000	29,732,000	3,104,000	389,000	2,926,000	2,000	394,000	868,000
National Total, % by Disposal Method	Solid Wastes = 26.4 %	21%	2.2%	0.3%	2%	0.001%	0.3%	0.6%

[a]Table only includes states with reported data. Therefore, the estimated total volume of drilling waste is less than the 148.7 million barrels discussed in the text. The estimated total volume reported in the text, 148.7 million barrels, includes drilling waste volumes estimated for states for which no responses to drilling related questions were received on the 1995 survey.

only modeled single oil or gas wells rather than risks from clusters of wells [U.S. Department of the Interior, Environmental Protection Agency, Office of Solid Waste and Emergency Response, 1987]. This means the risk modeling does not account for risks from the same waste type being disposed the same way in close proximity. Therefore, the risk model does not consider the cumulative environmental effect from thousands of wells in a geographic area, e.g., the Permian Basin [United States Department of Energy, n.d.].

In a presentation by the Oklahoma Corporation Commission (OCC) and the Association of Central Oklahoma Governments (ACOG) sometime around 2013, the OCC and ACOG stated that they had many water well pollution complaints over 20 years that were both suburban and rural [Billingsley and Harrington, n.d.]. This presentation details the types of contamination found in groundwater in Oklahoma, the sources of the contamination and the locations. It was clearly stated that some of this contamination came from drilling waste. Two example cases were shown regarding groundwater contamination. One case was within a gated community northwest of Oklahoma City, Oklahoma, and the second case was within a subdivision in Edmond, Oklahoma.

In another presentation given in 2014 by the North Dakota Geological Survey, the results of six buried drilling waste reserve pit studies were presented [Murphy, 2014]. This presentation clearly stated that leachate is being generated from buried drilling fluids. The degree of groundwater impact depends upon the site characteristics. The type of contamination being monitored was salts.

Unconventional oil and gas wells ushered in a new era in fossil fuel production in the United States that increased drilling. The result is that more drilling waste is being created. Therefore, in 2015, the United States Department of Energy's National Energy Technology Lab (NETL) published a paper where they studied leaching characteristics from the Marcellus shale [Stuckman et al., 2015]. The paper was clear that there is limited information on the environmental impacts from drilling waste being disposed in reserve pits and MSW landfills. The drilling cuttings studied were two samples of water-based cuttings, one sample of oil-based cuttings and one sample of an outcropping of the Marcellus shale. It was shown, through testing, that there were multiple trace elements in the samples that leach under different leaching scenarios and that total dissolved solids leach first with potential long-term releases from barium, cobalt, nickel, copper, lead and zinc. The paper states that leaching tests using deionized water and rainfall showed less leaching than acetate-based leaching tests that would mimic anoxic conditions such as those in landfills. The paper suggested further study of the environmental impacts on total dissolved solids on leaching quality and conducting further research on long-term release of certain elements under different disposal scenarios.

In 2018, the National Institutes of Health (NIH) in the United States published a literature review of environmental and health effects of people living near oil production sites [Johnston et al., 2019]. In this paper, NIH specifically reviewed literature on drilling fluids, wastewater and oilfield waste pits. For both water and oil-based drilling waste, the papers state the fluids can be discharged directly into oil field pits with the potential to leach into the soil or groundwater. It is further noted that elevated levels of certain metals have been seen in drilling fluids and that concentrations of polycyclic aromatic hydrocarbons (PAH) existed in agricultural fields next to pits. There are numerous mentions of salts, metals and hydrocarbons in this paper and how these contaminants are being found in surface water, groundwater and soil from oil field waste, including drilling waste.

In the same year as the NIH report, an article from Wyoming stated that groundwater had been contaminated from hydraulic fracturing, disposal of petrochemicals in unlined reserve pits and inadequately constructed gas wells [Thuermer Jr., 2018]. The contamination in the groundwater, the Wind River Formation aquifer, is methane gas and petrochemicals. The area that has the contaminated water is around Pavillion, Wyoming, and covers approximately 12 square miles. One researcher, a former USEPA hydrologist, said that the aquifer most likely cannot be cleaned up. One of the two former USEAP scientists, Dominic DiGiulio, said drilling waste was placed in some of the 40 unlined pits used in the field. Mr. DiGiulio further said that the drilling waste was a diesel-based fluid and that it contained between 55% and 79% diesel.

Even though many states in the United States do not collect volumes/tons or the chemical compounds in oil field waste, this information is collected by the Pennsylvania Department of Protection (PADEP). Therefore, researchers in 2018 at Resources for the Future (RFF) looked at some of this data to determine if solid waste from shale gas should be regulated as hazardous [Swiedler et al., 2019]. The vast majority, 96.6% by weight, of the solid waste reviewed by RFF from the PADEP was drill cuttings. The paper noted that their review was timely because USEPA was in the process of reviewing whether oil and gas waste should be regulated under RCRA Subtitle C. The conclusion by the RFF paper was that average concentrations and leachate from the oil field waste were not high enough to be regulated as hazardous, but there were some samples that were high enough to exceed RCRA thresholds for toxicity. The paper suggested that solid oil field waste should be regulated under RCRA but that RCRA standards omit some potentially harmful chemicals in solid wastes that should be included in the regulations. Finally, this paper, which focused on landfill disposal of solid drilling waste, stated that low levels of contamination are problematic because it is aggregated in the disposal method – landfills. Basically, this means that there are concerning chemicals in drilling waste and that even low-level concentrations are a concern when placed in close proximity.

In their 2019 report, USEPA stated that there was limited information on the leaching or volatilization of oil field wastes [U.S. Department of the Interior and Recovery, 2019]. However, the report reviewed USEPA's 1987 report on reserve pit disposal of drilling waste and stated that the 200 year limit used in 1987 likely underestimated potential long-term risks, and the groundwater model used for contaminants likely underestimated there mobility [U.S. Department of the Interior and Recovery, 2019]. The review of the 1987 report goes on to state that there are several uncertainties with the 1987 conclusions regarding drilling solids. These uncertainties are mainly associated with waste types, management practices, constituents (contaminants) and release pathways, and they further state that they did not attempt to update the model results due to a lack of data [U.S. Department of the Interior and Recovery, 2019].

After the USEPA report, an article was written in September 2019 regarding damaged lands in North Dakota from oil brine spills and old evaporation pits [Reader, n.d.]. The salts from these spills and oil pits are the same type of salts found in drilling waste. One farmer interviewed for the article said that his land was worthless unless a buyer assumes the liability. This view was further illustrated in the article by a quote from the chief executive officer, Claude Sem, of Farm Credit Services of North Dakota to the House Natural Resources Committee in 2015. Mr. Sem stated, "Farm Credit is concerned about taking security in any property that has or potentially has hazardous materials on the premises, such as from a saltwater spill or oil spill. In fact, we would be cautious of even taking collateral near the hazardous or potentially hazardous site. It is and has been the policy of FCSND to not take security in property that is or potentially will be environmentally damaged. FCS discourages credit applications on property that is contaminated and would, and will, deny approval if credit was requested."

Overall, the literature shows that drilling waste contains multiple contaminants that can cause harm depending on the disposal method. When buried, highly mobile salts are the first to leach, e.g., chlorides. Hydrocarbons and heavy metals also leach but more slowly. Of particular interest is barium. Even though barium is usually in the form of barite, which is virtually insoluble, it becomes mobile under anoxic conditions that are found in landfills and buried waste in onsite pits – the most used method of disposal. Barium contamination of groundwater from drilling waste and its effects on human health and the environment are documented in multiple locations [Pragst et al., 2017].

17.2.11 Corporate Frameworks for Sustainability Reporting of Drilling Waste

There are several corporate reporting frameworks that address drilling waste, and the most prevalent ones are IPIECA, the Sustainability Accounting Standards Board (SASB) and the Global Reporting Initiative (GRI). SASB Standards are intended to focus on environmental, social and governance (ESG) issues that have a financially material impact on a company, while GRI Standards focus on economic, environmental and social impacts in relation to sustainable development for all stakeholders [SASB, 2020]. Basic reporting principles of waste from most reporting frameworks involve reporting the type of waste generated, the tons or volumes of a waste stream generated and the management methods used for each waste stream.

IPIECA is an oil and gas industry association advancing environmental and social performance that was formed at the request of the United Nations [IPIECA, n.d.-a]. One of the things this organization provides is guidance frameworks to the oil and gas industry on non-financial reporting [IPIECA, n.d.-b]. In IPIECA's guidance "Sustainability reporting guidance for the oil and gas industry Module 4 Environment," it mentions "drilling muds and cuttings" and "drill mud and cuttings" under the "Materials management" section [IPIECA, n.d.-c]. In this section, it states that drilling muds and cuttings are excluded from core reporting, C3, and additional reporting, A1 [IPIECA, n.d.-c]. The guidance does state though companies **can** [emphasis added] report the excluded wastes under A3. Further, the statement under A3 says to report certain wastes separately, including "large-volume wastes, such as drill mud and cuttings," but it does not state to report the type of drilling waste, tons or volumes of drilling waste generated, or management methods used [IPIECA, n.d.-c].

SASB, which merged with the International Integrated Reporting Council (IIRC) in July 2021 to form the Value Reporting Foundation, provides standards for companies to communicate how sustainability issues impact long-term enterprise value [SASB, 2020]. SASB has developed industry standards for 77 different industries including "Extractives & Minerals Processing" of which "Oil & Gas – Exploration & Production" and "Oil & Gas – Services" are a subcategory [SASB, n.d.-a]. For each industry, the sustainability issues listed under "Environment" include disclosure topics "Water & Wastewater Management," "Ecological Impacts," and "Waste & Hazardous Waste Management" [SASB, n.d.-d]. Disclosure topics for "Oil & Gas – Exploration & Production" specifically include "Water & Wastewater Management" and "Ecological Impacts." For water issues, it states that contamination can come from multiple sources including "well fluids." For ecology impacts, it states that examples of habitat loss can come from disposal of drilling and associated wastes [SASB, n.d.-b]. However, the disclosure topic of "Waste & Hazardous Materials Management" is listed under the industry subcategory "Oil & Gas – Services" [SASB, n.d.-c]. For this disclosure topic, "Waste & Hazardous Materials Management," SASB states that drilling fluids pose a risk and also state this for the disclosure topic "Ecological Impacts." Since the exploration and production company is usually the legally responsible party for managing drilling waste when it is not taken to a third party, it seems prudent for drilling waste to be listed under "Oil & Gas – Exploration & Production," too, rather than just oil and gas services. However, it is clear that SASB considers drilling waste to be a financially material topic.

GRI is an international reporting organization that provides a common language for organizations to report their impacts [Global Reporting Initiative, n.d.-b]. Guidance provided by GRI covers a wide range of topics. In 2021, GRI launched its first Sector Standard which was for oil and gas [Global Reporting Initiative, n.d.-b]. Even though the GRI Standards state that an organization must determine what topics it considers material, the Oil and Gas

Sector Standard clearly states that waste is a likely material topic including drilling waste (muds and cuttings) because it is a high volume waste which "can contain chemical additives, hydrocarbons, metals, naturally occurring radioactive material (NORM) and salts" [Global Reporting Initiative, n.d.-a]. This document further details that the likely material topics for waste should be reported by waste type, tonnages or volume, amount diverted from disposal and amount directed to disposal [Global Reporting Initiative, n.d.-a].

Even though IPIECA does not specifically state that oil and gas companies should report on drilling waste or how it should reported, both SASB and GRI have clear guidance on drilling waste because it is a financial and environmentally material topic.

17.3 METHOD

The majority of the literature review came from sources within the United States or were about information on the United States. Drilling waste is clearly impactful in the United States because this region has the greatest well density of any country to date, the United States has a large number of locations left to drill and the laws in the United States are very favorable to continued oil and gas exploration. Also, even with the large number of drilled wells and wells to potentially drill, USEPA knows very little about the waste being produced from drilling and how it is affecting the environment. Therefore, Waste Analytics LLC began collecting public information and then deriving data on drilling waste in the United States for all publicly traded oil and gas companies. Waste Analytics' proprietary data includes the waste types and volumes generated, the waste management methods used and the environmental liabilities created by company. The research questions in this paper are answered using information from the literature review and information from Waste Analytics. Even though the rest of this paper will focus on the United States, there is clear indication that drilling waste is impactful regardless of where it is generated.

17.4 DISCUSSION

17.4.1 Is Drilling Waste an Externality?

Externalities are costs that are not included in the production of goods and can be either positive or negative. These benefits or costs are shifted to third parties, e.g., landowners. Most negative externalities are usually associated with environmental issues that are in the form of pollution. Also, it is known that pollution is usually the result of improperly managed wastes. Pollution can occur if the waste contains contaminants that harm human health or the environment, if the waste is a large-volume waste, if waste management methods vary significantly in protection and cost, or a combination of all these.

Water-based drilling waste is both liquid and solid. The base fluid, freshwater or saltwater, can change depending upon the section of the wellbore drilled. This means after the surface section of the well is drilled with freshwater to protect groundwater, all other drilling waste may contain some type of chemical additive. Also, just because the majority of the base fluid may be one fluid does not mean that other fluids are not present, e.g., a water-based drilling fluid may have a significant portion of oil (diesel) added for lubricity.

Oil-based drilling waste is usually solid because the oil-based drilling fluid is recycled due to its expense. The oil-based solids contain portions of the base oil, usually diesel onshore in the United States, saltwater, barite and other additives.

The literature supports that drilling waste contains salts, metals, hydrocarbons and emerging contaminants, e.g., PFAS. These contaminants are in both water-based and oil-based drilling waste. All of these contaminants can become mobile in the environment through leaching. Highly soluble salts, e.g., chlorides, usually move first. Heavy metals, hydrocarbons and emerging contaminants move more slowly. Even barium becomes mobile in anoxic environments [U.S. Department of the Interior, Environmental Protection Agency, Office of Solid Waste and Emergency Response, 1987].

Table 17.8 is from information in the Keller paper. Clearly, the amounts of contaminants per well that can be introduced in the environment are very large.

The reports presented by USEPA and the API show that drilling waste is mainly liquid waste. In fact, the last survey by the API in 1995 indicated that 73.6% of drilling waste was liquid and 26.4% of drilling waste was solid. This survey was done well before the beginning of unconventional wells being drilled in the United States. However, USEPA used this same liquid to solid relationship for their 2019 report. Waste Analytics' review of drilling waste from 2017 through 2020 indicates changes in the relationship of liquid waste to solid waste as shown in Table 17.9. In fact, the percentages of liquids and solids have basically reversed.

As USEPA pointed out in their 2019 report, unconventional wells have increased dramatically since 2000. It is reasonable to assume that changes have occurred in drilling waste that reflects the technological needs of drilling wells horizontally. In many cases for recent unconventional wells, the horizontal section of the wellbore exceeds one mile in length. Wellbores that turn 90 degrees and extend laterally for over one mile require shale inhibition to reduce formation swelling and to provide lubricity. This change is reflected in the types of drilling waste generated by base fluid. In API's 1995 survey, they stated that 98% of drilling waste by base fluid was water-based fluid

TABLE 17.8 Drilling waste contaminant concentrations and total contaminant weights per well [Keller, 2017]

Constituent	Concentrations Observed in Solid Drilling Waste (mg/kg)	Weight per Well Generated (lbs)
Chlorides	2000 to 200,000	4,000 to 170,000
Total Petroleum Hydrocarbons	Trace to 220,000	Trace to 168,000
Arsenic	Trace to 211	Trace to 530
Barium	Trace to 400,000	5,000 to 500,000
Cadmium	Trace to 16	Trace to 35
Chromium	Trace to 160	15 to 350
Lead	Trace to 270	Trace to 600
Mercury	Trace to 1.9	Trace to 4
Selenium	Trace to 27	Trace to 57
Silver	Trace to 7.2	Trace to 15

TABLE 17.9 Comparison of liquid drilling wastes and solid drilling wastes from API's1995 survey and an average of Waste Analytics' data from 2017 to 2020

	API's Average from 1995 Survey	Waste Analytics' Average from 2017–2020
Liquid Drilling Wastes	73.6%	28.5%
Solid Drilling Wastes	26.4%	71.5%

TABLE 17.10 Comparison of water-based solids and oil-based solids from an averageof Waste Analytics' data from 2017 to 2020

	Water-Based Solids	Oil-Based Solids
Waste Analytics' Average from 2017 to 2020	62.25%	37.75%

(freshwater and saltwater) and that 2% was oil-based fluid (oil and synthetic). Once again, USEPA did not collect new information for their 2019 report. However, from Waste Analytics' data on different waste types by base fluid, a dramatic change has occurred in the mud types used as shown in Table 17.10. There is far more usage of oil-based muds than assumed from API's 1995 survey.

TABLE 17.11   Waste management method percentages from Waste Analytics'
data from 2017 to 2020

Method	Waste Analytics' Average Percentage from 2017 to 2020
Burial	31.9%
Injected	14.9%
Land Applied	9.8%
Land Farmed	1.9%
Landfilled	39.9%
Recycled	0.1%
Road Applied	1.5%

From API's 1995 survey, liquid drilling waste was mainly managed by evaporation (47%) and solid drilling waste was usually buried in onsite reserve pits (21%) with only 2.001% of solid drilling waste going to a commercial facility or industrial/MSW landfill. Waste Analytics' analysis in Table 17.11 shows that burial is still often used, but landfilling has dramatically increased.

Burial involves leaving the drilling waste in reserve pits onsite. This is a very low-cost option. Landfilling involves loading the waste on trucks, transporting the waste and paying a fee for disposal at the landfill. Clearly, landfilling costs more than onsite burial. In USEPA's 1987 report, the vast cost differences are detailed where the national average for onsite burial is $2.04 per barrel and offsite disposal at a landfill with a synthetic liner and leachate collection system is $15.52 per barrel [U.S. Department of the Interior, Environmental Protection Agency, Office of Solid Waste and Emergency Response, 1987]. These differences occur in states where you can choose between these methods, e.g., Texas. This same USEPA report states most landfills with liners and leachate collection systems provide greater environmental protection.

The combination of the literature review and Waste Analytics' data shows that drilling waste contains contaminants that can harm human health and the environment; it is a large-volume waste stream on a per-well-drilled basis; there are management methods with different environmental protections being used; and there are large cost differences in these management methods. Therefore, drilling waste is an externality.

17.4.2   Is Drilling Waste a Material Issue?

In order to determine if the externalities created from drilling waste are large enough to be financially material and/or environmentally material, the impact must be determined. For financial materiality, it is important to look at near-term costs and long-term costs. For environmental materiality, it is important to look at the breadth of potential environmental impact to a community and region.

As has been shown by the studies conducted by USEPA, third-party sources and Waste Analytics, there are varied approaches to managing drilling waste from onsite burial without liners to offsite disposal facilities, e.g., landfills. The near-term costs for different management methods vary greatly. Using the figures from USEPA in their 1987 report, landfilling is approximately 655% more expensive than onsite burial without a liner. Of course, some states require drilling waste to be disposed in landfills. However, in other states, e.g., Texas, an oil and gas company can bury onsite or dispose of their waste in a landfill. If you assumed that 6,000 barrels of drilling waste are generated per well and using USEPA's 1987 report figures, it would cost $12,240 per well to bury drilling waste onsite versus $93,120 per well to landfill it. Using USEPA's 1987 disposal cost estimates with the United States Energy Information Administration's (USEIA) 2015 onshore horizontal drilling cost average of $2.2 M per well (U.S. Department of Energy, U.S. Energy Information Administration, 2016, p. 7), drilling waste disposal can vary from 0.5% for onsite burial to 4.2% for landfilling as a percentage of drilling cost. This is a wide disparity in near-term costs. If given a choice, as in Texas, why would any oil and gas company choose to incur costs to go to a landfill that is approximately 655% more expensive than onsite burial if it is legal? Oil and gas companies must be making these disposal method choices based on knowledge of environmental impacts.

Since drilling waste contains contaminants that are mobile and drilling waste is buried in reserve pits, it can be concluded that potential pollution is occurring from buried drilling waste. Therefore, it is reasonable to assume that if an oil and gas operator chooses the near-term, low-cost option of onsite burial, there should be some corresponding liability booked to the financial statements for long-term pollution. Waste Analytics has not found any liability booked to the balance sheets of any publicly traded oil and gas companies for buried drilling waste in the United States. This means this is probably an off-balance sheet liability. Most likely, these figures would give investors "pause" which is the definition of materiality according to SCOTUS.

To understand the financial impacts, Waste Analytics has calculated the estimated environmental liabilities created from drilling waste. In 2019, from the publicly traded oil and gas companies Waste Analytics reviewed, the drilling waste Estimated Environmental Liability (EEL) was $1.6B. Since the drilling waste usually remains where it was originally disposed, the yearly EEL is cumulative. Waste Analytics reviewed the cumulative EEL from 2017 to 2020 as a ratio of publicly traded oil companies' revenue for 2020 and found that it ranged from 0.01% to 23.37%. The single item materiality rule for total revenue from the Norwegian Research Council states something is material if it exceeds 1% of total revenue. Waste Analytics review showed that approximately 54% of the companies exceeded 1% of total revenue in cumulative EEL. This directly shows that drilling waste liabilities are financially material. In fact, this figure would be even higher if other costs associated with drilling waste pollution were included, i.e., sampling, testing, soil pollution and groundwater pollution clean up.

One of the most actively drilled areas in the United States is the Permian Basin. USEIA reported in 2021 that there were 27,540 active wells in the Midland Basin – a part of the Permian Basin entirely within Texas [United States Department of Energy, n.d.]. It is known that salt formations must be drilled through in this area in order to get to the oil and gas producing zones [Hovorka, 1997]. The drilling fluid used to drill these salt sections is saltwater mud. The resulting drilling waste can be assumed to contain the upper end of the chlorides per well of 200,000 mg/kg or 170,000 pounds [Keller, 2017] due to chemical saturation. This means that for just the active wells in the Midland Basin, 5.5 megagrams or 2.3 million tons of chlorides have been generated in drilling waste. Waste Analytics review of this waste stream from 2017 through 2020 shows that at least half has been buried resulting in the burial of 2.8 megagrams or 1.15 million tons of chlorides. This figure does not account for the previously plugged and abandoned wells in the Midland Basin that also generated salty drilling waste.

Oil and gas operators do use synthetic liners in reserve pits in many instances which could be assumed to mitigate leaching. However, if the waste is buried in the reserve pit, it must be closed with the addition of something to stiffen the drilling waste, e.g., soil, lime kiln dust and fly ash. The drying and burial process tears the liners. The intent of the liner is to prevent fluid loss during drilling but not to contain the buried waste. Therefore, the contaminants are free to leach during precipitation after closure.

The Midland Basin has three main aquifers: the Edwards-Trinity Plateau, the Pecos Valley and the Ogallala. They are all used in varying degrees for municipal water supplies and for agriculture. Since chlorides are highly mobile, they are markers for leaching from pollution sources. Any contamination of freshwater sources is of concern to stakeholders because water is critical for human habitation and the environment. Chloride contamination is well known in the Midland Basin, and oil and gas activity is sited as one of the sources.

The effects of groundwater contamination from drilling waste on a community are seen in Pavillion, Wyoming. Researchers have directly tied part of the contamination in the Wind River Formation aquifer to oil-based (diesel) drilling waste and stated that the likelihood the aquifer can be cleaned up is low. Unfortunately, the residents in the affected area who relied on the aquifer's water must now have their water trucked in. While this may be possible for a small isolated area, it is impossible for larger geographic areas with more people and industry, e.g., the Midland Basin.

Drilling waste is a financially material issue as shown by work from Waste Analytics. Obviously, oil and gas companies consider the issue financially important or they would not make decisions that are drastically cheaper in the short term. This waste stream is also environmentally material due to its ability to leach into groundwater resources. The known contaminants in drilling waste that all leach under various conditions and the widespread use of burial as a management method indicate that all stakeholders in these areas are affected.

17.4.3   Is Drilling Waste Addressed in Corporate Sustainability Reporting?

The United Nations states that wastes that affect human health and the environment should be reported by companies. Regulatory agencies usually require companies to report industrial and hazardous wastes' amounts and how they are disposed. Reporting of these wastes by the type, amount and management methods used is recommended by international reporting frameworks if they are material, e.g., SASB and GRI.

In May of 2021, Waste Analytics reviewed the latest sustainability reporting of 69 publicly traded oil and gas companies that drilled wells in the United States. The review of each companies' reporting was based on six questions listed below.

1. Does the company publish an ESG/Sustainability report?

2. Does the company provide a performance metrics table?

3. Does the company have a section regarding waste and waste management?

4. Does the company include waste metrics in their performance metrics table?

5. Does the company include metrics on drilling waste?

6. Does the company disclose specific disposal methods for drilling waste?

This review also included companies that mentioned IPIECA, SASB and GRI in their reporting. The results are in Table 17.12.

From this review, it is clear that most companies do have some form of sustainability report and that most have a performance metrics table. However, less than half mention waste or waste management in their report and even less include waste in their performance metrics. A much smaller number of companies include metrics specific to drilling waste, and only two of the

TABLE 17.12   Oil and gas corporate reporting from Waste Analytics' data

Topic	Number of Companies	Percentage of Total
Companies reviewed	69	100%
Companies that published ESG/Sustainability reports	49	71%
Companies that provided a performance metrics table(s)	44	64%
Companies that had a section regarding waste/waste management	28	41%
Companies that included waste metrics in the performance metrics table(s)	22	32%
Companies that included metrics specific to drilling wastes	3	4%
Companies that disclosed specific disposal methods used for drilling wastes	2	3%
Companies that mentioned GRI in their report	32	46%
Companies that mentioned SASB in their report	39	67%
Companies that mentioned IPIECA in their report	25	36%

companies disclose specific management methods for drilling waste. Interestingly though, 46% of the companies reviewed mention GRI in some way and a full 67% mention SASB. Therefore, the concept of waste reporting, including drilling waste, is known by most of these companies. Further, waste in general is mentioned in reporting by 41% of the companies. This means these companies either do not consider drilling waste a material issue or they are choosing to avoid reporting on drilling waste.

Given that IPIECA considers drilling waste a large-volume waste, GRI and SASB specifically mention drilling waste in their frameworks, USEPA states that drilling waste is a large-volume waste that has the potential to leach into groundwater, and there are numerous reports concerning contaminants in drilling waste and how contaminants in drilling waste leach, it is striking that oil and gas companies provide such limited information on the topic. Most of the companies that do provide metrics usually report the types and amounts of oil and produced water spills. These spills should be reported, but the volume of spills in relation to the amounts of drilling waste generated is very small. Often, the contaminants in the materials spilled are the same types of contaminants in drilling waste. Why should a company report on the types and amounts of spills but not report on a far larger waste stream that is known to create harm? This lack of transparency is extremely concerning and clearly indicates that this is a negative externality.

17.4.4 What Are the Potential Impacts If Drilling Waste Is Not Addressed?

Potential impacts from drilling waste are pollution-related issues that affect the environment that then affect human health and the financial system. All of these impacts are interrelated.

Environmental impacts from drilling waste are contaminated soil and groundwater from buried drilling waste. Literature is very clear that drilling waste contains salts, metals, hydrocarbons and emerging contaminants, e.g., PFAS. The evidence also supports that these contaminants have the capacity to move in the environment. It is also clear that drilling waste occurs in areas with increasing populations and areas of intense agriculture as is shown by looking at well density in the United States. This is not a problem in just one geographic region but is in multiple states, e.g., California, Oklahoma, Texas, Colorado, Pennsylvania, West Virginia, and New York.

USEPA anticipated that leaching from reserve pits that are unlined in areas with permeable soils has a high potential for migration into groundwater and soil [U.S. Department of the Interior, Environmental Protection Agency, Office of Solid Waste and Emergency Response, 1987]. They further state that these types of reserve pits can continue to cause problems after closure when drilling waste begins to leach into surrounding soil. The contaminants mentioned include chlorides, sodium, barium, chromium and arsenic.

As an example, in Oklahoma, contamination from drilling waste was shown in the OCC presentation of 2013 [Billingsley and Harrington, n.d.]. This presentation is clear that barium was found in water wells and that barium is found in drilling mud. Barium was present in water wells less than 25 feet deep and water wells greater than 25 deep. These contaminated water wells were from various parts of the state meaning they were not isolated to one area. The presentation also documents other contaminants in the water wells from oil and gas production including salts. The OCC points out that groundwater is important because approximately 40% of all water used in Oklahoma comes from groundwater, it provides water to more than 200 Oklahoma cities and towns, 295,000 Oklahomans use groundwater from domestic wells and 73% of all irrigation for agriculture is from groundwater. Obviously, groundwater is an important resource that is being impacted across Oklahoma by oil and gas.

The main paths of human health to be impacted from drilling waste are consumption of groundwater, consumption of foods irrigated with the groundwater, or foods grown in impacted soils. The NIH has detailed that soils contaminated from drilling fluids can harm human health through direct ingestion, crops, dermal contact and inhalation of contaminated soil particles. They also state that migration of these contaminants into groundwater affects communities [Johnston et al., 2019]. The literature review by NIH also discusses these impacts for both organic and inorganic contamination. Many of the chemicals found in oil and gas waste in this review indicated that they are endocrine disruptors which decreased pituitary hormone levels, increased body weights, disrupted development of ovarian follicles and altered uterine and ovary organ weights for females [Johnston et al., 2019]. It further states males also are subject to reproductive issues including testes weights, serum testosterone, body weights and decreased sperm counts.

In another study in Sudan, barium and lead were found in groundwater that originated from drilling waste and produced water pits, specifically from the use of barite in drilling mud [Pragst et al., 2017]. The researchers found both barium and lead in hair samples of the local community. Also, the researchers noted a rise in livestock deaths. The paper goes on to detail that lead exposure causes developmental neurotoxicity, reproductive dysfunction and toxicity to kidneys, blood and endocrine systems. They also note that barium is known for its acute toxicity, and the reported health effects include cardiovascular and kidney disease, and other issues associated with neurological and mental disorders.

Not only are environmental and human health issues being caused by drilling waste and how it is managed, but financial impacts are documented, too. These financial impacts relate to impacted land values and the potential for impacted housing values.

As previously noted in the article from North Dakota, the CEO of Farm Credit Services of North Dakota said that his lending agency would not take security in any property affected by a brine or oil spill. The landowner in the article further stated that unless someone wanted to take the liability of the contamination on his land his property is worthless. This means that contamination on property in the form of buried pits creates issues for the value of the land. Remediation of buried drilling waste would be cost prohibitive for landowners. Further, if contamination has migrated from their property to adjacent landowners, they may have legal obligations to compensate adjacent landowners for trespass. The result is that the landowner's asset becomes a liability.

In the OCC presentation, two case studies were shown that were previously discussed where saltwater plumes were found in housing developments under homes. These studies clearly showed groundwater contamination moving under these homes. In fact, some of these homes had been built directly over former pits. Even if remediation of the groundwater under these homes was possible, the value of these homes could be called into question. If the homeowners chose to sell their homes, it is doubtful the homes would be valued at the same amount as though they were not sitting on contamination. This means that the asset would be seriously devalued. Also, if the homeowner held a mortgage on the property, they would be paying principal and interest on an asset that could not be recovered. Since homes are the largest asset most citizens in the United States hold, their net worth might be negative. The only option for the landowner might be bankruptcy. The result is that the banking industry could be holding toxic assets.

All of these impacts, environmental, human health and financial, are potentially widespread due to number of wells that have been drilled in the United States. This means the effects have societal implications that are not well known.

17.5 CONCLUSION

It has been shown that drilling waste is a large-volume waste that has multiple contaminants that can move in the environment. Drilling waste management practices that allow these contaminants to enter soil and groundwater are still in use. USEPA identified these potential concerns in its 1987 report.

It is also clear that there are large differences in the costs of onsite burial and landfilling yet there are no corresponding liabilities booked by oil and gas companies when onsite burial is used. Waste Analytics' EEL calculation further supports that drilling waste would be a financially material issue to oil and gas companies if properly accounted. Simply put, drilling waste is a negative externality that is material especially when onsite burial is used.

At the moment, third parties bear the burden of the impacts from drilling waste. The widespread past and current drilling activity in the United States indicates that drilling waste is a national problem and not isolated to one geographic region. This means society is subsidizing the oil and gas industry in the United States regarding the management of drilling waste.

International reporting frameworks support that waste, including drilling waste, should be reported by corporations. This reporting should include the types of drilling waste generated the amounts of drilling waste generated and the management methods used for the drilling waste. As of 2021, no publicly traded oil and gas company drilling in the United States met all of these reporting standards. It is critical for this waste stream to be reported so that solutions can be discussed for this widespread issue. Unfortunately, there is a lack of required government reporting. Therefore, corporations must be held accountable for this material externality by the stakeholders of all involved including stockholders. Large asset managers and asset owners should begin to engage oil and gas companies regarding this waste stream because it has the potential to harm other investments. Without addressing this issue, human health will be affected, the environment harmed and assets written down. Currently, the only beneficiaries are oil and gas companies. It is unreasonable for individuals, communities, regions and other industries to bear the burden of issues created by drilling waste when they were not able to participate in the financial gain and are required to pay for its damage.

Areas for suggested further study on this topic are as follows:

1. Human health impacts from drilling waste disposal practices by region.

2. Detailed groundwater impacts from drilling waste by region.

3. Detailed soil impacts from drilling waste by region.

4. How liabilities created from drilling waste should be accounted for by oil and gas companies.

5. Determination of environmental protections from various centralized disposal facilities for oil and gas waste.

REFERENCES

American Coal Ash Association. (n.d.). Fly ash use in concrete increases slightly as overall coal ash recycling rate declines. https://acaa-usa.org/wp-content/uploads/coal-combustion-products-use/Coal-Ash-Production-and-Use.pdf.

American Petroleum Institute (2000). Overview of exploration and production waste volumes and waste management practices in the united states based on api survey of onshore and coastal exploration and production operations for 1995 and api survey of natural gas processing plants for 1995. ICF Consulting.

Bagnoli, M., Godwin, T., and Watts, S. G. (2021). On the real effects of changes in definitions of materiality. *Available at SSRN 3973688.*

Bartels, W. (2014). Sustainable insight: the essentials of materiality assessment. https://assets.kpmg/content/dam/kpmg/pdf/2014/10/materiality-assessment.pdf.

Billingsley, P. and Harrington, J. (n.d.). Preventing new groundwater pollution from old oilfield areas' oklahoma corporation commission and central oklahoma governments. https://cese.utulsa.edu/wp-content/uploads/2017/06/IPEC-2013-PREVENTING-GROUNDWATER-POLLUTION-FROM-OLD-OILFIELD-AREAS.pdf. Accessed: 2022-01-02.

Global Reporting Initiative. (n.d.-a). GRI 11: Oil and Gas Sector 2021, Global Sustainability Standards Board.

Global Reporting Initiative. (n.d.-b). Our mission and history. https://www.globalreporting.org/about-gri/mission-history.

Hale, A. (2017). Overview of OBM cuttings remediation methods and technologies. https://www.aade.org/application/files/2815/7365/8670/Arthur_Hale_Aramco-_Overview_of_OBM_Cuttings_Remediation_Methods_and_Technologies_11-162017.pdf.

Helbling, T. (2020). Externalities: prices do not capture all costs. *International Monetary Fund, Finance & Development.*

Hovorka, S. (1997). Salt Cavern Studies – Regional Map of Salt Thickness in the Midland Basin, Contract Number DE-AF22-96BC14978. Technical report, U.S. Department of Energy, Bureau of Economic Geology, The University of Texas.

Institute, C. F. (n.d.). Materiality thresholds in audits. https://corporatefinanceinstitute.com/resources/knowledge/accounting/materiality-threshold-in-audits. Accessed: 2022-01-22.

IPEICA (2009). Drilling fluids and health risk management.

IPIECA. (n.d.-a). About us. https://www.ipieca.org/about-us. Accessed: 2022-01-29.

IPIECA. (n.d.-b). Performance reporting. https://www.ipieca.org/our-work/sustainability/performance-reporting. Accessed: 2022-01-29.

IPIECA. (n.d.-c). Sustainability reporting guidance for the oil and gas industry module 4 environment. https://www.ipieca.org/media/5115/ipieca_sustainability-guide-2020_mod4-env.pdf. Accessed: 2022-01-19.

IPIECA. (n.d.-d). Worldwide drilling density. https://www.researchgate.net/ figure/World-oil-and-gas-well-distribution-and-density-courtesy-of-IHS-Energy_fig16_329707783/. Accessed: 2022-01-22.

Jha, P. K., Mahto, V., and Saxena, V. (2014). Emulsion based drilling fluids: an overview. *International Journal of ChemTech Research*, 6(4):2306–2315.

Johnston, J. E., Lim, E., and Roh, H. (2019). Impact of upstream oil extraction and environmental public health: a review of the evidence. *Science of the Total Environment*, 657:187–199.

Keller, R. P. (2017). Evaluating Drivers of Liability, Risk, and Cost While Enhancing Sustainability for Drilling Waste. Paper presented at the American Association of Drilling Engineers National Technical Conference and Exhibition, Houston, Texas.

Lodungi, J. F., Alfred, D., Khirulthzam, A., Adnan, F., and Tellichandran, S. (2016). A review in oil exploration and production waste discharges according to legislative and waste management practices perspective in malaysia. *International Journal of Waste Resources*, 7(1):260.

Murphy, E. (2014). Reserve Pit and Brine Pond Studies in North Dakota. *Energy Development and Transmission Committee Senator Rich Wardner, Chair Minot, ND April*, 8.

Nations, U. (n.d.). Goal 12: Ensure sustainable consumption and production patterns, goal 12 targets. https://www.un.org/sustainabledevelopment/ sustainable-consumption-production/. Accessed: 2022-01-07.

Norwegian Environment Agency. (n.d.). PFAS in mining and petroleum industry – use, emissions and alternatives.

Pragst, F., Stieglitz, K., Runge, H., Runow, K.-D., Quig, D., Osborne, R., Runge, C., and Ariki, J. (2017). High concentrations of lead and barium in hair of the rural population caused by water pollution in the thar jath oilfields in south sudan. *Forensic Science International*, 274:99–106.

Programme, U. N. E. (n.d.). Goal 12: Sustainable consumption and production goal 12: sustainable consumption and production. https://www. unep.org/explore-topics/sustainable-development-goals/why-do-sustainable-development-goals-matter/goal-12/. Accessed: 2022-01-08.

Ramirez Jr, P. (2009). *Reserve Pit Management: Risks to Migratory Birds. Cheyenne (WY): US Fish and Wildlife Service.*

Reader, H. P. (n.d.). The North Dakota Salt Lands. https://hpr1.com/index. php/feature/news/the-north-dakota-salt-lands. Accessed: 2022-01-10.

Reis, J. (1993). An Overview of the Environmental Issues Facing the Upstream Petroleum Industry. In *SPE Annual Technical Conference and Exhibition*. Society of Petroleum Engineers.

SASB. (n.d.-a). Find your industry, extractives & minerals processing. https://www.sasb.org/standards/materiality-finder/find. Accessed: 2022-01-29.

SASB. (n.d.-b). Oil & gas – exploration & production, disclosure topics. https://www.sasb.org/standards/materiality-finder/find/?industry[]= EM-EP. Accessed: 2022-01-29.

SASB. (n.d.-c). Oil & gas – services, disclosure topics. https://www.sasb. org/standards/materiality-finder/find/?industry[]=EM-SV. Accessed: 2022-01-29.

SASB. (n.d.-d). The research scope of sasb standards encompasses a range of sustainability issues, environment. https://www.sasb.org/standards/ materiality-finder/?lang=en-us. Accessed: 2022-01-29.

SASB (2020). Sasb standards & other esg frameworks. https://www.sasb. org/about/sasb-and-other-esg-frameworks. Accessed 2022-01-29.

Scott, B. (2010). Opportunities and obstacles to reducing the environmental footprint of natural gas development in the Uinta Basin' Colorado Law Scholarly Commons. https://scholar.law.colorado.edu/cgi/viewcontent. cgi?article=1004&context=reducing-environmental-footprint-of-natural-gas-development-in-uintah-basin.

Stuckman, M., Thomas, C., Lopano, C., and Hakala, A. (2015). Leaching characteristics of drill cuttings from unconventional gas reservoirs. In *Unconventional Resources Technology Conference, San Antonio, Texas, 20-22 July 2015*, pages 1301–1308. Society of Exploration Geophysicists, American Association of Petroleum.

Swiedler, E. W., Muehlenbachs, L. A., Chu, Z., Shih, J.-S., and Krupnick, A. (2019). Should solid waste from shale gas development be regulated as hazardous waste? *Energy Policy*, 129:1020–1033.

Thuermer Jr., A. (2018). Pavillion water experts fault leaky gas, unlined pits. https://wyofile.com/pavillion-water-experts-fault-leaky-gas-wells-unlined-pits/. Accessed: 2022-01-28.

Tuncan, A., Tuncan, M., and Koyuncu, H. (1996). Reuse of stabilized petroleum drilling wastes as sub-bases. *AA Balkema.*

United States Department of Energy, E. I. A. (n.d.). Today in energy drilling and completion improvement support improvement in permian basin hydrocarbons production. https://www.eia.gov/todayinenergy/detail.php?id=50016. Accessed: 2022-01-08.

U.S. Department of the Interior, E. P. A. (2018). Inventory of U.S. greenhouse gas emissions and sinks 1990–2016: Abandoned oil and gas wells. https://www.epa.gov/sites/default/files/2018-01/documents/2018_complete_report.pdf. Accessed: 2022-01-09.

U.S. Department of the Interior, E. P. A. O. o. R. C. and Recovery (2019). Management of exploration and production wastes: factors informing a decision on the need for regulatory action.

U.S. Department of the Interior, Environmental Protection Agency, Office of Solid Waste and Emergency Response (1987). Report to Congress, Management of Wastes from the Exploration, Development, and Production of Crude Oil, Natural Gas, and Geothermal Energy. 1.

CHAPTER 18

ESG Scores and Price Momentum Are Compatible: Revisited

Matus Padysak

Quantpedia.com

CONTENTS

18.1 INTRODUCTION

It is indisputable that the term sustainable investing is one of the most discussed topics among financial practitioners and academics as well. After all, being more sustainable has become a significant part of our lives (e.g., recycling, renewable energies, minimizing individual carbon footprint, etc.). Nowadays, socially responsible and sustainable investing is deeply connected with the term "ESG" score, where E stands for environmental, S for social and G for governance ratings of firms. Even though the ESG score is quite a novel term in the markets, responsible investing is much older. The origins probably did not have any financial motivation at all. Screening based on Christian values or Shariah-compliant investments is much older than the ESG. Additionally, not investing in firms with widely recognized environmental scandals such as pollution is also much older.

Naturally, the emergence of ESG has opened questions about how ESG is related to financial performance and risks. However, the answer is not that

DOI: 10.1201/9781003293644-18

straightforward, and it depends on the ESG data, investment universe and, most importantly, on the approach to ESG investing.

The greatest agreement among research seems to be in the proposition that ESG helps to minimize risks. According to [Freedman and Stagliano, 1991], responsible firms are less vulnerable to government-imposed fines. [Ashwin Kumar et al., 2016] identified that implementing ESG lowers the vulnerability to companies' reputational, political and regulatory risk, leading to lower cash flow and profitability volatility. Lower risk is also present among mutual funds, as shown by [Nofsinger and Varma, 2012], especially during crises. More responsible mutual funds outperform other mutual funds during bear markets. Lower risk was also identified in [Padysak, n.d.]. In the individual stocks, the ESG score is negatively related to the volatility (higher ESG equals lower volatility) and positively associated with the maximal drawdowns during five subperiods in 2010–2020.

[Verheyden et al., 2016] studied whether excluding the worst ESG stocks would reduce profitability. The results show that the performance is not reduced with the negative screening approach, and in fact, the risk-adjusted return is slightly larger than the benchmark. A similar approach was used by [Jondeau et al., 2021]. The paper shows that the portfolio can become significantly more environmentally friendly with a much lower carbon footprint by excluding a few most significant polluters. Furthermore, the performance difference compared to the benchmark is insignificant. As a result, the strategy can be used to construct passive portfolios (with relatively infrequent yearly rebalancing) that are more "Paris-aligned".

Secondly, the ESG scores are frequently utilized in ESG level strategies that consist of going long (short) top (bottom) ESG stocks. There seems to be a consent that the ESG level strategies tend to be profitable ([Nagy et al., 2016], [Dorfleitner et al., 2013] and [Hanicova and Vojtko, 2020]). The literature also recognizes ESG momentum strategies that identify the improvements in ESG scores and transforms these changes into investment decisions ([Nagy et al., 2016] and [Hanicova and Vojtko, 2020]).

Another strand of literature examines whether it is possible to integrate the ESG scores into traditional investment styles such as value, growth, or momentum. [Kaiser, 2020] explores how to blend value, growth and momentum investing with ESG investing by considering both ranks of the investment style and ESG metric. The author shows that it is possible to make the investments more sustainable for both the U.S. and European investment universe without sacrificing performance. [Coqueret et al., 2021] also examined the combination of traditional styles: size, value, momentum, investment and profitability with ESG scores to build portfolios. The results also show that implementing ESG does not necessarily lead to lower out-of-sample performance. The research by [Padysak, n.d.] focused on integrating ESG and momentum investing. The study was based on a widely recognized mathematical 0-1 knapsack problem, which is often explained by a situation of a robber that has limited space (capacity) in the knapsack and aims to rob as much as possible (maximize his utility). One could expand this problem to the financial markets, and the research has shown that it is possible to maximize the ESG score of the momentum portfolio or maximize the ESG portfolio's momentum score. As a result, the momentum portfolio with maximized ESG score was more sustainable and less risky, which can be truly practical since high risk and crashes are often regarded as the Achilles' heel of momentum investing.

On the other hand, the ESG portfolio can be much more profitable when the momentum style is included. Especially the risk-reducing ability of ESG scores can be a vital addition to momentum investing. The research about the momentum trend has come a long way since the original paper of [Jegadeesh and Titman, 1993]. [Barroso and Santa-Clara, 2015] suggest a risk-managed momentum, where the momentum is divided by the volatility to help to mitigate the adverse risks. [Butt et al., 2020] use a similar approach and suggest a variance scaled momentum in emerging markets to reduce exposure to market and liquidity factors, which boosts the performance in EM investment universe. [Fan et al., 2020] suggest a generalized risk-adjusted momentum to alleviate downside risks recognized as momentum crashes. In ESG investing, it is compelling that the ESG inclusion into momentum style investing had similar benefits as risk-adjusted momentum while being more sustainable. The critical question is whether this relationship still holds.

This research builds on [Padysak, n.d.] and could be understood as a continuation of the study. Firstly, we prolong the sample and thus provide an out-of-sample test of the proposed ESG-MOM (MOM-ESG) maximizing strategy. The prolonged sample includes a major (but short-lived) Covid financial crisis and can be characterized as the period of even greater ESG adoption when the ESG and socially responsible investing went mainstream. Secondly, we examine a slightly different methodology. The previous paper has focused on a 0-1 knapsack problem meaning that each stock is either chosen as an investment or not, and the optimal solution was found by stochastic optimization. This research examines the bounded knapsack problem, which can invest in any stock with a weight between zero and one. While it might seem that such a parametrization would lead to an excessive number of stocks, in practice, the condition that the overall momentum must be high for the ESG portfolio (or overall ESG must be high for the momentum portfolio) ensures that the number of stocks is comparable to picking the top ESG/MOM stocks without any optimization.

The rest of this chapter is structured as follows. In Section 18.2 we describe the ESG data and explain how these data are being filtered. In Section 18.3 we provide the theoretical basis for the ESG-MOM or MOM-ESG knapsack portfolios. Section is related to their practical implementation and, most importantly, to test their performance through portfolio sorts. We conclude the chapter with a summary discussion and conclusions set out in Section 18.4.

18.2 DATA

Monthly ESG data are sourced from OWL ESG (formerly OWL Analytics). The company provides data, indexes and evaluation metrics related to sustainable investing. The delivered dataset is formatted as the comma-separated values. Thus, it is readily made for analysis in data analysis software such as R or Python.

Although each dimension of the ESG is available, we are interested in a scaled Total ESG score between zero and one. ESG data were matched with price data, and the universe was filtered such that stocks were from the U.S. region. Small firms and firms without tickers were omitted. Furthermore, stocks for which the price series ended sooner than the ESG series were also omitted. This leaves us with 636 stocks from April 2010 to January 2022. Factor portfolios were obtained from Kenneth R. French's data library.

18.3 METHODOLOGY

Assume there are $S = 1, 2, 3, ..., n$ objects. Each object has at least two characteristics: weight and price. The object's weight could be characterized as the following vector: $w = (w_1, ..., w_n)^T$, and the price could be characterized as follows: $p = (p_1, ..., p_n)^T$. The knapsack problem could be easily explained as a real-life situation when one wants to choose several objects from the available set, such that the price of objects is maximized, and it fits into the knapsack. In other words, one aims to maximize the price of the chosen objects such that the weight constraints are not violated. Furthermore, if the vector $q = (q_1, ..., q_n)^T$ denotes the quantities of each object that is put into a knapsack, the problem could be formulated as follows:

$$\max q^T p \tag{18.1}$$

$$s.t. q^T w \leq W \text{ and } q_i \geq 0$$

where W is the weight capacity of the knapsack. If the only condition is that the $q_i \geq 0$, it is an unbounded knapsack problem; if the $q_i \in \{0, 1\}$, it is a 0-1 knapsack problem as in [Padysak, n.d.]. In this study, we examine the bounded knapsack problem where the bounds are zero and one to ensure that no stock has a too large weight in the portfolio, and as a result, the portfolio is concentrated. Consequently, the optimization problem is defined as follows:

TABLE 18.1 An example of ESG data from OWL ESG

Company Name	ISIN	shareClassFIGI	Region	Global Sector Count	Global SubSector Count	Global Industry Count	Region Sector Count	Region SubSector Count	Region Industry Count	E1	E1 Regional Sector Rank
RSA Insurance Group plc	GB00BKMKR23	BBG001S7TGP7	EURO	579	117	117	164	36	36	81	8
W. R. Berkley Corporation	US0844231029	BBG001SSP463	NAMR	579	117	117	225	63	63		
Illinois Tool Works Inc.	US4523081093	BBG001SSSDX0	NAMR	485	24	44	182	8	20	37	30
CME Group Inc	US12572Q1058	BBG001S86547	NAMR	579	307	73	225	107	32		
The Mosaic Company	US61945C1036	BBG001S7UN1	NAMR	271	97	97	111	39	99	37	29
CARBO Ceramics Inc.	US1407811Q58	BBG001SDMN12	NAMR	164	155	56	80	77	34		
AMN Healthcare Services	US0017441O17	BBG001S6Q193	NAMR	168	94	36	96	62	25		
ArthroCare Corporation	US0431361OO7	BBG001S7DZQ1	NAMR	168	94	58	96	62	37		
The Swatch Group Ltd	CH0012255151	BBG001SCPBCS	EURO	429	106	25	119	33	9	46	56
Thor Industries	US88S1G01018	BBG001SSWPHG	NAMR	429	106	17	188	44	10		
SL Green Realty Corp.	US78440X1019	BBG001S9Z1V2	NAMR	579	133	76	76	76	76	33	29
Regency Centers Corporation	US7585849 1032	BBG001S7H752	NAMR	579	133	76	225	52	45	33	51
New York Community Bancorp	US6494451031	BBG001SD3222	NAMR	579	307	234	225	107	75	33	51
Cincinnati Bell Inc.	US1718711062	BBF001S7QQX9	NAMR	118	118	21	76	76	76		
Canadian Utilities Limited	CA1367178326	BBG001SSYFP7	NAMR	118	97	21	51	51	10		
Zep	US98944B1089	BBG001SSW713	NAMR	271	97	97	111	39	39		
Contact Energy Ltd	N2CEN E0001S6	BBG001S673F3	PAC	118	118	21	20	20	4		
NYSE Euronext	US6294911010	BBG001S7SYY9	NAMR	579	307	73	225	107	32	35	36
Brookdale Senior Living Inc.	US1124631045	BBG001SCHC58	NAMR	168	94	36	96	62	25		
Procter & Gamble Company	US7427181091	BBG001SSV4L9	NAMR	205	40	40	80	19	19	52	10
American Science and Engineering	US0294291077	BBG001SSNWC6	NAMR	273	178	27	154	100	21		
AvalonBay Communities	US0534841012	BBG001S7J2HB	NAMR	579	133	76	225	52	45		
Charles River Laboratories International	US1598641074	BBG001S7Q2f1	NAMR	168	74	16	96	34	13		
Penn Virginia Corporation	US7078821 9G0	BBG001S73HN3	NAMR	164	155	92	80	77	39		
IStar Financial Inc.	US45031U1016	BBG001SDG572	NAMR	579	133	76	225	52	45	33	51
Hospitality Properties Trust	US44106M1027	BBG001S81XC2	NAMR	579	133	76	225	52	45		
Frontier Communications	US35906A1088	BBG001S81ZY3	NAMR	579	133	76	76	76	76		
Liberty Property Trust	US5311721048	BBG001S7F282	NAMR	579	133	76	225	52	45	49	18

$$\max q^T p \tag{18.2}$$

$$s.t. q^T w \leq W, q_i \geq 0 \text{ and } q_i \leq 1$$

If we aim to **maximize the ESG score of the momentum portfolio**: the vector p consists of individual ESG scores of the stocks, w consists of momentum ranks (we opt for the traditional 12-month momentum with the past one month skipped), and the W is defined to target 10% of the highest momentum stocks. Therefore, the weight cap can be defined as the sum of the 64 (given the number of stocks of 635) lowest ranks (in a system where the highest value gets the lowest rank): $W = \Sigma_{i=1}^{64} i$

If we aim to **maximize the momentum of the ESG portfolio**: the vector p consists of individual momentum ranks of the stocks (however, in this case, the high momentum equates to a high rank), and w consists of individually modified ESG scores such that we subtract one from each score and take an absolute value of them (to be in line with the idea of the weight budget); the W is defined to target 10% of the highest ESG stocks (lowest modified ESG scores). Therefore, the weight cap can be defined as the sum of the modified ESG scores that are lower than the first decile. If we define the modified ESG scores as e, the

$$W = \Sigma_{i, e_i \leq q(e, 0.1)} e_i \tag{18.3}$$

where $q(e, 0.1)$ is the 10% percentile of e.

As both cases have requirements that are represented by linear relationships, we utilize linear programming to find the optimal solutions. We opt for package lpSolve in R.

18.4   INVESTIGATION AND RESULTS

In the previous section, we have provided the theoretical basis for the ESG-MOM or MOM-ESG knapsack portfolios. This section is related to their practical implementation and, most importantly, to test their performance through portfolio sorts.

We perform optimizations each month to find weights of individual stocks for portfolios that maximize ESG in the momentum portfolio or momentum in the ESG portfolio using the most recent monthly data. Based on the weights, we form portfolios over the subsequent months for an out-of-sample test of the performance. To scale the investment to one, the individual weight is defined as $w_{stock,i} = \frac{q_i}{\Sigma q_i}$. We remark that the maximizing portfolios are long-only (since the weights are bounded to 0 and 1). Therefore, we compare the knapsack portfolios with long-only momentum and ESG benchmarks. The long-only momentum strategy buys the top 10% momentum stocks and is a traditional

TABLE 18.2 Performance comparison of MOM/ESG strategies. Ret is the annualized return, Sd is the annualized volatility (standard deviation), Max DD is the maximum drawdown, ESG is the average ESG score of the portfolio over the sample period and N is the average number of stocks included in the portfolio over the sample. The sample period spans from 1.4.2010 to 3.1.2022.

	Ret	Sd	Max DD	Ret/Sd	ESG	N
MOM-ESG	22.54	22.71	−26.74	0.99	0.65	53.23
ESG-MOM	17.01	16.01	−22.43	1.06	0.941	58.51
MOM	23.62	22.88	−27	1.03	0.396	63
ESG	17.84	16.14	−23.95	1.10	0.95	64

long-only momentum strategy. The long-only ESG strategy is based on the ESG level strategies and buys the top 10% ESG stocks. Both benchmarks are rebalanced monthly, the same as the knapsack strategies

We denote the strategies as follows: MOM-ESG is a strategy that maximizes the ESG score of the momentum portfolio, ESG-MOM is a strategy that maximizes the momentum of the ESG portfolio, MOM is the momentum portfolio and ESG is the ESG level strategy.

While the previous results showed that the momentum portfolios are less risky, the prolonged sample does not exhibit this pattern. Still, the optimized momentum portfolios are much more sustainable (as measured by the ESG scores). The performance difference is small, which is in line with other literature, which shows that ESG inclusion might not be the most profitable financial strategy. Still, when thoughtfully implemented, it does not hurt the performance. Although the riskiness of the MOM-ESG strategy is only slightly lower than that of the MOM strategy, the performance gap between them is lower compared to the previous research.

For the ESG portfolio, the momentum optimization does not seem to be worth it. The ESG score is slightly lower, and so is the performance compared to the ESG portfolio. However, it is noteworthy that the pure ESG portfolio is the least volatile, which is also in line with other literature. Furthermore, the ESG portfolio is not only less volatile than the momentum portfolio, which is not surprising but also compared to the equally weighted portfolio of all stocks in the sample (18.76%).

As the next step, we examine whether the portfolios are not driven by the common equity factors such as the Fama and French 5 and momentum factors. However, we must expect that the momentum exposure of the momentum tilted portfolios will be significant. We provide results also for the long-only ESG portfolio, for which the factor analysis can be a vital by-product of our study.

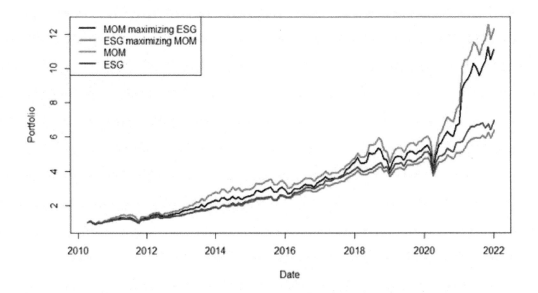

Figure 18.1   Performance comparison of MOM/ESG strategies

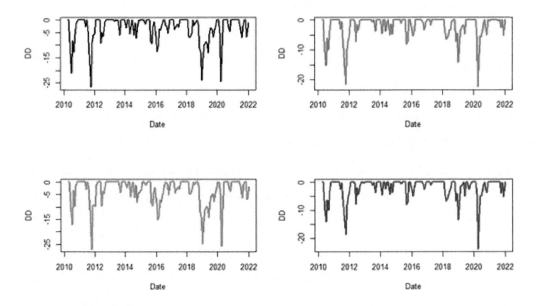

Figure 18.2   Drawdown charts of MOM/ESG strategies. The colors match those in Figure 18.1

TABLE 18.3 Factor analysis of MOM/ESG strategies

Panel A: Fama French 3 factor model	α	MKT	SMS	HML	RMW	CMA	MOM
MOM-ESG	0.007	1.046	0.752	0.011			
	(2.033)	(12.803)	(5.484)	(0.097)			
ESG-MOM	0.003	0.949	0.118	0.077			
	(1.649)	(20.702)	(1.536)	(1.198)			
MOM	0.007	1.069	0.823	−0.088			
	(2.176)	(13.572)	(6.224)	(−0.8)			
ESG	0.004	0.96	0.148	0.126			
	(2.167)	(22.656)	(2.079)	(2.12)			
Panel B: Fama French 3 factor model + Momentum factor							
MOM-ESG	0.005	1.167	0.801	0.305			0.609
	(1.658)	(15.713)	(6.629)	(2.753)			(6.418)
ESG-MOM	0.002	0.996	0.137	0.191			0.237
	(1.649)	(22.22)	(1.881)	(2.863)			(4.137)
MOM	0.005	1.184	0.869	0.189			0.576
	(1.821)	(16.424)	(7.414)	(1.758)			(6.246)
ESG	0.004	0.971	0.152	0.153			0.056
	(2.056)	(22.169)	(2.138)	(2.346)			(1.005)
Panel C: Fama French 5 factor model + Momentum factor							
MOM-ESG	0.005	1.181	0.735	0.265	−0.193	0.175	0.592
	(1.716)	(15.646)	(5.497)	(2.033)	(−1.163)	(0.852)	(6.187)
ESG-MOM	0.002	0.993	0.177	0.186	0.115	−0.0181	0.244
	(1.131)	(21.735)	(2.189)	(2.353)	(1.152)	(−0.145)	(4.207)
MOM	0.006	1.185	0.798	0.222	−0.207	−0.038	0.576
	(2.01)	(16.156)	(6.137)	(1.752)	(−1.284)	(−1.284)	(6.246)
ESG	0.003	0.963	0.213	0.159	0.177	−0.07	0.068
	(1.813)	(21.759)	(2.716)	(2.078)	(1.817)	(−0.583)	(1.213)

As expected, the long-only strategies have significant market exposure, and the momentum tilted strategies have sizeable and significant momentum factor exposure. Among the other factors, both momentum and ESG strategies (and also their optimized variants) seem to be significantly related to the size factor, and there is also a not negligible value factor relationship. The most recent factors – profitability and investment seem to have an insignificant effect on the strategies. Finally, the alpha is relatively tiny and not highly significant (albeit depending on the required significance level, we could conclude that the strategies are statistically significant). However, this only underlines the known facts from the other literature. The ESG itself can be profitable and does not harm the performance, but especially over the longer sample, without the recent "hype," it is not a miracle. Furthermore, not only the inclusion of ESG is not harmful to performance, but the sustainability is much larger.

18.5 CONCLUSIONS

This research builds on the previous research by [Padysak, n.d.]. Although some fascinating results from the previous research were not repeated, we show that ESG and momentum are still more than compatible. We build momentum portfolios that aim to maximize ESG scores or ESG portfolios that seek to maximize the momentum. Mathematically, the portfolios are derived from the knapsack optimization problem, which can be solved as a problem of linear programming in its bounded version. For comparison, we provide long-only ESG and momentum strategy portfolios.

The prolonged sample shows that it still holds that momentum and ESG style investing can be combined. The momentum strategy itself is widely recognized as profitable, but it is not that "ESG friendly" as measured by ESG scores. The ESG of the momentum portfolio could be much more attractive and significantly larger with our optimization procedure without a sizeable decrease in performance. Furthermore, we show that the data still support the notion that the ESG can help to reduce the risk, e.g., measured by volatility. However, this effect does not significantly materialize in momentum tilted portfolios, even though it is present in pure ESG level strategy.

Lastly, we also provide a factor analysis for all the aforementioned strategies. Each system has a significant and large market beta, which was expected, as these strategies are long-only. In addition, the momentum-related strategies also have significant momentum exposure. Among the other factors, the most sizeable result is that both ESG and momentum strategies (and their combinations) have considerable exposure to the size factor.

Lastly, although we present results for approximately the top 10% of stocks of the respective investment universes, the study should be understood as a general framework for implementing other investing factors. Based on the individually defined weight cap and maximizing the preferred style in the ESG portfolio or ESG in the preferred style portfolio, one should alter the approach to fit individual needs.

REFERENCES

Ashwin Kumar, N., Smith, C., Badis, L., Wang, N., Ambrosy, P., and Tavares, R. (2016). ESG factors and risk-adjusted performance: A new quantitative model. *Journal of Sustainable Finance & Investment*, 6(4):292–300.

Barroso, P. and Santa-Clara, P. (2015). Momentum has its moments. *Journal of Financial Economics*, 116(1):111–120.

Butt, H. A., Kolari, J. W., and Sadaqat, M. (2020). Revisiting momentum profits in emerging markets. *Pacific-Basin Finance Journal*, 65:101–486.

Coqueret, G., Stiernegrip, S., Morgenstern, C., Kelly, J., Frey-Skott, J., and Osterberg, B. (2021). Boosting esg-based optimization with asset pricing characteristics. *Available at SSRN 3877242.*

Dorfleitner, G., Utz, S., and Wimmer, M. (2013). Where and when does it pay to be good? A global long-term analysis of esg investing. In *26th Australasian Finance and Banking Conference.*

Fan, M., Kearney, F., Li, Y., and Liu, J. (2020). Momentum and the cross-section of stock volatility. *Available at SSRN 3977553.*

Freedman, M. and Stagliano, A. (1991). Differences in social-cost disclosures: a market test of investor reactions. *Accounting, Auditing & Accountability Journal,* 4(1).

Hanicova, D. and Vojtko, R. (2020). Backtesting ESG Factor Investing Strategies. *Available at SSRN 3633303.*

Jegadeesh, N. and Titman, S. (1993). Returns to buying winners and selling losers: implications for stock market efficiency. *The Journal of Finance,* 48(1):65–91.

Jondeau, E., Mojon, B., and Pereira da Silva, L. A. (2021). Building benchmarks portfolios with decreasing carbon footprints. *Swiss Finance Institute Research Paper,* pages 21–91.

Kaiser, L. (2020). ESG integration: value, growth and momentum. *Journal of Asset Management,* 21(1):32–51.

Nagy, Z., Kassam, A., and Lee, L.-E. (2016). Can ESG add alpha? An analysis of ESG tilt and momentum strategies. *The Journal of Investing,* 25(2):113–124.

Nofsinger, J. R. and Varma, A. (2012). Individuals and their local utility stocks: preference for the familiar. *Financial Review,* 47(3):423–443.

Padysak, M. (n.d.). ESG Scores and Price Momentum Are More Than Compatible. *Available at SSRN: https://ssrn.com/abstract=3650163 or http://dx.doi.org/10.2139/ssrn.3650163.*

Verheyden, T., Eccles, R. G., and Feiner, A. (2016). ESG for all? The impact of ESG screening on return, risk, and diversification. *Journal of Applied Corporate Finance,* 28(2):47–55.

(V)

DIRECTORY OF ALTERNATIVE DATA VENDORS

ALEXANDRIA TECHNOLOGY

About company: Alexandria Technology (www.alexandriatechnology.com) employs advanced machine learning techniques to replicate the decisions of domain experts when classifying critical information in unstructured content. Alexandria's Contextual Text Analytics (ACTA) engine was first used to classify DNA and now decodes the subject matter and sentiment in financial news.

The company's unique approach allows for relevant classification of virtually any unstructured content set, in any language, which currently includes native Japanese. In its current form, analyzing financial news, ACTA matches the decision of a research analyst with extremely high accuracy for multiple asset classes in real time.

Head office location: New York, United State

Other sites: London, United Kingdom and Los Angeles, California

Chief executive officer or Managing Director: Dan Joldzic CEO, Chris Kantos MD

Key services provided by the company:

The Alexandria analytical engine classifies financial news for entities, events and sentiment for multiple asset classes including global equities, economies, commodities, currencies and government debt instruments. This breakthrough technology avoids the accuracy and flexibility problems inherent in standard word- and rule-based approaches, while classifying articles at much lower latencies.

While the engineering goal is to match the analyst, Alexandria's research finds their data is predictive of price and volatility, and uncorrelated to traditional factors.

Contact: dan.joldzic@alexandriatechnology.com +16467591442; chris@alexandriatechnology.com +447731769537

BLOOMBERG L.P.

About company: Bloomberg L.P. is a privately held financial, software, data and media company. Data.Bloomberg.com, a ready-to-use data website where clients can easily discover, access and use all of Bloomberg's market-leading, high quality data. This includes reference data, regulatory data, pricing data, ESG data and quantitative data, all available in a standardized format via multiple delivery channels such as SFTP, REST API and in the cloud. Data.Bloomberg.com is also the place where clients can generate custom datasets, including requesting historical data, and manage their data inventory, all through one easy to use web interface.

Head office location: New York, USA

Key services provided by the company: Bloomberg developed and built his own computerized system to provide real-time market data, financial calculations and other financial analytics.

Bloomberg GO allows subscribers to access the Bloomberg Professional service to monitor and analyze real-time financial data, search financial news, obtain price quotes and send electronic messages through the Bloomberg Messaging Service. Bloomberg connects alternative data providers to the most influential global network of financial services professionals on our enterprise data distribution platform. Our data experts can help firms set up data for distribution and integration into client workflows, helping improve the efficiency and customer experience of their products.

Contact: Americas +1 212 318 2000 EMEA +44 20 7330 7500 Asia Pacific +65 6212 1000

CHINASCOPE

About company: ChinaScope (www.chinascope.com) is a leading financial technology company focusing on extraction and interpretation of unstructured and semi-structured data related to China. We are committed to providing intelligent data solutions for decision making to financial institutions, banks, enterprises and governments.

ChinaScope offers a full range of data feed services including SAM supply chain data, SmarTag news analysis, customer-supplier relationships, fundamental financials and corporate actions. In addition, we provide scenario-based intelligent data solutions. In a world that is becoming more and more intertwined economically and geopolitically, making sense of content coming out of China in the context of daily decisions becomes an inescapable reality. ChinaScope complements our clients' existing decision-making apparatus by providing high-quality distilled information on China.

Head office location: Shanghai, China

Other sites: Beijing, Shenzhen, Nanjing, Chongqing, Jinan, Hong Kong

Chief Executive Officer: Tom Liu

Chief Data Officer: Lucas Lu

Head of Quantitative Research: Song Lu

Key services provided by the company:

Data product 1: China Segment Analysis & Mapping (SAM)
ChinaScope's proprietary Segment Analysis & Mapping (SAM) data categorizes 20k public companies by standardized product segments they are operating in. The data classification system is mapped to the GICS industry classification, effectively creating 12 layers of GICS subcategories, which include 6k+ standard product nodes. Product nodes are further categorized into upstream and downstream links, ultimately forming industry-level value chain networks.

Data product 2: China News Analytics (SmarTag) ChinaScope provides a news analytics data product (SmarTag) by using ChinaScope proprietary NLP technology. The NLP parses 1,600+ Chinese language news sites, extracts meta data and classifies relevant information into 7 categories: Company, Industry, Event, Region, People, Product, Themes. In addition, ChinaScope delivers sentiment scores to both the news article and the entity mentioned in the news respectively (positive, negative, neutral, adding up to 100

Data product 3: China Customer-Supplier Relationships Customer-Supplier Relationships data describes the direct customer and supplier relationships of China A-Share companies. The data is extracted from financial filings of A-Share companies by extracting entity names from disclosure of major customers and suppliers and from counterparties of accounts receivables and accounts payables. The dataset covers relationships centered on all listed companies on the Shanghai Stock Exchange and the Shenzhen Stock Exchange.

Contact: liu.tom@chinascope.com; lucas.lu@chinascope.com; song.lu@chinascope.com

EXANTE DATA

About company: Exante Data services the world's most sophisticated market participants using a combination of in-depth analysis and human conceptual thinking. We use unique data to go deeper than others. We strive to focus on topics that are away from the consensus or before (ex ante!) the consensus. We think of our client relationships as long-term collaborations.

Head office location: Manhattan, New York

Other sites: https://moneyinsideout.exantedata.com/

Chief executive officer or Managing Director: Jens Nordvig

Marketing and sales director: Kaye Gentle

Key services provided by the company:

- A comprehensive Data and Analytics Platform with >100k series, proprietary models and self-updating charts. Data are organized thematically.

- Macro Strategy which includes market commentary, capital flow analysis, trade updates and political risk notes. We focus on topics that truly matter to markets, as opposed to maintenance research.

- Our Global Team of Analysts that are available on Bloomberg chat and via regular video calls. Our senior strategists and advisors are based in NYC, London, Norway, Spain, Colombia and the US East Coast.

- COVID tracking, forecasting and mobility. This as a big element of our macro strategy in 2022-2021, but our COVID focus currently is mostly centered on China – mobility and alt data.

Contact:
info@exantedata.com
Kaye Gentle (kaye.gentle@exantedata.com)
Chief of Staff — Global Client Relationship Manager
(332) 220-3500

FACTEUS

About company: Facteus is on a mission to help democratize data insights by safely unlocking valuable data sets and increasing data literacy through technology. Through our synthetic data engine, MIMIC™, we safely unlock more data sets for research and insights. Our alternative data research, predictive analytics and insights platform, Quantamatics™, provides a turnkey solution for streamlining data sourcing, management, data science and insights delivery into a wide range of research and investment workflows.

Head office location: Beaverton, Oregon

Other sites: N/A

Chief executive officer or Managing Director: CHRIS MARSH

Marketing and sales director: NOEL BOSCO

Key services provided by the company: Analytics, PCI Security, Enterprise Reporting, Big Data, Business Intelligence, Data Anaysis, Statistical Analysis, Machine Learning, Data Monetization, Financial Data, Alternative Data and Investment Research

Contact: info@facteus.com

FACTSET

About company: FactSet creates flexible, open data and software solutions for tens of thousands of investment professionals around the world, providing instant access to financial data and analytics that investors use to make crucial decisions.

For 40 years, through market changes and technological progress, our focus has always been to provide exceptional client service. From more than 60 offices in 23 countries, we're all working together toward the goal of creating value for our clients, and we're proud that 95% of asset managers who use FactSet continue to use FactSet, year after year. As big as we grow, as far as we reach, and as successful as we become, we stay connected to our clients and to each other.

Head office location: Norwalk, Connecticut

Other sites: N/A

Chief executive officer or Managing Director: Phil Snow

Marketing and sales director: Christina Sradj

Key services provided by the company: Fintech, portfolio management, investment management, investment banking, wealth management, trading, risk and compliance, client reporting, OMS, EMS, data delivery, professional services, open technology, performance and attribution, risk management and analysis, and financial data.

Contact: support@factset.com

Christina Sradj - csradj@factset.com

GTCOM Technology Corporation

About company:

Headquartered in Santa Clara, CA, GTCOM Technology Corporation (GTCOM US) is a fintech research and data analytical company. With advanced NLP technologies and strong research methodology using Alternative Data, we help clients identify proper data sources, categorize and tag unstructured data, create and evaluate datasets and deliver in-depth value in the form of dataset, dashboards and analytic reports for the financial and capital markets industries.

Head office location: Santa Clara, CA, USA

Chief executive officer or Managing Director: Allen Yan

Marketing and sales director: Raymond Jiao

Key services provided by the company:

Comprehensive alternative data solution from China, including

- Sentiment Data

- Geolocation Data

- E-Commerce Data

- Recruitment Data

- ESG Indicators

Other services we provide China-related data sourcing; alternative data-based index development; research reports.

Contact: BD@gtcom-us.com

INFOTRIE FINANCIAL SOLUTIONS

About company:
InfoTrie is an Alternative Data Specialist & an AI-powered company since 2012 covering millions of heterogeneous sources and converting them into unique datasets.

We have offices in Europe and Asia and are trusted by Asset managers, Hedge Funds, Researchers, FinTechs as well as large corporations & financial institutions across the globe.

Our decades of experience enable us to run a wide range of pre-built, out-of-the box solutions and to offer targeted feeds leveraging our expertise in Financial Engineering, Artificial Intelligence and Natural Language Processing (NLP).

Head office location: Singapore

Chief Executive Officer: Frederic Georjon

Executive Director: Suresh Kumar

Our Alternative Data Catalogue:

- Sentiment Data

- Job Postings

- E-Commerce Analytics

- SEC and Regulatory Filings

Key services provided by the company:

- We enable businesses to extract and validate massive unstructured datasets into meaningful information through a simple integration and flexible delivery methods.

- We globally monitor 100,000s of publicly traded & private companies, people, topics and asset classes to provide seamless enriched datasets via our standard iFeed API.

- Bespoke datasets and taxonomies.

Contact: contact@infotrie.com

MARKETPSYCH DATA

About company: MarketPysch Data provides natural language processing services (NLP) and news and social media-derived sentiment and macroeconomic data through Refinitiv. Refinitiv MarketPsych analytics provide metadata and sentiment feeds on millions of companies, 100+ currencies, 200+ commodities, 100,000+ cities, states and countries. The data is delivered in real-time, minutely, hourly and daily formats, and it spans from 1998 to the present. The data is derived from tens of thousands of global business and finance-related news and social media sources. The aggregated Refinitiv MarketPsych Analytics delivers quantified time series of dozens of granular categories of emotion (fear, anger, joy), expectations (optimism, earningsForecast, interestRateForecast), political risk (governmentInstability) and macroeconomic (inflation, unemployment) factors, among many others. MarketPsych began work on financial text analytics in 2004. MarketPsych aims to provide the global standard in financial media analytics.

Head office location: Singapore

Chief executive officer or Managing Director: Richard L. Peterson

Marketing and sales director: NLP software and data products are provided independently and through Refinitiv.

Key services provided by the company: The company's primary business is the creation and distribution of media analytics software including sentiment data, NLP engines and processing tools.

Contact: +1.323.389.1813

NIKKEI FTRI

About company:
Nikkei FTRI is a research institute that is 100% owned by Nikkei Inc. and has been a leading company in a credit risk management field in Japan with a number of clients such as banks, asset managers and government agencies. Our products SMACOM and FTRI News Dolphin are investment-decision tools based on alternative data as well as traditional data.

Head office location: Tokyo, Japan

Other sites: N/A

Chief executive officer or Managing Director: Tomoyuki Miyagi

Marketing and sales director: Akiko Song, Yoshifumi Yamashita

Key services provided by the company:

News Analysis

News analysis provides tag data and positive/negative scores estimated from Japanese reliable news sources from Nikkei, NQN and QUICK. This service has achieved a highly accurate analysis by applying natural language processing technology of Nikkei FTRI and is well predictive of subsequent cross-sectional returns in Japan.

SMACOM
SMACOM provides unique scores analyzed by Nikkei FTRI and helps professional institutional investors with their investment decision-making. The scores are based on alternative data as well as traditional data such as news, corporate disclosures, financial statements, macro/microeconomic data and various alternative data, some of which are exclusive to us.

Contact:
Phone: +81-3-6273-7743
Email: global_div@ftri.co.jp

NOWCAST INC.

About company:

Nowcast Inc. is the leading alternative data and actionable insight provider in Japan, we work to make Japanese alternative data accessible to clients no matter where they are from. We focus on developing and providing high-quality datasets through our strong Japanese data source network and data science capabilities. Our current focus in providing consumer transaction and geolocation datasets, in which we cover over 500 listed and unlisted Japanese companies and REITs. Nowcast data and insights are used by investment professionals, government institutes, think tanks, corporates from across the world.

Head office location: Tokyo, Japan

Other sites: N/A

Chief executive officer or Managing Director: Masashi Tsujinaka

Marketing and sales director: Alex Occhipinti

Key services provided by the company:

- Consumer transaction and geolocation data including credit card and Point of Sales data. This data allows investors, governments and corporates to gain insights at both the macro- and microlevel through a broad product suite aimed at meeting the unique needs of our client base.

- Analyst insight reports developed by our analysts to add in-depth context that can explain the trends in our data and help better support your analysis

- Data mastering for corporates to maximise the potential of their dataset in assisting with management, marketing and other key business decisions

Contact:

- Masashi Tsujinaka, CEO - E: tsujinaka@nowcast.co.jp M: +8180-5633-6412

- Alex Occhipinti, Sales – E: alexander.occhipinti@nowcast.co.jp M: +8170-8944-2396

RAVENPACK

About the company: Since 2003, RavenPack has been one of the leading data analytics providers in financial services, allowing firms to quickly extract value and insights from large amounts of unstructured text data. RavenPack's products allow companies to enhance returns, reduce risk and increase operational efficiency. The company's clients include some of the most successful and most sophisticated hedge funds, banks and asset managers in the world.

Head office location: Marbella, Spain

Other sites: New York, NY

Chief executive officer or Managing Director: Armando Gonzalez

Marketing director: Max Colas

Sales director: Malcolm Bain

Key services provided by the company:

RavenPack Edge, the company's flagship product, enables the systematic inclusion of unstructured and hard-to-find data into existing knowledge workflows. This helps finance professionals and businesses make more informed decisions, stay ahead of competition and mitigate emerging risks.

RavenPack Edge delivers structured analytics on published content from high-quality news sources and job feeds, including gated content from Dow Jones Newswires, Barron's, The Wall Street Journal, MarketWatch, FactSet, LinkUp, Benzinga and over 40,000 more web and social media sources. It tracks over 12 million entities, such as companies, individuals, currencies, products, services or themes. Leveraging Nearly 20 Years Of Research, RavenPack Edge combines proprietary Natural Language Processing (NLP) technology powered by heuristic and Machine Learning (ML) engines and 20+ Years of high-quality, fully normalized archive with billions of documents.

Contact: mbain@ravenpack.com

REFINITIV

About company:

Refinitiv, an LSEG (London Stock Exchange Group) business, is one of the world's largest providers of financial markets data and infrastructure. With $6.25 billion in revenue, over 40,000 customers and 400,000 end users across 190 countries, Refinitiv is powering participants across the global financial marketplace. We provide information, insights and technology that enable customers to execute critical investing, trading and risk decisions with confidence. By combining a unique open platform with best-in-class data and expertise, we connect people to choice and opportunity – driving performance, innovation and growth for our customers and partners.

Head office location: New York City, US (Operational); London, England, UK (Corporate)

Chief Executive Officer: David W. Craig

Key services provided by the company:

- Business consultancy – Dynamic business consultancy capabilities, facilitated by trusted experts across the globe. Working in close partnership with our clients, the team works to understand customers' workflows and business requirements, making sure our proposed solutions are aligned with their strategic vision.

- Solutions engineering – Technical delivery of installations, moves and changes (IMAC) across the full range of Refinitiv products. The team delivers additional value to customer engagements by leveraging specialist knowledge and deep expertise of Refinitiv's products to ensure an accurate and timely delivery.

- Technical consultancy – Specialized technical consulting activities delivering solutions projects. Widely recognized technical experts, this team works to ensure Refinitiv products are correctly embedded into our customers' environments. They manage all aspects of complex implementations or migrations.

- Customer technical authority – Subject matter experts working with stakeholders across the business to ratify the design and pre-sale for our more complex implementations, complete the detailed solution design post-sale and remain on point with the customer and the delivery teams to provide technical expertise continuity, meeting the customer requirements and supporting the wider business strategy.

- Project management – Professional certification with extensive product expertise define this team. They assist in managing initiatives where scale, scope, complexity or sensitivity requires superior organization and coordination. Working with our customers, they combine their knowledge of Refinitiv products with vast experience in delivering proven and sustainable end-to-end solutions.

- Custom solutions – Custom solutions are high quality applications developed based on customer's specific needs. Our highly skilled team addresses particular challenges by building adapters to Refinitiv APIs, developing workflow solutions and envisioning futuristic solutions aligned to Refinitiv strategy.

Contact: lemuel.brewster@refinitiv.com, tarek.fleihan@refinitiv.com, silke.marsh@refinitiv.com, stuti.singh@refinitiv.com

Index